T0142836

Studies in Systems, Decision and Control

Volume 98

About this Series

The series “Studies in Systems, Decision and Control” (SSDC) covers both new developments and advances, as well as the state of the art, in the various areas of broadly perceived systems, decision making and control- quickly, up to date and with a high quality. The intent is to cover the theory, applications, and perspectives on the state of the art and future developments relevant to systems, decision making, control, complex processes and related areas, as embedded in the fields of engineering, computer science, physics, economics, social and life sciences, as well as the paradigms and methodologies behind them. The series contains monographs, textbooks, lecture notes and edited volumes in systems, decision making and control spanning the areas of Cyber-Physical Systems, Autonomous Systems, Sensor Networks, Control Systems, Energy Systems, Automotive Systems, Biological Systems, Vehicular Networking and Connected Vehicles, Aerospace Systems, Automation, Manufacturing, Smart Grids, Nonlinear Systems, Power Systems, Robotics, Social Systems, Economic Systems and other. Of particular value to both the contributors and the readership are the short publication timeframe and the world-wide distribution and exposure which enable both a wide and rapid dissemination of research output.

More information about this series at http://www.springer.com/series/13304

Carsten Last

From Global to Local Statistical Shape Priors

Novel Methods to Obtain Accurate Reconstruction Results with a Limited Amount of Training Shapes

Carsten Last
Institut für Robotik und Prozessinformatik
Technische Universität Braunschweig
Braunschweig
Germany

ISSN 2198-4182 ISSN 2198-4190 (electronic)
Studies in Systems, Decision and Control
ISBN 978-3-319-85169-3 ISBN 978-3-319-53508-1 (eBook)
DOI 10.1007/978-3-319-53508-1

Printed on acid-free paper

This Springer imprint is published by Springer Nature
The registered company is Springer International Publishing AG
The registered company address is: Gewerbestrasse 11, 6330 Cham, Switzerland

For my wife Carolin.
Without you I could not have done this.

Preface

The present thesis originates from my scientific work as employee at the *Institut für Robotik und Prozessinformatik, Technische Universität Braunschweig*, Germany. However, the completion of this thesis would not have been possible without the help of others.

At first, I would like to thank Prof. Dr.-Ing. Friedrich M. Wahl for directing me towards the interesting topic of medical image segmentation and giving me the freedom to develop my own ideas within this field. He provided me with excellent technical resources and a great working environment, and he gave me the opportunity to present my ideas at various national and international conferences. I would also like to thank Prof. Dr. Thomas Vetter for his effort to review this thesis as a second referee and for providing the morphable face model as well as the additional face scans that are used for evaluation.

My thanks go especially also to Dr.-Ing. Simon Winkelbach for his continuous support in developing and discussing new ideas, but as well for proofreading this thesis. His input had a great influence on the final outcome. I also have to thank the whole staff and all students of the *Institut für Robotik und Prozessinformatik* for creating such a nice and friendly working atmosphere.

Additionally, I would like to thank the German Research Foundation (DFG) for funding parts of my research within the project *Roboterunterstützte, erwartungskonforme Endoskopführung in der endosalen Chirurgie*. In this regard, I would like to thank Prof. Dr. med. Dr. h.c. Friedrich Bootz from the *Klinik und Poliklinik für Hals-Nasen-Ohrenheilkunde/Chirurgie* at the *Universitätsklinikum Bonn* for providing me with the paranasal sinuses database that is being used within this thesis. Also, I would like to thank my former colleagues Dr.-Ing. Ralf Westphal, Markus Rilk, and Dr. med. Klaus Eichhorn for the fruitful and friendly cooperation within this project.

Special thanks go to my family for their continuous moral and financial support throughout all my years of study and to my wife Carolin for the countless hours that she spent discussing and proofreading this thesis.

Braunschweig, Germany Carsten Last
December 2016

Contents

List of Figures

List of Tables

List of Algorithms

List of Algorithms

Chapter 1
Basics

We assume that the reader of this thesis has a sound understanding of digital image processing methods. Some good textbooks about digital image processing are e.g. [45, 56, 131]. The purpose of this chapter is to provide the reader with additional basics that are needed for the understanding of this thesis. We will first give some fundamental mathematical naming conventions in Sect. 1.1. Afterwards, in Sects. 1.2 and 1.3, we will continue the mathematical introduction by briefly repeating some basics from vector calculus and by introducing the calculus of variations. With these mathematical tools at hand, we will subsequently present the so-called *level set methods* in Sect. 1.4, an elegant way to represent curves and surfaces. Finally, in Sect. 1.5, we will put it all together in order to obtain a powerful approach to image segmentation problems which plays an important role throughout this thesis.

1.1 Naming Conventions

We start the introductory chapter with some elementary mathematical naming conventions that are important for the understanding of this thesis.

Vectors, or vector-valued functions, are always interpreted as row-vectors. They are denoted by small or capital letters overlain with an arrow. So, a vector $\vec{a} \in \mathbb{R}^n$ or $\vec{A} \in \mathbb{R}^n$ is given as

$$\vec{a} = (a_1, \ldots, a_n) \qquad \text{or} \qquad \vec{A} = (A_1, \ldots, A_n)\,, \tag{1.1}$$

respectively.

Matrices are denoted by bold capital letters. So, a matrix $\mathbf{A} \in \mathbb{R}^{m \times n}$ is given as

$$\mathbf{A} = \begin{pmatrix} A_{11} & \ldots & A_{1n} \\ \vdots & \ddots & \vdots \\ A_{m1} & \ldots & A_{mn} \end{pmatrix}. \tag{1.2}$$

C. Last, *From Global to Local Statistical Shape Priors*, Studies in Systems, Decision and Control 98, DOI 10.1007/978-3-319-53508-1_1

The **Inner Product** of two vectors $\vec{a} \in \mathbb{R}^n$ and $\vec{b} \in \mathbb{R}^n$, or also *scalar product* or *dot product*, is defined as

$$\langle \vec{a}, \vec{b} \rangle = \vec{a}\,\vec{b}^T = (a_1, \dots, a_n) \begin{pmatrix} b_1 \\ \vdots \\ b_n \end{pmatrix} = a_1 b_1 + \dots + a_n b_n \,. \tag{1.3}$$

The **Outer Product** of two vectors $\vec{a} \in \mathbb{R}^n$ and $\vec{b} \in \mathbb{R}^n$, or also *tensor product*, is defined as

$$\vec{a} \otimes \vec{b} = \vec{a}^{\,T}\,\vec{b} = \begin{pmatrix} a_1 \\ \vdots \\ a_n \end{pmatrix} (b_1, \dots, b_n) = \begin{pmatrix} a_1 b_1 & a_1 b_2 & \dots & a_1 b_n \\ a_2 b_1 & a_2 b_2 & \dots & a_2 b_n \\ \vdots & \vdots & \ddots & \vdots \\ a_n b_1 & a_n b_2 & \dots & a_n b_n \end{pmatrix} . \tag{1.4}$$

1.2 Vector Calculus

In this section, we repeat some basics from vector calculus that occur in the mathematical derivations from the upcoming chapters. Additionally, these basics are crucial for the understanding of the calculus of variations which we are going to introduce in the next section.

The vector calculus is a branch of mathematics that deals with the differentiation and integration of vector fields in two or more dimensions. We are especially interested in real-valued vector fields, the so-called scalar fields, i.e. functions that map from a multi-dimensional domain $\Omega \subset \mathbb{R}^d$, where typically $d \in \{2, 3\}$, to the field of real numbers $\mathbb{R}$. Such a scalar field $I : \Omega \subset \mathbb{R}^d \to \mathbb{R}$ may represent for example two-dimensional or three-dimensional image information. For $d = 2$ we interpret the function I as an *image* and for $d = 3$ we interpret I as a *volumetric image*, respectively. Other scalar fields in this thesis are the level set function which will be introduced in Sect. 1.4 and the weight fields which we will introduce in Chap. 3.

1.2.1 Pixels and Voxels

In two dimensions, the elements of the domain Ω are often called *pixels* and in three dimensions they are often called *voxels*. The term *pixel* is a short form for *picture element* and the term *voxel* is a short term for *volume element*, respectively. In this thesis, we are considering two-dimensional (2D) as well as three-dimensional (3D) image information. So, everywhere were both terms *pixel* as well as *voxel* are appropriate, we will use the general term *element* in order to avoid confusion.

1.2.2 The Gradient and Necessary Condition for a Minimum

A common task in ordinary vector calculus is to find the minima of a scalar field. A necessary condition for a minimum in a certain point is that the gradient of the scalar field equals zero in the sought-after point. For a scalar field $h : \mathbb{R}^d \rightarrow \mathbb{R}$, i.e. h is a real-valued function of d variables, the gradient is defined as [23, Chap. 13.2.2]

$$\nabla h(x_1, \ldots, x_d) = \left(\frac{\partial h}{\partial x_1}, \ldots, \frac{\partial h}{\partial x_d} \right), \tag{1.5}$$

where $\frac{\partial h}{\partial x_i}$ denotes the partial derivative of the function h with regard to the variable x_i. Then, the necessary condition for a minimum can be written as [23, Chap. 18.2.5]

$$\nabla h(x_1, \ldots, x_d) = \vec{0}_d\,, \tag{1.6}$$

where $\vec{0}_d$ is a d-dimensional row-vector that consists only of zeros.

1.2.3 Directional Derivative

The directional derivative of a scalar field h describes the rate of change of the multivariate function $h : \mathbb{R}^d \rightarrow \mathbb{R}$ in the direction of a vector $\vec{p} \in \mathbb{R}^d$. It can be obtained as the inner product of the gradient with the normalized vector $\vec{p}$ [23, Chap. 13.2.2.2]:

$$\nabla_{\vec{p}}\, h = \langle \nabla h, \frac{\vec{p}}{||\vec{p}||} \rangle\,. \tag{1.7}$$

So, the directional derivative is equal to the signed length of the projection of the gradient onto the vector $\vec{p}$. In the case that $\vec{p}$ is a unit vector, it reduces to

$$\nabla_{\vec{p}}\, h = \langle \nabla h, \vec{p} \rangle\,. \tag{1.8}$$

1.2.4 Descent Direction

For a scalar field $h : \mathbb{R}^d \rightarrow \mathbb{R}$, a *descent direction* $\vec{p} \in \mathbb{R}^d$ in a point $\vec{x}$ is a direction along which the directional derivative is negative [23, Chap. 18.2.1.2]:

$$\langle \nabla h(\vec{x}), \vec{p} \rangle < 0\,. \tag{1.9}$$

One specific descent direction is the negative gradient $-\nabla h(\vec{x})$ of the function h, because when we insert $-\nabla h(\vec{x})$ into Eq. (1.9), we obtain

$$\langle \nabla h(\vec{x}), -\nabla h(\vec{x}) \rangle = -||\nabla h(\vec{x})||^2 \leq 0 \,. \tag{1.10}$$

So, as long as the negative gradient differs from zero, it indicates a descent direction of the function h. In fact, the negative gradient always points in the direction of steepest descent, i.e. the direction in which the function values decrease the most [23, Chap. 18.2.5.1].

1.2.5 Descent Methods

A common way to find a local minimum of a scalar field h is to start with an initial guess $\vec{x}^0$ for the minimum, choose a descent direction $\vec{p}$ in the point $\vec{x}^0$, and take repeated steps in the descent direction:

$$\vec{x}^{k+1} = \vec{x}^k + \gamma^k \, \vec{p}^k \,, \tag{1.11}$$

where $k \in \mathbb{N}$ denotes the k-th iteration and $\gamma^k \in \mathbb{R}$ is the step size in each iteration. Equation (1.11) is iterated until a stationary point of the function h is reached which most likely indicates a local minimum as we are moving in a descent direction of the function h.[1] In the case that the negative gradient is chosen as a descent direction, Eq. (1.11) becomes

$$\vec{x}^{k+1} = \vec{x}^k - \gamma^k \, \nabla h(\vec{x}^k) \,, \tag{1.12}$$

which is called *gradient descent method* or also *method of steepest descent* [23, Chap. 18.2.5.1].

A necessary condition for a minimum of a scalar field has been given in Eq. (1.6). This means that the gradient descent method should stop at the desired local minimum as the gradient becomes zero. However, in practical implementations the gradient will never exactly be zero due to noise in the data or numerical quantization errors. So, it is a common practice to either iterate Eq. (1.11) for a fixed number of iterations or to check whether two consecutive solutions $\vec{x}^k$ and $\vec{x}^{k+1}$ lie within a predefined convergence tolerance ϵ and stop iterating Eq. (1.11) in the case that the following termination condition is fulfilled:

$$||\vec{x}^{k+1} - \vec{x}^k||^2 < \epsilon \,. \tag{1.13}$$

1.3 Calculus of Variations

The vector calculus, introduced in Sect. 1.2, enables us to find the minima or maxima of a vector field. As we will see later on, in this thesis we will additionally deal with the question which function h minimizes or maximizes a functional E. A functional

[1] The second possibility would be a saddle point.

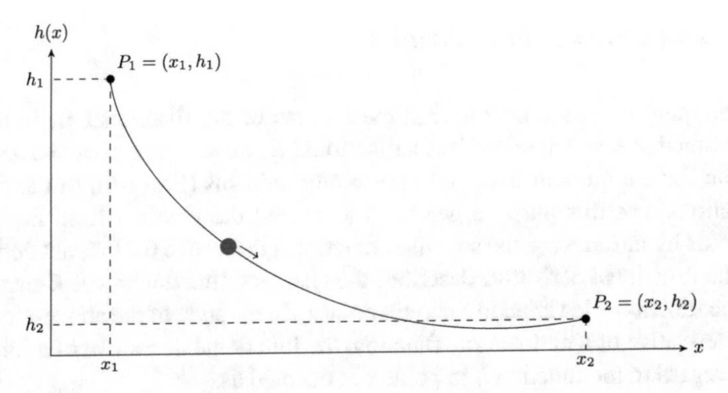

Fig. 1.1 Depiction of the brachistochrone curve on which the *blue* body reaches the point P_2 in shortest time when starting with zero velocity in point P_1

is a mapping that maps a function h, defined inside a function space Γ, to the field of real numbers $\mathbb{R}$ so that $E : \Gamma \to \mathbb{R}$ and $E : h \mapsto E(h)$. In order to answer this question, we need the calculus of variations.

The calculus of variations originates from the so-called *brachistochrone problem* where one searches for the curve along which a body has to move in order to reach a given end point $P_2 = (x_2, h_2)$ from a given start point $P_1 = (x_1, h_1)$ in shortest time when the body is accelerated only by constant gravity g and the friction is neglected (see Fig. 1.1). In this case, the function h describes the curve and the functional E describes the time that the body needs to travel from point P_1 to point P_2 along the curve h [87]:

$$E(h) = \int_{x_1}^{x_2} \sqrt{\frac{1 + h'(x)^2}{2g(h_1 - h(x))}}\, dx\,. \tag{1.14}$$

The curve that minimizes this functional is called a *brachistochrone curve*, hence the name *brachistochrone problem*. Another example for a functional is the entropy of a function. It is defined as [55]

$$E(h) = -\int h(x) \ln h(x)\, dx\,. \tag{1.15}$$

In this thesis, we will later introduce functionals that are of the form

$$E(h) = \int_{\Omega} F(h, \frac{\partial h}{\partial x_1}, \ldots, \frac{\partial h}{\partial x_d}, x_1, \ldots, x_d)\, d\vec{x}\,. \tag{1.16}$$

In Eq. (1.16), the functional E maps a function $h : \mathbb{R}^d \to \mathbb{R}$ with multiple arguments, i.e. $h : \vec{x} \mapsto h(\vec{x})$ with $\vec{x} = (x_1, \ldots, x_d)$, to the field of real numbers $\mathbb{R}$ by integrating a term F over a multi-dimensional domain $\Omega \subset \mathbb{R}^d$. The term F thereby depends on the function h, its partial derivatives $\frac{\partial h}{\partial x_1}, \ldots, \frac{\partial h}{\partial x_d}$ and its arguments $x_1, \ldots, x_d$.

1.3.1 The Functional Derivative

Now, the question arises how to find the minima of the functional E. In order to find the function h which minimizes a functional E, we can generalize the necessary condition for a minimum from ordinary vector calculus (Eq. (1.6)) to the calculus of variations. For this purpose, we need to extend the gradient from Eq. (1.5) to functionals by introducing the so-called functional derivative (or Fréchet derivative) [55]. The functional derivative describes the change of the functional E that occurs when the function h is varied in a certain point x. In the style of the above-mentioned partial derivative of a multivariate function, the functional derivative of a functional E with regard to the function h in point x is denoted as

$$\frac{\partial E(h)}{\partial h(x)} . \tag{1.17}$$

For a functional that is of the form

$$E(h) = \int F(h, h', x)\, dx , \tag{1.18}$$

i.e. the functional E maps an univariate function $h : \mathbb{R} \to \mathbb{R}$ to the field of real numbers $\mathbb{R}$ by integrating over a term F that depends on the function h, its derivative h' and its argument x, the functional derivative can be obtained as [23, 87, Chap. 10.3.2]

$$\frac{\partial E(h)}{\partial h(x)} = \frac{\partial F(h, h', x)}{\partial h} - \frac{\mathrm{d}}{\mathrm{d}x} \frac{\partial F(h, h', x)}{\partial h'} . \tag{1.19}$$

By setting Eq. (1.19) to zero, we obtain a necessary condition for a minimum of the functional E from Eq. (1.18):

$$\frac{\partial F(h, h', x)}{\partial h} - \frac{\mathrm{d}}{\mathrm{d}x} \frac{\partial F(h, h', x)}{\partial h'} = 0 . \tag{1.20}$$

Equation (1.20) is called the *Euler differential equation of the calculus of variations* [23, Chap. 10.3.2], or also *Euler-Lagrange equation* as it has the same structure as the *Lagrange equations of the second kind* in classical mechanics [87]. An exact solution of the *Euler-Lagrange equation* exists only for some simple problems [23, Chap. 10.5]. In most cases, one uses numerical methods to obtain the desired solution. The most straightforward way to find a minimum of the functional E is to perform functional gradient descent [13, 23, Chap. 10.5]:

$$h^{k+1}(x) = h^k(x) - \gamma^k \frac{\partial E(h^k)}{\partial h^k(x)} , \tag{1.21}$$

where $k \in \mathbb{N}$ denotes the k-th iteration and $\gamma^k \in \mathbb{R}$ is the step size in each iteration.

1.3.2 Differentiation Rules

For the functional derivative most of the differentiation rules from ordinary vector calculus apply [58]. In this thesis, we need the following three properties:

1. *The functional derivative is linear*: Let $F : \Gamma \to \mathbb{R}$ and $G : \Gamma \to \mathbb{R}$ be two functionals that map from a function space Γ to the set of real numbers $\mathbb{R}$ so that a function $h \in \Gamma$ is mapped to $F(h)$ and $G(h)$, respectively. Then, for two real constants $a \in \mathbb{R}$ and $b \in \mathbb{R}$, it holds that [96]

$$\frac{\partial(aF(h) + bG(h))}{\partial h(x)} = a\,\frac{\partial F(h)}{\partial h(x)} + b\,\frac{\partial G(h)}{\partial h(x)}\,. \tag{1.22}$$

2. *The product rule applies*: Let again $F : \Gamma \to \mathbb{R}$ and $G : \Gamma \to \mathbb{R}$ be two functionals that map from a function space Γ to the set of real numbers. Then, for the product of the two functionals the derivative is given as [58]

$$\frac{\partial(F(h)\,G(h))}{\partial h(x)} = \frac{\partial F(h)}{\partial h(x)}\,G(h) + F(h)\,\frac{\partial G(h)}{\partial h(x)}\,. \tag{1.23}$$

3. *The chain rule applies*: Let $F : \Gamma \to \mathbb{R}$ be a functional given on some function space Γ. Now, we consider an element $G : x \to G(x)$ in Γ which is again a functional depending on some function $k : y \to k(y)$. We further denote this as $G(k)(x)$. Then, via $k \to F(G(k))$, F becomes a functional that depends on the function k itself and the functional derivative of F with regard to k at point y is given as [49]

$$\frac{\partial F(G(k))}{\partial k(y)} = \int \frac{\partial F(G)}{\partial G(x)}\frac{\partial G(k)(x)}{\partial k(y)}\,dx\,. \tag{1.24}$$

Also, when we replace the functional $G(k)(x)$ by an ordinary function $g(k)$, the integral vanishes and we get [58]

$$\frac{\partial F(g(k))}{\partial k(y)} = \frac{\partial F(g)}{\partial g(k(y))}g'(k(y))\,. \tag{1.25}$$

1.3.3 Functionals of Functions with Multiple Arguments

A very useful extension of Eq. (1.19) allows us to determine the functional derivative of the functional E that has been defined in Eq. (1.16), i.e. a functional that maps a multivariate function $h : \mathbb{R}^d \to \mathbb{R}$ to the field of real numbers $\mathbb{R}$ by integrating a term F over a multi-dimensional domain $\Omega \subset \mathbb{R}^d$, where the term F depends on the function h, its partial derivatives $\frac{\partial h}{\partial x_1}, \ldots, \frac{\partial h}{\partial x_d}$ and its arguments $x_1, \ldots, x_d$. In this case, the functional derivative is given as [23, 87, Chap. 10.4.2]

$$\frac{\partial E(h)}{\partial h(\vec{x})} = \frac{\partial F}{\partial h} - \sum_{u=1}^{d} \frac{\partial}{\partial x_u} \frac{\partial F}{\partial \left(\frac{\partial h}{\partial x_u}\right)}. \tag{1.26}$$

1.3.4 Functionals of Multiple Functions

Another useful extension of Eq. (1.19) considers functionals E that depend on multiple scalar functions $h_1, \ldots, h_m$, with $h_i : \mathbb{R} \to \mathbb{R}$:

$$E(h_1, \ldots, h_m) = \int F(h_1, \ldots, h_m, h_1', \ldots, h_m', x)\, dx\,. \tag{1.27}$$

In this case, the partial functional derivative of the functional E with regard to the function h_i is obtained with the help of Eq. (1.19) to

$$\frac{\partial E(\vec{h})}{\partial h_i(x)} = \frac{\partial F(h_1, \ldots, h_m, h_1', \ldots, h_m', x)}{\partial h_i} - \frac{\mathrm{d}}{\mathrm{d}x} \frac{\partial F(h_1, \ldots, h_m, h_1', \ldots, h_m', x)}{\partial h_i'}, \tag{1.28}$$

where $\vec{h} = (h_1, \ldots, h_m)$ are all scalar functions combined in one vector-valued function $\vec{h} : \mathbb{R}^d \to \mathbb{R}^m$. Then, the functional derivative of the functional E with regard to the vector valued function $\vec{h}$ is given as the vector of the partial functional derivatives [23, 87, Chap. 10.3.5]:

$$\frac{\partial E(\vec{h})}{\partial \vec{h}(x)} = \left(\frac{\partial E(\vec{h})}{\partial h_1(x)}, \ldots, \frac{\partial E(\vec{h})}{\partial h_m(x)} \right), \tag{1.29}$$

and the necessary condition for a minimum $\left(\frac{\partial E(\vec{h})}{\partial \vec{h}(x)} = \vec{0}_m\right)$ leads to a system of m Euler-Lagrange equations, one for each function h_i:

$$\begin{gathered}
\frac{\partial F(h_1, \ldots, h_m, h_1', \ldots, h_m', x)}{\partial h_1} - \frac{\mathrm{d}}{\mathrm{d}x} \frac{\partial F(h_1, \ldots, h_m, h_1', \ldots, h_m', x)}{\partial h_1'} = 0 \\
\vdots \\
\frac{\partial F(h_1, \ldots, h_m, h_1', \ldots, h_m', x)}{\partial h_m} - \frac{\mathrm{d}}{\mathrm{d}x} \frac{\partial F(h_1, \ldots, h_m, h_1', \ldots, h_m', x)}{\partial h_m'} = 0\,.
\end{gathered} \tag{1.30}$$

1.3.5 The Functional Differential

The generalization of the directional derivative from ordinary vector calculus (c.f. Sect. 1.2.3) to the calculus of variations is called the functional differential (or Gâteaux derivative). It gives the rate of change of the functional $E(h)$ along a function $f \in \Gamma$:

$$dE(h; f) = \int \frac{\partial E(h)}{\partial h(x)} f(x)\, dx \tag{1.31}$$

1.4 Level Set Methods

Now, having defined the mathematical basics, another important point is the mathematical representation of moving curves and surfaces. This is important because later on in this thesis we will model the high-level shape information with the help of closed curves or surfaces, respectively. The most simple way is to represent curves and surfaces in a Lagrangian framework, i.e. as a set of connected points (c.f. Sect. 2.1). However, this simple description has some severe drawbacks when the curve or surface evolves over time. For example, one has to explicitly deal with self-intersections and topology changes.

In order to deal with these problems, the so-called *level set methods* have been introduced by Osher and Sethian in 1988 as a new way to describe moving interfaces [93]. Instead of describing moving curves and surfaces in a Lagrangian framework, Osher and Sethian proposed to describe them in an Eulerian framework as the *zero level set* of a higher-dimensional embedding function, the so-called *level set function*. Their technique is used in a variety of different applications, e.g. fluid dynamics, materials science, computational geometry, robot motion planning, and image processing [115]. We will introduce the level set function in Sect. 1.4.1. Afterwards, in Sect. 1.4.2, we will introduce the famous *level set equation* which can be used to evolve the level set function over time.

1.4.1 Level Set Function

One can represent a closed curve $C : [a, b] \to \Omega \subset \mathbb{R}^2$ with $C(a) = C(b)$ via the set of points in which a higher-dimensional embedding function $\Phi : \Omega \to \mathbb{R}$ equals zero:

$$C = \{\vec{x} \mid \Phi(\vec{x}) = 0\}\ . \tag{1.32}$$

The set defined in Eq. (1.32) is called the *zero level set* of the *level set function* Φ and hence C is also called *zero level curve* of the function Φ. Equation (1.32) also applies for closed surfaces in three dimensions or closed hypersurfaces in even higher dimensions.

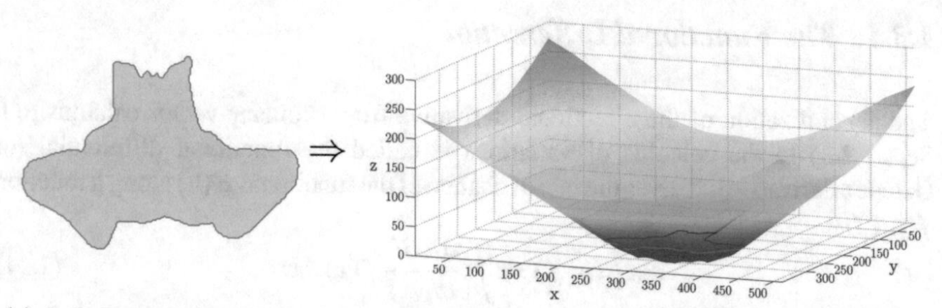

(a): Left: Boundary (red curve) and interior (blue region) of the paranasal sinuses. Right: Corresponding implicit level set representation with highlighted zero level set (red curve).

(b): Depiction of a zero level curve (black) undergoing topology changes when the red level set function is modified. The zero level plane is depicted in blue.
Source: Modified from Wikipedia (public domain).

Fig. 1.2 Exemplary level set representations of different curves

In general, one can represent a $(d-1)$-dimensional hypersurface C as the zero level set of a d-dimensional embedding function Φ (where $d=2$ for curves and $d=3$ for surfaces) [23, Chap. 17.1.4.1]. It is common to use the *signed distance function* as higher-dimensional embedding function. The signed distance function is defined in any element $\vec{x}$ of the domain Ω as the shortest Euclidean distance to the nearest point $\vec{p}$ on a closed curve C. Positive distances are used for elements $\vec{x}$ that reside outside the closed curve C and negative distances are used for elements $\vec{x}$ that reside inside the closed curve C, respectively[2]:

$$\Phi(\vec{x}) = \operatorname{sign}(\vec{x}) \min_{\vec{p} \in C} \|\vec{x} - \vec{p}\| , \tag{1.33}$$

where $\operatorname{sign}(\vec{x})$ takes the value -1 if the point $\vec{x}$ lies inside the closed curve C and $+1$ otherwise. An example for a signed distance representation of a closed curve can be seen in Fig. 1.2a. A major advantage of such an implicit zero level representation of a moving curve, in contrast to e.g. an explicit parametric curve representation, is that topology changes of the moving curve are handled implicitly by the level set function (c.f. Fig. 1.2b).

For all $\vec{x} \in C$, the gradient $\nabla\Phi(\vec{x})$ of the level set function Φ is always normal to the zero level curve C [115, Eq. (1.8)]. This is because the zero level set represents an isocontour of the level set function, and the gradient of a function is always

[2]Please note that the reverse definition is also widespread, i.e. positive distances are used for elements inside the closed curve and negative distances are used for elements outside the curve, respectively. However, the derivations from the following sections are based on the first mentioned definition.

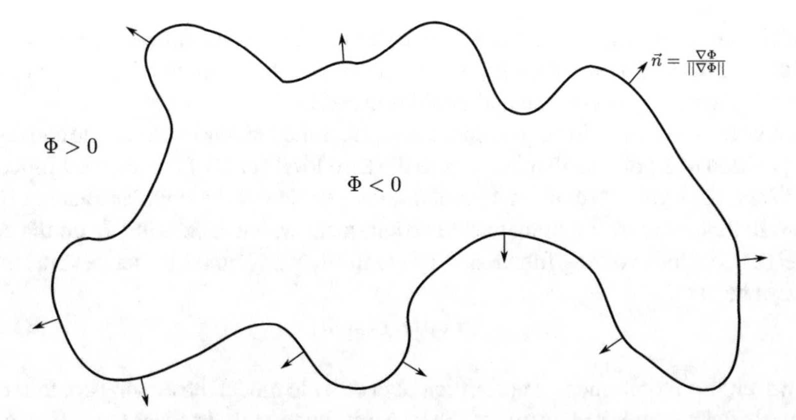

Fig. 1.3 Zero level set (*black curve*) and normalized gradient vectors (*black arrows*) of the level set function Φ. The normalized gradient vectors represent the outward pointing normal vectors of the curve

orthogonal to the isocontours of the function as it points in the direction of maximal function value increase. This can be seen when we assume that an isocontour C is given as a parameterized curve $C : \mathbb{R} \rightarrow \Omega$ with curve parameter α. As the values of the level set function Φ on the isocontour C are constant, the derivative $\frac{\mathrm{d}}{\mathrm{d}\alpha}\Phi(C(\alpha))$ of Φ along the curve is always zero. By the chain rule we get

$$\frac{\mathrm{d}}{\mathrm{d}\alpha}\Phi(C(\alpha)) = \langle\nabla\Phi(C(\alpha)), \frac{\mathrm{d}}{\mathrm{d}\alpha}C(\alpha)\rangle = 0\,. \tag{1.34}$$

This implies that the result of the dot product between the gradient of the level set function $\nabla\Phi(C(\alpha))$ and the tangent vector of the curve $\frac{\mathrm{d}}{\mathrm{d}\alpha}C(\alpha)$ is always zero. Thus, the gradient of the level set function has to be orthogonal to the tangent vector of the curve in each point on the curve.

As the gradient $\nabla\Phi$ always points in the direction of maximal function value increase, $\vec{n} = \frac{\nabla\Phi}{||\nabla\Phi||}$ represents the outward pointing normal when the values of the level set function inside the closed curve C are negative, and $\vec{n}$ represents the inward pointing normal when the values of the level set function inside the closed curve C are positive, respectively. This property is shown in Fig. 1.3. When Φ is a signed distance function, we additionally have $||\nabla\Phi|| = 1$ so that $\vec{n} = \nabla\Phi$ [115, Chap. 11.3].

1.4.2 Level Set Evolution

In this section, we want to answer the question how to evolve the level set function Φ over time in order to model moving contours. At first, one has to expand the domain of the level set functions by an additional time dimension in order to being able to model a time-varying level set function. By doing this, such a time-varying level set function is defined as $\Phi : \Omega \times \mathbb{R}_0^+ \to \mathbb{R}$, where $\mathbb{R}_0^+$ (the set of positive real numbers including zero) represents the additional time axis.

Now, to derive a motion equation that describes the moving contour, one considers the position of a point $\vec{x}_0$ that belongs to the zero level set C of the level set function Φ. When the level set function Φ evolves, the position of the point $\vec{x}_0$ changes over time. In this process, the fundamental requirement which each point $\vec{x}_0$ on the zero level set C of the evolving function Φ has to fulfill at any time t is that its value must always be zero:

$$\Phi(\vec{x}_0(t), t) = 0 . \tag{1.35}$$

However, this requirement is not sufficient in order to model the moving contour as it neglects the remaining elements of the level set function Φ. In order to obtain a consistent motion equation, the points $\vec{x}_l$ on every other level set $C_l = \{\vec{x}_l \mid \Phi(\vec{x}_l) = l\}$ have to move as well in an *appropriate* way.[3] So, Eq. (1.35) generalizes to

$$\Phi(\vec{x}_l(t), t) = l , \; \forall l \in \mathbb{R} . \tag{1.36}$$

Now, in order to find out how the level set function Φ evolves over time, one differentiates Eq. (1.36) with regard to t. By the chain rule, one obtains the following partial differential equation which describes the temporal variation of the level set function Φ:

$$\langle \nabla\Phi(\vec{x}(t), t), \frac{\mathrm{d}}{\mathrm{d}t}\vec{x}(t)\rangle + \frac{\mathrm{d}}{\mathrm{d}t}\Phi(\vec{x}(t), t) = 0 . \tag{1.37}$$

By moving the first term on the left hand side of Eq. (1.37) to the right hand side, it can be rewritten as

$$\frac{\mathrm{d}}{\mathrm{d}t}\Phi(\vec{x}(t), t) = -\langle \nabla\Phi(\vec{x}(t), t), \frac{\mathrm{d}}{\mathrm{d}t}\vec{x}(t)\rangle . \tag{1.38}$$

The second term in the inner product on the right hand side of Eq. (1.38) represents the change in location of a point $\vec{x}$ at time t. It can be interpreted as a time-varying velocity $\vec{v}(\vec{x}(t))$. When one assumes that the change in location of a point $\vec{x}$ depends only on the location of the point, but not on the time t, the time-varying velocity $\vec{v}(\vec{x}(t))$ can be simplified to a time-independent velocity $v(\vec{x})$.

Then, the right hand side of Eq. (1.38) is given as the dot product between the elements of a velocity field $\vec{v} : \Omega \to \mathbb{R}$ and the gradient $\nabla\Phi$ of the level set function. So, one can simplify Eq. (1.38) to

$$\frac{\mathrm{d}}{\mathrm{d}t}\Phi(\vec{x}(t), t) = -\langle \nabla\Phi(\vec{x}(t), t), \vec{v}(\vec{x})\rangle . \tag{1.39}$$

[3] Appropriate means for example that the remaining elements have to move in a way such that the level set function remains the signed distance function [115]. We will come back to this point in Sect. 1.4.3.

Now we have obtained a description of how the level set function Φ evolves under the influence of a velocity field $\vec{v} : \Omega \to \mathbb{R}$. This equation is also called *transport equation* as it is as well used to model how physical quantities are transported under the influence of the velocity field $\vec{v}$ [16]. When we expand Eq. (1.39) to

$$\frac{\mathrm{d}}{\mathrm{d}t}\Phi(\vec{x}(t),t) = -||\nabla\Phi(\vec{x}(t),t)||\,\langle\frac{\nabla\Phi(\vec{x}(t),t)}{||\nabla\Phi(\vec{x}(t),t)||},\vec{v}(\vec{x})\rangle\,, \tag{1.40}$$

it can be seen that the second term on the right hand side of Eq. (1.40) is the directional derivative of the level set function Φ along the velocity vector $\vec{v}$, i.e. the rate of change of the level set function along the vector $\vec{v}$ (c.f Eq. (1.7)). The other way around, this also means that only the fraction of the velocity which acts parallel to the normal of the evolving level curves has an influence on the level set evolution. This is not surprising as an evolution tangential to the normal would have no influence on the location of the level curves. Consequently, when we give the dot product from Eq. (1.39) in its magnitude/phase representation:

$$\frac{\mathrm{d}}{\mathrm{d}t}\Phi(\vec{x}(t),t) = -||\nabla\Phi(\vec{x}(t),t)||\,||\vec{v}(\vec{x})||\cos(\nabla\Phi(\vec{x}(t),t),\vec{v}(\vec{x}))\,, \tag{1.41}$$

the magnitude of the velocity normal to the level curves is given as

$$F(\vec{x}) = ||\vec{v}(\vec{x})||\cos(\nabla\Phi(\vec{x}(t),t),\vec{v}(\vec{x})) \tag{1.42}$$

and Eq. (1.39) can be further simplified to

$$\frac{\mathrm{d}}{\mathrm{d}t}\Phi(\vec{x}(t),t) = -||\nabla\Phi(\vec{x}(t),t)||\,F(\vec{x})\,. \tag{1.43}$$

This is the famous *level set equation* as defined by Osher and Sethian in [93]. It describes how the level curves of the level set function Φ evolve when a velocity with magnitude F is applied in normal direction.

1.4.3 Level Set Methods for Image Segmentation Problems

The level set framework can be used for image segmentation tasks. One of the earliest approaches has been proposed by Malladi et al. [86]. They defined the velocity

$$F = g\,(1-\beta\,\kappa) \tag{1.44}$$

so that the level set Eq. (1.43) becomes

$$\begin{aligned}\frac{\mathrm{d}}{\mathrm{d}t}\Phi &= -g\,(1-\beta\,\kappa)\,||\nabla\Phi|| \\ &= -g\,||\nabla\Phi|| + g\,\beta\,\kappa\,||\nabla\Phi||\,,\end{aligned} \tag{1.45}$$

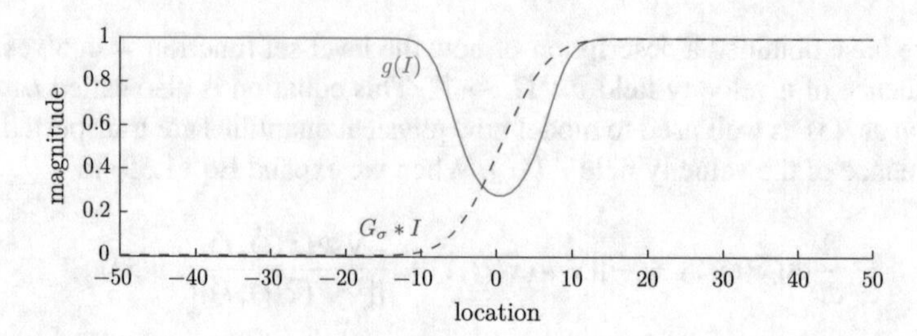

Fig. 1.4 Example of an edge-indicator function g: The smoothed (normalized) image values $G_\sigma * I$ are shown as a *dashed blue line* and the edge indicator function g is shown in *solid red*, respectively

where g is an edge indicator function, κ denotes the curvature of the zero level curve, and $\beta \in \mathbb{R}$ is a scalar weighting factor. The edge indicator function g is defined as

$$g(I) = \frac{1}{1 + ||\nabla(G_\sigma * I)||^2}, \tag{1.46}$$

where $I : \Omega \to \mathbb{R}$ denotes the image and G_σ is a Gaussian smoothing kernel with standard deviation σ.[4] An example of the edge-indicator function g can be seen in Fig. 1.4. The edge-indicator function approaches zero in image regions with large gradient magnitudes, which most likely indicate edges in the image, and it is close to one otherwise. This means that the first term on the right-hand side of Eq. (1.45) drives the zero level curve outwards with unit speed and it slows down in regions which are likely to contain image edges. Ideally, the zero level set expansion should stop at the image edges.

The second term on the right hand side of Eq. (1.45) describes a curve that moves under its curvature κ, which is also called *Euclidean curve shortening flow*. The curvature can be obtained as the divergence of the unit normal vector of the curve [115, Chap. 1.3]:

$$\kappa = \operatorname{div}\left(\frac{\nabla\Phi}{||\nabla\Phi||}\right) = \langle\nabla, \frac{\nabla\Phi}{||\nabla\Phi||}\rangle. \tag{1.47}$$

For $d = 2$, Eq. (1.47) reads [115, Eq. (1.9)]

$$\kappa = \frac{\frac{\partial^2\Phi}{\partial x^2}(\frac{\partial\Phi}{\partial y})^2 - 2\frac{\partial\Phi}{\partial x}\frac{\partial\Phi}{\partial y}\frac{\partial^2\Phi}{\partial xy} + \frac{\partial^2\Phi}{\partial y^2}(\frac{\partial\Phi}{\partial x})^2}{((\frac{\partial\Phi}{\partial x})^2 + (\frac{\partial\Phi}{\partial y})^2)^{\frac{3}{2}}}. \tag{1.48}$$

A curve that moves under its curvature tends to deform to a circle and then shrink to a single dot until it finally disappears (see Fig. 1.5). So, the *Euclidean curve shortening flow* acts as a regularizer which ensures that the zero level set stays as short as possible. Because of the multiplication with the edge indicator function g, the shrinking should stop when the moving contour approaches an image edge.

[4]The operator '$*$' in Eq. (1.46) denotes a convolution.

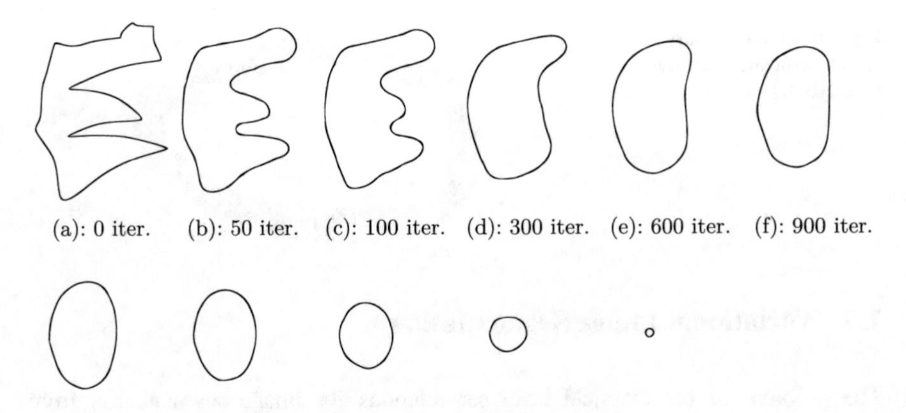

Fig. 1.5 Curve moving under the Euclidean curve shortening flow

Applying this motion equation not only shortens but also smoothes the curve, because large oscillations quickly disappear. The scalar weighting factor β can thereby be used to tune the influence of the regularization term on the final segmentation result. The larger the weighting factor β, the smoother the resulting curve.

What has not been mentioned so far is that the velocity F from Eq. (1.45) makes sense only for the zero level set of Φ as it determines how the zero level curve should evolve based on the available image information. However, as mentioned in Sect. 1.4.2, the derivation of the level set method assumes that the velocity field F is defined on the whole domain Ω. One way to address this issue is to construct a so-called *extension velocity* $\hat{F}$ which is defined as [86]

$$\hat{F}(\vec{x}) = \begin{cases} F(\vec{x}) & \text{, if } \vec{x} \in C \\ F(\vec{y}) & \text{, otherwise}, \end{cases} \tag{1.49}$$

where $\vec{y}$ denotes the nearest point in the zero level set:

$$\vec{y} = \arg\min_{\vec{p}} ||\vec{x} - \vec{p}||, \text{ with } \vec{p} \in C\,. \tag{1.50}$$

Another more common approach is to update the level set function Φ only in a small vicinity (usually a few pixels) around the zero level set – called the *Narrow Band* (NB) (see Fig. 1.6) – and to assume that the speed function F is valid in this region. When using this *narrow band level set method*, one has to deal with the fact that the level set function Φ remains unchanged outside the narrow band. This is commonly solved in such a way that the level set is evolved for a couple of timesteps. Then, before the zero level set leaves the narrow band, the level set function is re-initialized to a signed distance function [86]. This process is repeated until the final segmentation result has been obtained.

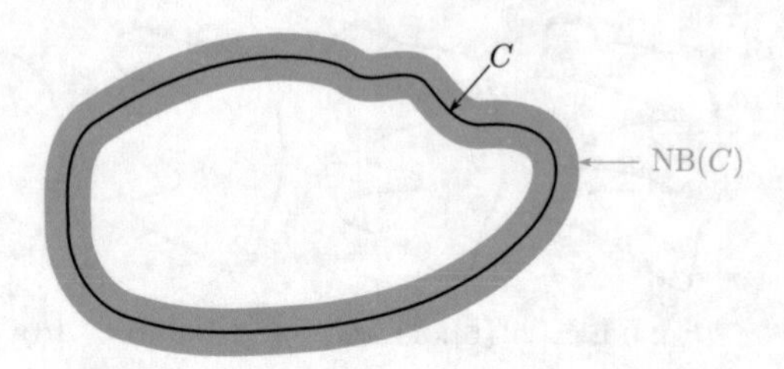

Fig. 1.6 Zero level curve C and corresponding narrow band $\mathrm{NB}(C)$

1.5 Variational Image Segmentation

The problem of the classical level set methods for image segmentation from Sect. 1.4.3 is that the used speed functions are heuristically motivated and they lack a sound mathematical foundation. For the discussed example from Sect. 1.4.3, this has the practical effect that the zero level curve tends to overrun weak image edges as the proposed edge-indicator function g never exactly reaches zero for these edges. In order to address the issue of heuristically motivated speed functions, *variational level set methods* have been proposed where the level set evolution is defined via the negative functional derivative of an appropriate energy functional $E(\Phi)$ [40]:

$$\frac{\mathrm{d}}{\mathrm{d}t}\Phi(\vec{x}) = -\frac{\partial E(\Phi)}{\partial \Phi(\vec{x})} . \tag{1.51}$$

The functional $E(\Phi)$ exactly describes the segmentation problem. This has the advantage that the solution of the problem can be obtained via functional gradient descent (c.f. Eq. (1.21)):

$$\Phi^{k+1}(\vec{x}) = \Phi^{k}(\vec{x}) - \gamma^{k}\frac{\partial E(\Phi^{k})}{\partial \Phi^{k}(\vec{x})} , \tag{1.52}$$

where k denotes the iteration and γ^k denotes the stepsize in each iteration, respectively.

In the following two subsections, we will shortly introduce two famous variational level set approaches, the *Geodesic Active Contours* approach and the *Active Contours Without Edges* approach, respectively, as they will become important later throughout this thesis.

1.5.1 Geodesic Active Contours

The *Geodesic Active Contours* level set evolution has first been defined by Caselles et al. in 1997 [25] and has obtained a lot of attention since then. With currently 5183 citations according to Google Scholar (as of May 7, 2015), it is one of the most cited papers in the level set literature. It provides a sound mathematical foundation for a moving curve C that is attracted by object boundaries which are characterized by image edges (similar to the approach discussed in Sect. 1.4.3), and embeds this contour evolution in the level set framework from Sect. 1.4. In 2005, Li et al. further extended this approach by providing a variational formulation directly for the level set function Φ instead of the curve C. Their energy is given as [78]

$$E(\Phi) = \int_{\Omega} \delta_{\Phi(\vec{x})}\, g(I(\vec{x}))||\nabla\Phi(\vec{x})||\, d\vec{x}\,, \tag{1.53}$$

where δ_Φ denotes the Dirac delta function that is nonzero only at the zero-crossings of Φ, and g is the edge-indicator function defined in Eq. (1.46).

It can be seen that the energy is minimal when an as short as possible zero level set is located at object edges that are characterized by strong gradients. The functional derivative of Eq. (1.53) can be obtained as [78]

$$\frac{\partial E(\Phi)}{\partial \Phi(\vec{x})} = \delta_\Phi \left[-\langle \nabla g, \frac{\nabla\Phi}{||\nabla\Phi||} \rangle - g \operatorname{div}\left(\frac{\nabla\Phi}{||\nabla\Phi||} \right) \right] . \tag{1.54}$$

The first term on the right hand side of Eq. (1.54) is called *advection term* as it exerts a force on the zero level set that pulls it in the direction of the nearest image edge (c.f. Sect. 3.3.4.2), and the second term on the right hand side is the Euclidean curve shortening flow which is also part of Eq. (1.45).

In order to counteract the shrinking effect of the Euclidean curve shortening flow or to speed up the segmentation process, it is common to extend Eq. (1.53) with another term that acts as an *area constraint* [78]:

$$E(\Phi) = \int_{\Omega} \delta_{\Phi(\vec{x})}\, g(I(\vec{x}))||\nabla\Phi(\vec{x})||\, d\vec{x} + \nu \int_{\Omega} g(I(\vec{x})) H(-\Phi(\vec{x}))\, d\vec{x}\,, \tag{1.55}$$

where $\nu \in \mathbb{R}$ is a real weighting factor and H denotes the Heaviside step function.

The functional derivative of Eq. (1.55) can be obtained as [78]

$$\begin{aligned} \frac{\partial E(\Phi)}{\partial \Phi(\vec{x})} &= \delta_\Phi \left[-\langle \nabla g, \frac{\nabla\Phi}{\|\nabla\Phi\|} \rangle - g \operatorname{div}\left(\frac{\nabla\Phi}{\|\nabla\Phi\|} \right) - \nu g \right] \\ &= \delta_\Phi \left[-\langle \nabla g, \frac{\nabla\Phi}{\|\nabla\Phi\|} \rangle - g \left[\operatorname{div}\left(\frac{\nabla\Phi}{\|\nabla\Phi\|} \right) + \nu \right] \right] . \end{aligned} \tag{1.56}$$

It can be seen that the *area constraint* acts as a *balloon force* that inflates the moving contour when ν is smaller than zero and that deflates the moving contour when ν is greater than zero by subtracting or adding the constant value ν to the level set function Φ. Again, the multiplication with the edge indicator function g has the effect that the expansion or shrinkage is stopped when the moving contour approaches an image edge. So, choosing a value of ν smaller than zero can counteract the shrinkage effect of the Euclidean curve shortening flow and choosing a value of ν greater than zero can speed up the segmentation by amplifying the shrinkage effect when the zero

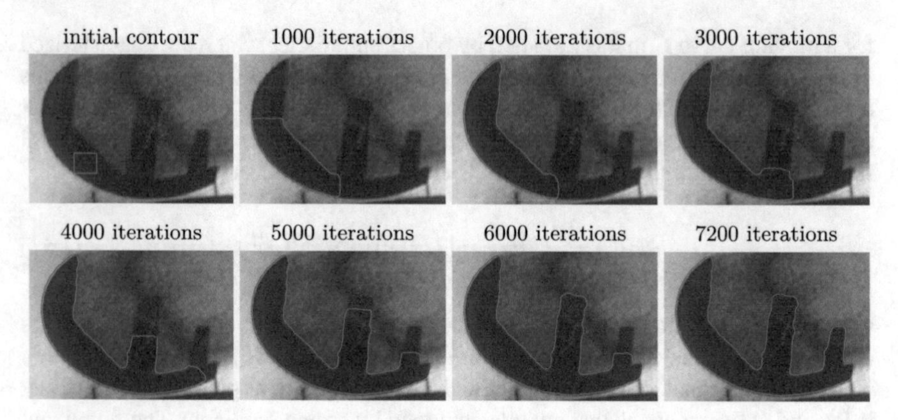

Fig. 1.7 Exemplary level set segmentation of an implanted knee prosthesis

level set is initialized outside the object of interest. For $\nu = 1$, the *area constraint* corresponds to the first term on the right hand side of Eq. (1.45).

Another common practice is to extend the energy from Eq. (1.56) with scalar weighting factors for both terms on the right hand side so that their influence can be controlled [65]. It can easily be shown that this extension results in a functional derivative of the form

$$\frac{\partial E(\Phi)}{\partial \Phi(\vec{x})} = \delta_\Phi \left[-\alpha \left\langle \nabla g, \frac{\nabla \Phi}{||\nabla \Phi||} \right\rangle - \beta g \operatorname{div}\left(\frac{\nabla \Phi}{||\nabla \Phi||} \right) - \nu g \right], \tag{1.57}$$

where the influence of the individual terms on the curve evolution can be managed with the three parameters $\alpha \in \mathbb{R}$, $\beta \in \mathbb{R}$, and $\nu \in \mathbb{R}$.

So, the variational derivation of a level set evolution equation for edge-based image segmentation has led to an equation that is similar to Eq. (1.45) but has one important difference: the *advection term*. This additional term exerts a force in the direction of the image edges so that an overrun of weak image edges is greatly reduced. We will pick up on this point later in Sect. 3.3.4.2.

An example for the segmentation of an implanted knee prosthesis [8] with the described approach can be seen in Fig. 1.7. The zero level curve is initialized as a square inside the object of interest. Then it is evolved with the help of Eq. (1.57) until it approaches the edges of the prosthesis. The standard deviation of the Gaussian smoothing kernel in the definition of g has been chosen to $\sigma = 1.0$ and the weighting factors have been chosen to $\alpha = \beta = 5.0$ and $\nu = 0.4$.

1.5.2 Active Contours Without Edges

So far, we have only discussed edge-based image segmentation approaches. However, with variational level set methods it is also possible to segment an image into two distinctive regions that must not necessarily be separated by large image gradients.

Such an approach is also called region-based segmentation as it relies only on the statistical information about the two regions but not on information about the boundary that separates them. The most famous region-based variational level set segmentation approach, with currently 7361 citations according to Google Scholar (as of June 30, 2015), is the *Active Contours Without Edges* approach by Chan and Vese [26].

The *Active Contours Without Edges* approach tries to segment an image into two regions with approximately constant intensities c_1 and c_2. The corresponding energy functional is given as

$$\begin{aligned} E(\Phi) = \lambda_1 \int_\Omega (I - c_1)^2 H(-\Phi)\, d\vec{x} + \lambda_2 \int_\Omega (I - c_2)^2 H(\Phi)\, d\vec{x} \\ +\mu \int_\Omega \delta_\Phi ||\nabla \Phi||\, d\vec{x} + \nu \int_\Omega H(-\Phi)\, d\vec{x}\,, \end{aligned} \tag{1.58}$$

where λ_1, λ_2, μ, and ν denote real weighting factors, and δ_Φ again denotes the Dirac delta function which is nonzero only at the zero-crossings of Φ. It is given as the derivative of the Heaviside function $H(\Phi)$.

The first term on the right hand side of Eq. (1.58) penalizes the total squared deviation from the image intensities inside the zero level curve to the constant intensity c_1 and the second term penalizes the total squared deviation from the image intensities outside the zero level curve to the constant intensity c_2. Hence, these two terms become minimal when the intensity values inside the closed curve mostly resemble c_1 and the intensity values outside the closed curve mostly resemble c_2, respectively.

The other two terms are again regularization terms that favor a short curve (third term) or a curve with a small interior (fourth term). Now, one can obtain the functional derivative of Eq. (1.58) as [26]

$$\frac{\partial E(\Phi)}{\partial \Phi(\vec{x})} = \delta_\Phi \left[-\lambda_1 (I - c_1)^2 + \lambda_2 (I - c_2)^2 - \mu \,\mathrm{div}\left(\frac{\nabla \Phi}{||\nabla \Phi||} \right) - \nu \right]. \tag{1.59}$$

Along the various terms on the right hand side of Eq. (1.59), the third and the fourth term are the already known *Euclidean curve shortening flow* and the *balloon force*, which both act as a regularizer on the moving curve. The first two terms are the so-called *fitting terms* as they drive the zero level contour inwards when the intensity on the zero level contour differs from c_1 and outwards when the intensity on the zero level contour differs from c_2, respectively.

Chapter 2
Statistical Shape Models (SSMs)

The automated extraction of objects from two-dimensional or three-dimensional image data is a complicated task that is not easy to solve. This is especially the case in medical image segmentation problems where the goal often is to extract highly variable anatomical structures from relatively low-quality image data [59]. The image data thereby originates for example from approaches that produce 2D images of the desired anatomy, like conventional X-ray imaging, or from newer approaches that produce 3D volumetric images of the anatomy under consideration, like e.g. *Computed Tomography* (CT) scans or *Magnetic Resonance Imaging* (MRI). Some exemplary segmentation problems will be discussed in Chap. 4. They include the automated extraction of the paranasal sinuses (Sect. 4.1) and the bones in the human knee (Sect. 4.2), which are both needed for surgical planning, but also the extraction of faces from range scan data (Sect. 4.3) is an increasingly popular application as it is needed for real time facial animation [79] or face recognition [20].

For problems of the above-mentioned kind, the available image or range information often does not suffice in order to extract the desired structures due to artifacts in the data, missing data, or poor contrast [59]. As a consequence, high-level information is needed in order to replace the missing information and to distinguish between correct and erroneous information. Two examples can be seen in Fig. 2.1. Without anatomical knowledge it is hard to identify the paranasal sinuses in the CT slice in Fig. 2.1a. However, it is easy to fill the holes of the disturbed range scan of a human face in Fig. 2.1b as all humans have a clear understanding of how a face should look like. Likewise, a medical expert (e.g. a surgeon) has no problem outlining the paranasal sinuses as he or she possesses the required anatomical knowledge.

Now, the question is how this high-level knowledge can be integrated into automatic object extraction approaches. This is where the so-called *Statistical Shape Models* (SSMs) come into play: When many hand-segmentations of the desired object class (e.g. the paranasal sinuses) or many complete range scans of the desired object class (e.g. human faces) are available, one can use this information in order to extract objects from the same object class in newly acquired data.

There exist many different variants of statistical shape models. For an extensive overview we refer the reader to the review article by Heimann and Meinzer [60].

C. Last, *From Global to Local Statistical Shape Priors*, Studies in Systems, Decision and Control 98, DOI 10.1007/978-3-319-53508-1_2

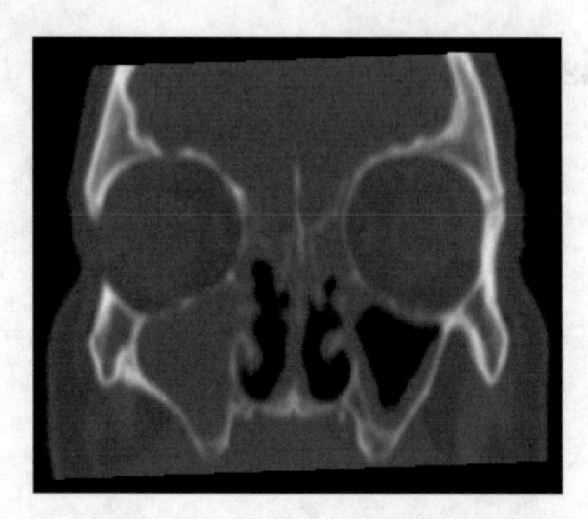

(a): CT cross-section of a human head.

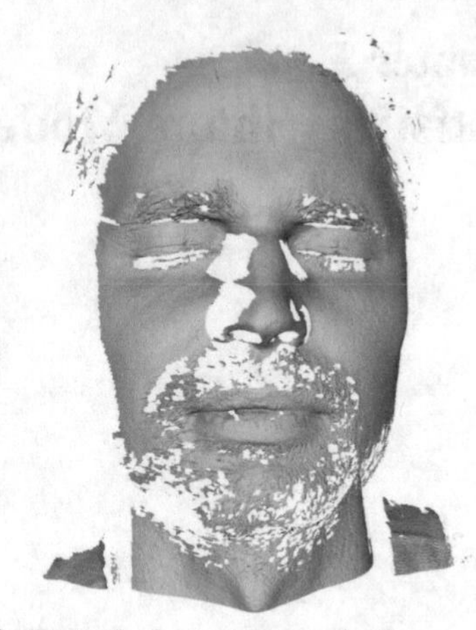

(b): Disturbed range scan of a human face.

Fig. 2.1 Exemplary CT cross-section (**a**) and range scan (**b**) of a human head

In the following sections, we will first concentrate on the so-called *Point Distribution Model* (PDM) that has been introduced by Cootes et al. in [32] as it is probably the best-known method in the area of statistical shape models [60]. Afterwards, in Sect. 2.3, we will discuss an extension of the PDM that works with implicit, level set-based shape representations, and we will deal with the problem of rigidly aligning the training shapes in Sect. 2.4. In Sect. 2.5, we will address the problem of limited training data, and in Sect. 2.5 we will finally mention some previous approaches by other researchers who tried to address this fundamental problem.

2.1 Definition of Shape

In the introduction to this chapter, we have explained that we are interested in introducing high-level information into object extraction problems by modeling the statistical shape properties of a particular object class. So, before we can continue, we need to define what we mean when we speak about the *shape* of an object. A common general definition of the term *shape* is as follows [64, p. 1, Definition 1.1]:

> Shape is all the geometrical information that remains when location, scale and rotational effects are removed from an object.[1]

This definition has been originally coined by Kendall [70] back in 1977.

More specifically, when we refer to the geometrical information of an object, we mean the silhouette of an object, i.e. the geometric outline of an object in 2D or 3D [37]. This geometric outline can be represented in a variety of different ways,

[1]Reprinted from [64, p. 1, Definition 1.1], © 2016 John Wiley and Sons Ltd.

like. e.g. splines [18] or fourier descriptors [122]. However, the two most popular representations are polygons (in 2D or equivalently polygon meshes in 3D) [60] and the zero level set of a higher-dimensional level set function that has been introduced in Sect. 1.4.1 [40].

A two-dimensional shape $\vec{C}$ that is represented by a polygon is completely defined by an ordered list of 2D points

$$\vec{C} = (x_1, y_1, \ldots, x_r, y_r), \tag{2.1}$$

together with the conventions that each point (x_i, y_i) is connected by a straight line to its successor (x_{i+1}, y_{i+1}) and that the point (x_r, y_r) is connected by a straight line to the point (x_1, y_1). This last convention may optionally be dropped in order to allow *open* shapes. Similarly, a three-dimensional shape $\vec{C}$ that is represented by a polygon mesh is completely defined by an ordered list of 3D points

$$\vec{C} = (x_1, y_1, z_1, \ldots, x_r, y_r, z_r), \tag{2.2}$$

together with a set of connectivity rules that define which points (x_i, y_i, z_i) have to be connected by straight lines in order to obtain a surface that is composed out of many polygons of a specific type (e.g. triangles or quadrilaterals). As in Eqs. (2.1) and (2.2) the shape $\vec{C}$ is directly defined by a list of points, we refer to these representations as *explicit* shape representations. Some examples can be seen in Fig. 2.2.

In contrast to the above-mentioned *explicit* shape representations, the zero level set of a higher-dimensional level set function is called an *implicit* shape representation as the shape C is indirectly obtained as the set of points for which a higher-dimensional embedding function $\Phi : \Omega \to \mathbb{R}$ equals zero. This has been defined in Eq. (1.32) as

$$C = \{\vec{x} \mid \Phi(\vec{x}) = 0\},$$

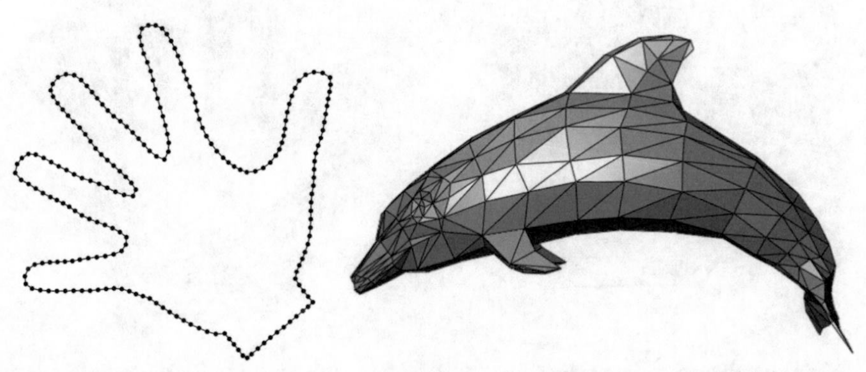

(a): Polygon representing a hand. (b): Triangle mesh representing a dolphin.

Fig. 2.2 Exemplary explicit shape representations of a hand (**a**) and a dolphin (**b**). *Source of subfigure* (b): Wikipedia (public domain)

where $\vec{x} \in \mathbb{R}^d$ and $\Omega \subset \mathbb{R}^d$. The variable d denotes the dimension of the object so that typical values for d are 2 or 3. As already mentioned in Sect. 1.4.1, it is common to use the signed distance function as the higher-dimensional embedding function for the shape C. However, other functions are also possible [41]. For an exemplary depiction of such an implicit shape see Fig. 1.2a.

2.2 An Explicit Linear Parametric Statistical Shape Model

Now that we have defined the term shape and provided two methods to represent a shape, we continue by introducing the *Point Distribution Model* (PDM) by Cootes et al. In order to construct the PDM, one needs first of all a set of (generally hand-labeled) training shapes $\{\vec{C}_1, ..., \vec{C}_n\}$ that are given in the explicit shape representation discussed above. All these shapes must belong to the same shape class, which means that all training samples represent different shapes of the same object, e.g. a hand (c.f. Fig. 2.3). Additionally, all the training shapes $\vec{C}_i$ have to be in correspondence to each other. This involves that all training shapes must be represented by the same amount of points and that all points have to be located at corresponding positions along all training shapes. Furthermore, we demand that all training shapes are aligned to each other with regard to rotation, translation, and scale as this information does not belong to the shape information of a specific shape class according to the definition in Sect. 2.1. How this alignment can be achieved will be discussed in Sect. 2.4. An example of 40 aligned training shapes for the hand shape class, where each shape

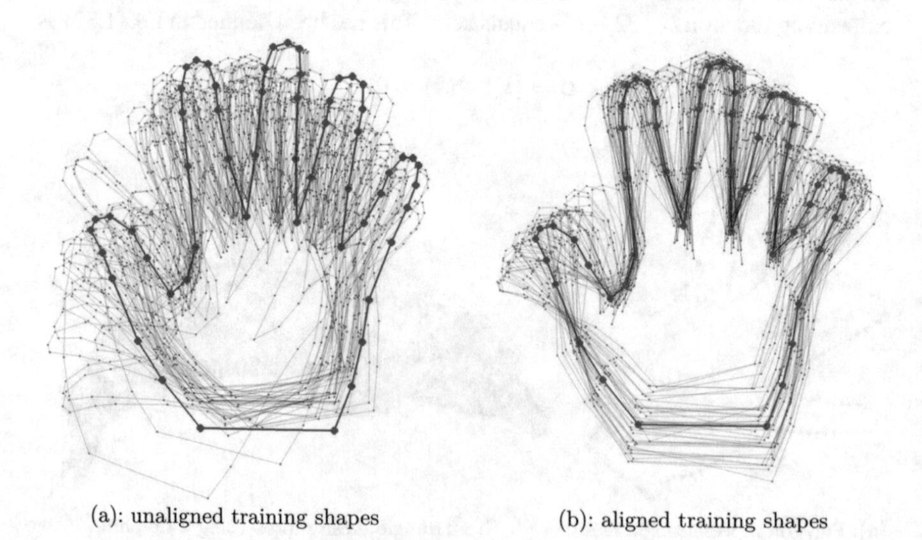

Fig. 2.3 Mean shape (*blue*) overlaid with the trainings shapes before (**a**) and after (**b**) they have been rigidly aligned with regard to rotation, translation, and scale. The shapes originate from the database that is described in [123]

is represented by 56 points, can be seen in Fig. 2.3b. The shapes originate from the database that is described in [123].

When all above-mentioned prerequisites are met, the simplest way to model new shapes is to allow linear combinations of the training shapes [19]:

$$\vec{C}_{\text{bary}}(\vec{w}) = \sum_{i=1}^{n} w_i \vec{C}_i\,, \text{ with } \sum_{i=1}^{n} w_i = 1\,. \tag{2.3}$$

In Eq. (2.3), the vector $\vec{w} = (w_1, \ldots, w_n)$ is called the parameter vector (or weight vector) of the model $\vec{C}_{\text{bary}}$. In this case, the parameters $(w_1, \ldots, w_n)$ are know as normalized *barycentric coordinates* [119], and the parameters $(\frac{1}{n}, \ldots, \frac{1}{n})$ denote the barycenter (or balance point) of the training shapes. The barycenter is identical to the sample mean of the training shapes:

$$\vec{C}_{\text{bary}}\left(\frac{1}{n}, \ldots, \frac{1}{n}\right) = \sum_{i=1}^{n} \frac{1}{n} \vec{C}_i = \frac{1}{n} \sum_{i=1}^{n} \vec{C}_i = \bar{\vec{C}}\,. \tag{2.4}$$

This representation allows to quickly generate new shapes from a set of training shapes. However, it has the drawbacks that linear dependencies between the training shapes are not considered and that it contains no information about the likelihood of a modeled shape $\vec{C}_{\text{bary}}(\vec{w})$.

In order to consider the statistical distribution of the training shapes, it is often reasonable to assume that all shapes of a specific shape class are distributed according to a multivariate Gaussian distribution with mean $\vec{\mu} \in \mathbb{R}^{dr}$ and covariance $\Sigma \in \mathbb{R}^{dr \times dr}$, where d is the dimension of the training shapes (i.e. $d \in \{2, 3\}$) and r are the number of points on each shape (c.f Eqs. (2.1) and (2.2)). The Gaussian distribution of likely shapes $\vec{C}$ is then given as

$$\vec{C} \sim \mathcal{N}(\vec{\mu}, \Sigma) = \frac{1}{\sqrt{(2\pi)^{dr}|\Sigma|}} \exp\left(-\frac{1}{2}(\vec{C} - \vec{\mu})\Sigma^{-1}(\vec{C} - \vec{\mu})^T\right). \tag{2.5}$$

In Eq. (2.5), an estimate of the mean $\vec{\mu}$ is given by the sample mean $\bar{\vec{C}}$ from Eq. (2.4) and an estimate $\hat{\Sigma}$ of the covariance matrix Σ can be obtained as detailed in appendix D. Now, one can assign a probability to each shape of the model from Eq. (2.3) via

$$P\left(\vec{C}_{\text{bary}}(\vec{w})\right) = \frac{1}{\sqrt{(2\pi)^{dr}|\hat{\Sigma}|}} \times \exp\left(-\frac{1}{2}\left(\vec{C}_{\text{bary}}(\vec{w}) - \bar{\vec{C}}\right)\hat{\Sigma}^{-1}\left(\vec{C}_{\text{bary}}(\vec{w}) - \bar{\vec{C}}\right)^T\right). \tag{2.6}$$

The drawback of Eq. (2.6) is that it contains a lot of redundant information. As mentioned above, the sample covariance matrix has the dimensions $dr \times dr$. However, only n training shapes have been used to estimate the sample covariance matrix, where typically $n \ll dr$. In this case, it follows from the construction of the sample covariance matrix that it has at most rank $n - 1$ [23, Eq. (4.26f)]. So, a multivariate Gaussian distribution with a dimension less or equal to $(n - 1)$ would suffice in order to model the data statistics.

This low-dimensional Gaussian distribution can be obtained by applying the so-called *Principal Component Analysis* (PCA) [66] to the training shapes. The PCA is mathematically defined as an orthogonal transformation that maps the data to a new coordinate system in which most variation of the data occurs along the first axis, the second-most variation occurs along the second axis, and so on. The axes of the new coordinate system can be obtained as the eigenvectors of the sample covariance matrix $\hat{\Sigma}$. They are called *principal axes* or also *main modes of variation* and we denote them as $\{\tilde{\vec{C}}_1, \ldots, \tilde{\vec{C}}_m\}$. It is $\tilde{\vec{C}}_i \in \mathbb{R}^{dr}$ and $m \leq (n - 1)$ as the sample covariance matrix has at most $(n - 1)$ nonzero eigenvalues. The eigenvalues $\{\sigma_1^2, \ldots, \sigma_m^2\}$ of the sample covariance matrix $\hat{\Sigma}$ correspond to the variances along the principal axes and are called *principal components*. They have to be sorted in descending order $\sigma_1^2 \geq \sigma_2^2 \geq \cdots \geq \sigma_m^2$ in order to be in accordance with the definition of the PCA (i.e. the most variation of the data occurs along the first principal axis etc.).

By stacking the eigenvectors $\tilde{\vec{C}}_i$ of the sample covariance matrix on top of each other, one obtains a matrix

$$\tilde{\mathbf{C}} = \begin{pmatrix} \tilde{\vec{C}}_1 \\ \vdots \\ \tilde{\vec{C}}_m \end{pmatrix}, \tag{2.7}$$

the rows of which define the linear subspace that is spanned by the training shapes. We denote this subspace as *space of feasible shapes*. The representations of the training shapes $\vec{C}_i$ in this low-dimensional space of feasible shapes are given by the following coordinate transformation:

$$\vec{w}_i = \left(\vec{C}_i - \bar{\vec{C}}\right) \tilde{\mathbf{C}}^T, \tag{2.8}$$

and the distribution of the training shapes in the m-dimensional subspace $\tilde{\mathbf{C}}$ is defined by the following multivariate Gaussian distribution:

$$P(\vec{w}) = \frac{1}{\sqrt{(2\pi)^m |\tilde{\Sigma}|}} \exp\left(-\frac{1}{2} \vec{w} \tilde{\Sigma}^{-1} \vec{w}^T\right), \tag{2.9}$$

where $\tilde{\Sigma}$ is a diagonal matrix composed of the m most important eigenvalues $\{\sigma_1^2, \ldots, \sigma_m^2\}$ of the sample covariance matrix $\hat{\Sigma}$:

$$\tilde{\Sigma} = \begin{pmatrix} \sigma_1^2 & 0 & \dots & 0 \\ 0 & \sigma_2^2 & \dots & 0 \\ \vdots & \vdots & \ddots & \vdots \\ 0 & 0 & \dots & \sigma_m^2 \end{pmatrix}. \tag{2.10}$$

The inverse transformation that transforms the training shapes from the low-dimensional subspace back to the original shape space also exists and is given as

$$\vec{C}_i = \bar{\vec{C}} + \vec{w}_i \tilde{\mathbf{C}}. \tag{2.11}$$

The space of feasible shapes must not necessarily be spanned by all eigenvectors of the sample covariance matrix. When we recall the definition of the PCA, we remind us that the linear PCA subspace is constructed in such a way that the more axes are added to the subspace, the more variance in the data can be explained. So, when all eigenvectors of the sample covariance matrix are used to define the linear PCA subspace, all variance in the training data is represented in the subspace. However, those eigenvectors which have only small eigenvalues most likely correspond to noise in the training data which should not be represented in the space of feasible shapes. So, one typically chooses only the $m < n - 1$ most important eigenvectors to define the space of feasible shapes. An example of the shape space can be seen in Fig. 2.4. All training shapes from Fig. 2.3 have been projected with the help of Eq. (2.8) onto the $m = 2$ most important modes of variation. The estimated Gaussian weight distribution is also shown.

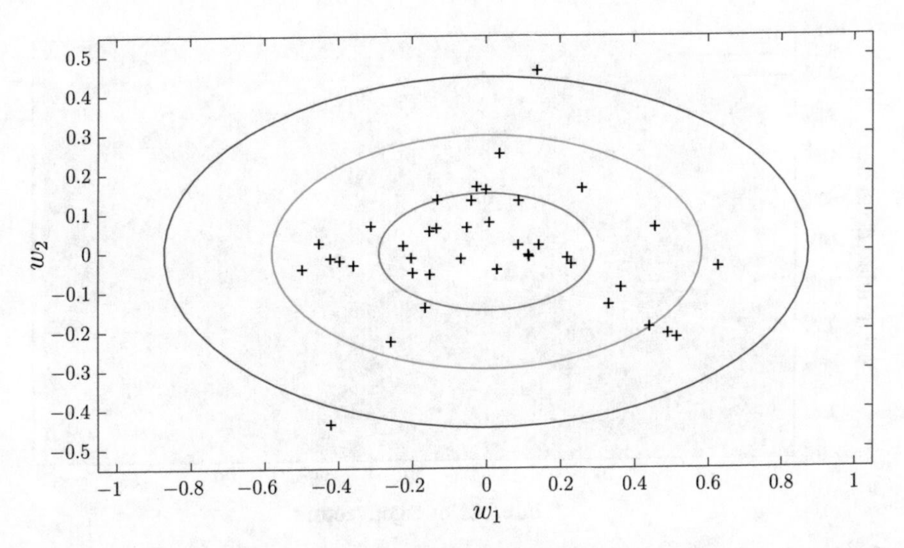

Fig. 2.4 Distribution of the training shapes (*black crosses*), projected onto the two main modes of variation. The ellipses at 1, 2, and 3 standard deviations of the corresponding Gaussian distribution are shown in *red*, *green*, and *blue*

In the case that fewer than available eigenvectors are used to define the shape space, Eq. (2.8) yields the orthogonal projection of the training shape C_i into the space of feasible shapes, i.e. the vector $\vec{w}_i$ denotes the point inside the low-dimensional space of feasible shapes which has the shortest euclidean distance to the original shape C_i:

$$\left\| \left(\bar{\vec{C}} + \vec{w}_i \tilde{\mathbf{C}} \right) - \vec{C}_i \right\| \rightarrow \min\ . \tag{2.12}$$

There exist different criteria to determine which eigenvectors are important. Each eigenvector $\tilde{\vec{C}}_i$ represents a certain amount of the total variance in the training shapes. It can be calculated with the help of the corresponding eigenvalue σ_i^2 to

$$\frac{\sigma_i^2}{\sum_{u=1}^{n-1} \sigma_u^2}\ , \tag{2.13}$$

where $\sum_{u=1}^{n-1} \sigma_u^2$ gives the total variance in the training shapes.

Consequently, one common way to define the shape space is to keep only as much eigenvectors until the accumulated variance of the m most important eigenvectors exceeds a fixed percentage (typically 90 to 98%) of the total variance in the training data [60]. The cumulative variance of the hand shapes can be seen in Fig. 2.5.

Another criterion is to search for an *elbow*, i.e. a characteristic drop, in the scree graph. In a scree graph, the abscissa gives the index i of the eigenvalue and the ordinate gives its value σ_i^2. As the eigenvalues are given in descending order, the

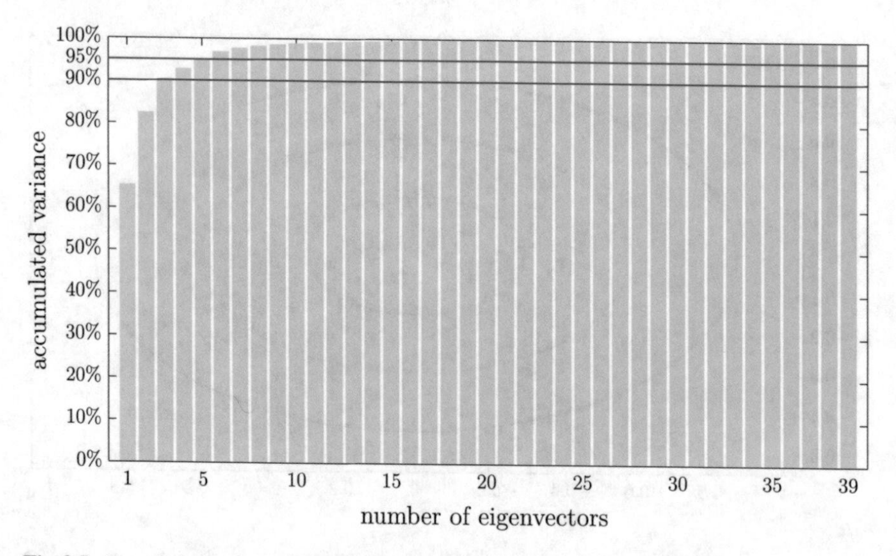

Fig. 2.5 Cumulative variance for a growing number of eigenvectors. Three eigenvectors are enough to account for 90% of the total variation and five eigenvectors capture roughly 95% of the total variation, respectively

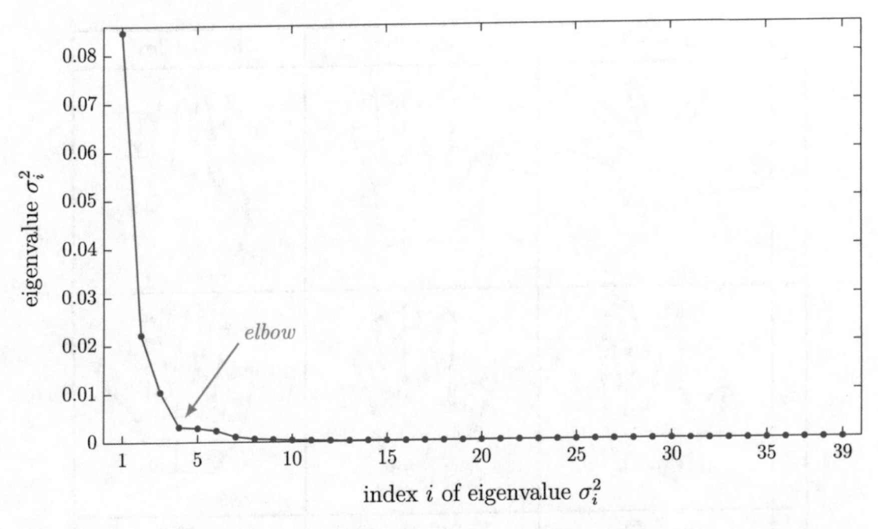

Fig. 2.6 Scree graph, showing the eigenvalues σ_i^2 in descending order of importance

scree graph is monotonically decreasing, where the decrease between the first few eigenvalues is typically larger than the decrease between the last few eigenvalues. Now, all eigenvalues left of and including the elbow are considered as important and the remaining eigenvalues are considered as unimportant. The basic idea of this is that the eigenvectors which correspond to noise in the training data are assumed to have approximately the same variance whereas the eigenvectors which correspond to meaningful variations are assumed to have different variances, respectively [66]. The scree graph for the hand shapes can be seen in Fig. 2.6.

Now, having identified the space of feasible shapes, the *Point Distribution Model* (PDM) by Cootes et al. is defined as

$$\vec{C}_{\text{PDM}}(\vec{w}) = \bar{\vec{C}} + \sum_{i=1}^{m} w_i \tilde{\vec{C}}_i \,, \tag{2.14}$$

i.e. all the modeled shapes $\vec{C}_{\text{PDM}}(\vec{w})$ are obtained as a linear combination of the m most important eigenvectors $\tilde{\vec{C}}_i$ of the sample covariance matrix (the *main modes of variation*) centered around the mean shape $\bar{\vec{C}}$. So, similar to the barycentric shape model in Eq. (2.3), the vector $\vec{w} = (w_1, \ldots, w_m)$ is the parameter vector (or weight vector) of the model $\vec{C}_{\text{PDM}}(\vec{w})$ as it controls the influence that each main mode of variation has on the modeled shape. The shape variations that occur when varying the five most important shape weights are shown in Fig. 2.7.

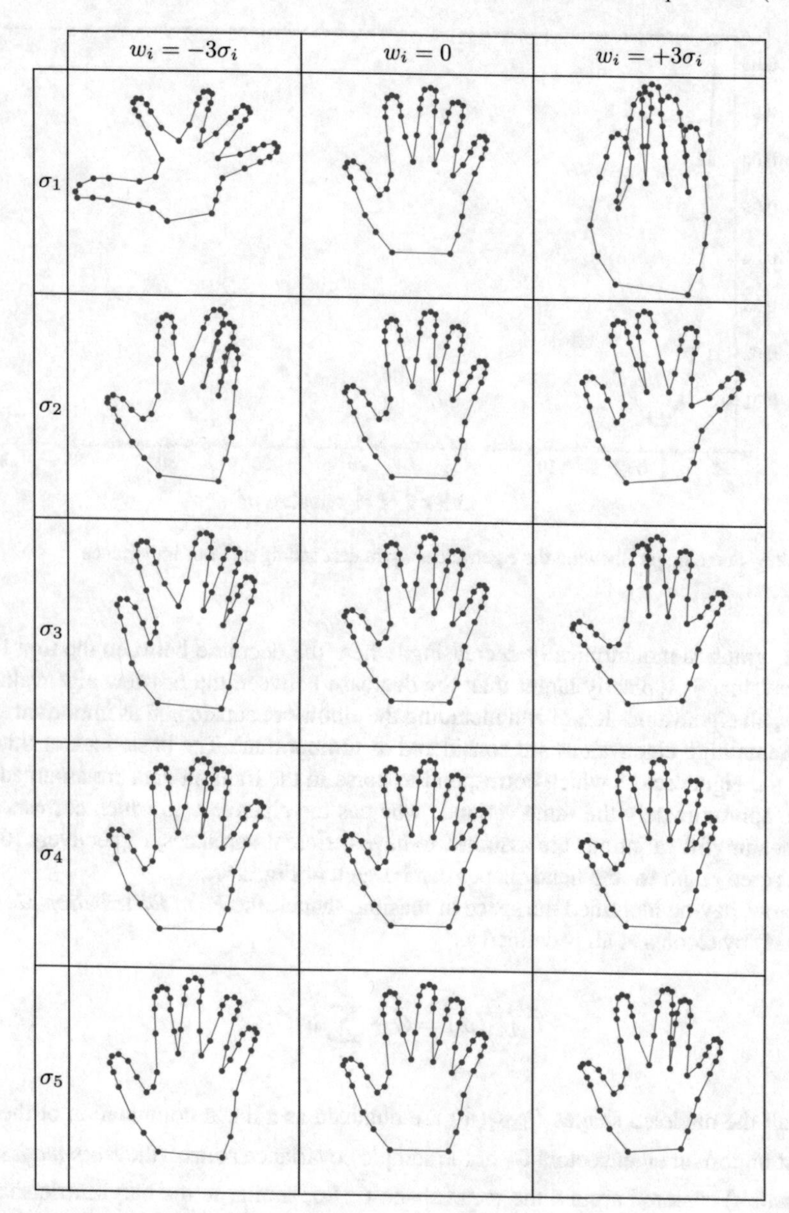

Fig. 2.7 Variation of the mean shape (*middle column*) by ± 3 standard deviations σ_i along the five most important eigenvectors (*left* and *right columns*)

In order to generate only likely shapes, i.e. shapes that correspond to the trained multivariate Gaussian distribution from Eq. (2.9), Cootes et al. additionally limit each weight w_i to be within three standard deviations of the Gaussian distribution:

$$-3\sigma_i \leq w_i \leq 3\sigma_i\,. \tag{2.15}$$

Another possibility is to constrain the weight vector $\vec{w}$ to lie inside the hyper-ellipsoid that defines the three standard-deviation boundary of the multivariate distribution [60]:

$$\sqrt{\sum_{i=1}^{m} \frac{w_i^2}{\sigma_i^2}} \leq 3\,. \tag{2.16}$$

Because of the facts that the shapes $\vec{C}_{\text{PDM}}(\vec{w})$ which are modeled by the PDM are obtained as linear combinations of the main modes of variation $\vec{\tilde{C}}_i$ and that a parameter vector (or weight vector) $\vec{w}$ is used to generate the shapes, the PDM is also called a *linear parametric statistical shape model* (linear parametric SSM). The construction of such a linear parametric SSM can be summarized as shown in Algorithm 2.1.

The m most important eigenvectors $\vec{\tilde{C}}_i$ of the sample covariance matrix together with the mean shape $\vec{\bar{C}}$ then define the low dimensional space of feasible shapes according to Eq. (2.14) and the corresponding eigenvalues σ_i^2 define the distribution of plausible shapes according to Eq. (2.9), respectively.[2]

2.3 An Implicit Linear Parametric Statistical Shape Model

Most of the currently used shape models are based on the PDM that has been defined in the last section [60]. This is most likely due to the fact that representing a shape as a set of points is very intuitive. Furthermore, the PDM is easy to understand and straightforward to implement. However, using points to describe a set of training shapes has also one big disadvantage: Dense point-correspondences need to be defined along the training shapes that are used to train the model. These dense point-correspondences may be determined manually or automatically. Yet, both approaches have their drawbacks. While the manual definition of point-correspondences is tedious and time-consuming, the automatic generation of point-correspondences is a difficult problem that is not easy to solve [60]. This contradicts the fact that wrong point-correspondences can cause strong artifacts in the shapes generated by the statistical shape model [98].

[2] Instead of explicitly setting up the covariance matrix, it is computationally more efficient to compute the *Singular Value Decomposition* (SVD) of the matrix $\breve{\mathbf{C}}$, because the right-singular vectors of the matrix $\breve{\mathbf{C}}$ are the eigenvectors of the matrix $\breve{\mathbf{C}}^T\breve{\mathbf{C}}$ [23, Chap. 4.6.3].

1. Start with a set of n rigidly-aligned (with regard to rotation, translation, and scale) training shapes, represented as dr-dimensional row vectors:

$$\{\vec{C}_1, ..., \vec{C}_n\}, \text{ with } \vec{C}_i \in \mathbb{R}^{dr}. \tag{2.17}$$

2. Compute the sample mean of the training shapes:

$$\bar{\vec{C}} = \sum_{i=1}^{n} \vec{C}_i. \tag{2.18}$$

3. Subtract the sample mean from all training shapes in order to obtain a new set of zero-mean training shapes:

$$\{\breve{\vec{C}}_1, ..., \breve{\vec{C}}_n\}, \text{ with } \breve{\vec{C}}_i \in \mathbb{R}^{dr}. \tag{2.19}$$

4. Stack all mean-subtracted shape vectors $\breve{\vec{C}}_i$ on top of each other in order to obtain the shape matrix $\breve{\mathbf{C}} \in \mathbb{R}^{n \times dr}$, with $n \ll r$ and $d \in \{2, 3\}$:

$$\breve{\mathbf{C}} = \begin{pmatrix} \breve{\vec{C}}_1 \\ \vdots \\ \breve{\vec{C}}_n \end{pmatrix}. \tag{2.20}$$

5. Estimate the sample covariance matrix $\hat{\Sigma} \in \mathbb{R}^{dr \times dr}$ of the mean-subtracted training shapes:

$$\hat{\Sigma} = \frac{1}{n-1} \breve{\mathbf{C}}^T \breve{\mathbf{C}}. \tag{2.21}$$

6. Compute the (at most) $n-1$ nonzero eigenvectors of the sample covariance matrix:

$$\{\tilde{\vec{C}}_1, ..., \tilde{\vec{C}}_{n-1}\}, \text{ with } \tilde{\vec{C}}_i \in \mathbb{R}^{dr}. \tag{2.22}$$

Algorithm 2.1: Necessary steps to obtain an explicit linear parametric SSM.

So, in order to deal with this problem, Leventon et al. were the first to propose a statistical shape model which does not require any point-correspondences at all [77]. They chose to represent the training shapes C_i implicitly as the zero level set of a signed distance function Φ_i:

$$C_i = \{\vec{x} \mid \Phi_i(\vec{x}) = 0\}, \tag{2.23}$$

where the signed distance functions Φ_i are defined as in Eq. 1.33 (c.f. Sect. 1.4).

In general, the signed distance functions Φ_i are given as functions that map from a continuous domain $\Omega \subset \mathbb{R}^d$, with $d \in \{2, 3\}$, to the field of real numbers $\mathbb{R}$. However, in practical implementations, one typically uses sampled versions Φ_i^s of the continuous level set functions Φ_i. These functions $\Phi_i^s : \Omega^s \subset \mathbb{N}^d \to \mathbb{R}$ are defined on a discrete grid Ω^s, where $\Omega^s = \{1, \ldots, j\} \times \{1, \ldots, k\}$ for $d = 2$ and $\Omega^s = \{1, \ldots, j\} \times \{1, \ldots, k\} \times \{1, \ldots, l\}$ for $d = 3$, respectively. See Fig. 1.2a for a 2D-example with $j = 512$ and $k = 366$.

Such two-dimensional discrete functions Φ_i^s can be represented as jk-dimensional row-vectors $\vec{\Phi}_i$ by successively appending the rows of the grid in a single vector:

$$\vec{\Phi}_i = \left(\Phi_i^s(1,1), \ldots, \Phi_i^s(j,1), \ldots, \Phi_i^s(1,k), \ldots, \Phi_i^s(j,k)\right). \tag{2.24}$$

This can be extended to three-dimensional discrete functions Φ_i^s when we first vectorize each of the l two-dimensional slices to a jk-dimensional row-vector as explained above and then successively append the thus obtained vectors to a jkl-dimensional row-vector:

$$\begin{aligned}\vec{\Phi}_i =& \big(\Phi_i^s(1,1,1), \ldots, \Phi_i^s(j,1,1), \ldots, \Phi_i^s(1,k,1), \ldots, \Phi_i^s(j,k,1), \ldots \\ & \Phi_i^s(1,1,l), \ldots, \Phi_i^s(j,1,l), \ldots, \Phi_i^s(1,k,l), \ldots, \Phi_i^s(j,k,l)\big).\end{aligned} \tag{2.25}$$

1. Start with a set of n rigidly-aligned (with regard to rotation, translation, and scale) training shapes, implicitly represented as the zero level sets of discretely sampled signed-distance functions:
$$\{\Phi_1^s, ..., \Phi_n^s\}, \text{ with } \Phi_i^s : \Omega^s \subset \mathbb{N}^d \rightarrow \mathbb{R}. \tag{2.26}$$
2. Convert the discrete functions Φ_i^s to jk- or jkl-dimensional row-vectors $\vec{\Phi}_i$ with the help of Eq. (2.24) (for $d = 2$) or Eq. (2.25) (for $d = 3$), respectively:
$$\{\vec{\Phi}_1, ..., \vec{\Phi}_n\}, \text{ with } \begin{cases} \vec{\Phi}_i \in \mathbb{R}^{jk} & \text{, if } d = 2 \\ \vec{\Phi}_i \in \mathbb{R}^{jkl} & \text{, if } d = 3. \end{cases} \tag{2.27}$$
3. Compute the sample mean of the training shapes:
$$\bar{\vec{\Phi}} = \sum_{i=1}^{n} \vec{\Phi}_i. \tag{2.28}$$
4. Subtract the sample mean from all training shapes in order to obtain a new set of zero-mean training shapes:
$$\{\breve{\vec{\Phi}}_1, ..., \breve{\vec{\Phi}}_n\}, \text{ with } \begin{cases} \breve{\vec{\Phi}}_i \in \mathbb{R}^{jk} & \text{, if } d = 2 \\ \breve{\vec{\Phi}}_i \in \mathbb{R}^{jkl} & \text{, if } d = 3. \end{cases} \tag{2.29}$$
5. Stack all mean-subtracted shape vectors $\breve{\vec{\Phi}}_i$ on top of each other in order to obtain the shape matrix $\breve{\mathbf{\Phi}}$, where $n \ll jk < jkl$:
$$\breve{\mathbf{\Phi}} = \begin{pmatrix} \breve{\vec{\Phi}}_1 \\ \vdots \\ \breve{\vec{\Phi}}_n \end{pmatrix}, \text{ with } \begin{cases} \breve{\mathbf{\Phi}} \in \mathbb{R}^{n \times jk} & \text{, if } d = 2 \\ \breve{\mathbf{\Phi}} \in \mathbb{R}^{n \times jkl} & \text{, if } d = 3. \end{cases} \tag{2.30}$$
6. Estimate the sample covariance matrix $\hat{\Sigma}$ of the mean-subtracted training shapes:
$$\hat{\mathbf{\Sigma}} = \frac{1}{n-1} \breve{\mathbf{\Phi}}^T \breve{\mathbf{\Phi}}, \text{ with } \begin{cases} \hat{\mathbf{\Sigma}} \in \mathbb{R}^{jk \times jk} & \text{, if } d = 2 \\ \hat{\mathbf{\Sigma}} \in \mathbb{R}^{jkl \times jkl} & \text{, if } d = 3. \end{cases} \tag{2.31}$$
7. Compute the (at most) $n - 1$ nonzero eigenvectors of the sample covariance matrix (see also footnote 2 on page 31):
$$\{\tilde{\vec{\Phi}}_1, ..., \tilde{\vec{\Phi}}_{n-1}\}, \text{ with } \tilde{\vec{\Phi}}_i \in \mathbb{R}^{dr}. \tag{2.32}$$

Algorithm 2.2: Necessary steps to obtain an implicit linear parametric SSM.

Now, with the help of Eqs. (2.24) and (2.25), one is able to transform the implicitly represented training shapes from Eq. (2.23) into a vectorial representation. This enables us to build a statistical shape model with the help of the implicit training shapes, exactly as explained in Sect. 2.2.

So, the construction of an implicit linear parametric SSM can be summarized as shown in Algorithm 2.2.

Like for the explicit SSM from Sect. 2.2, the m most important eigenvectors of the sample covariance matrix $\vec{\tilde{\Phi}}_i$ together with the mean shape $\vec{\bar{\Phi}}$ define the low dimensional space of feasible shapes according to

$$\vec{\Phi}_{\text{glob}}(\vec{w}) = \vec{\bar{\Phi}} + \sum_{i=1}^{m} w_i \, \vec{\tilde{\Phi}}_i \,, \tag{2.33}$$

with $\vec{w} = (w_1, \ldots, w_m)$, and the corresponding eigenvalues σ_i^2 define the distribution of plausible shapes according to Eq. (2.9), respectively. By inverting the steps which were necessary to vectorize the training shapes (Eq. (2.24) or Eq. (2.25), respectively), and by considering the level sets again as continuous functions, one obtains a shape model which is able to generate shapes that are implicitly represented through a higher-dimensional function $\vec{\Phi}_{\text{glob}}(\vec{w}) : \Omega \subset \mathbb{R}^d \to \mathbb{R}$:

$$\Phi_{\text{glob}}(\vec{w}) = \bar{\Phi} + \sum_{i=1}^{m} w_i \, \tilde{\Phi}_i \,. \tag{2.34}$$

As for the PDM from Sect. 2.2, all modeled shapes are obtained as linear combinations of the main modes of variation $\tilde{\Phi}_i$ and a parameter vector (or weight vector) $\vec{w}$ is used to generate the shapes, hence the term *implicit linear parametric statistical shape model*. Explicit representations of the generated shapes can be obtained as the zero level set of the higher dimensional function $\Phi_{\text{glob}}(\vec{w})$:

$$C_{\text{glob}}(\vec{w}) = \left\{ \vec{x} \mid \vec{\Phi}_{\text{glob}}(\vec{w})(\vec{x}) = 0 \right\} . \tag{2.35}$$

A schematic overview of the training process can be seen in Fig. 2.8. In the depicted setup, the training set consists of four shapes, each having a spatial resolution of 3×3 pixels, and the two main modes of variation are used to define the linear SSM subspace.

An example for the just described implicit linear parametric statistical shape model can be seen in Figs. 2.9, 2.10, 2.11, 2.12, 2.13 and 2.14. The statistical shape model that is shown describes the variation of the outer bony boundary which surrounds the nasal cavity and the paranasal sinuses. The used training shapes are depicted in appendix A (green line). A more detailed description of the dataset follows later in Sect. 4.1.1.

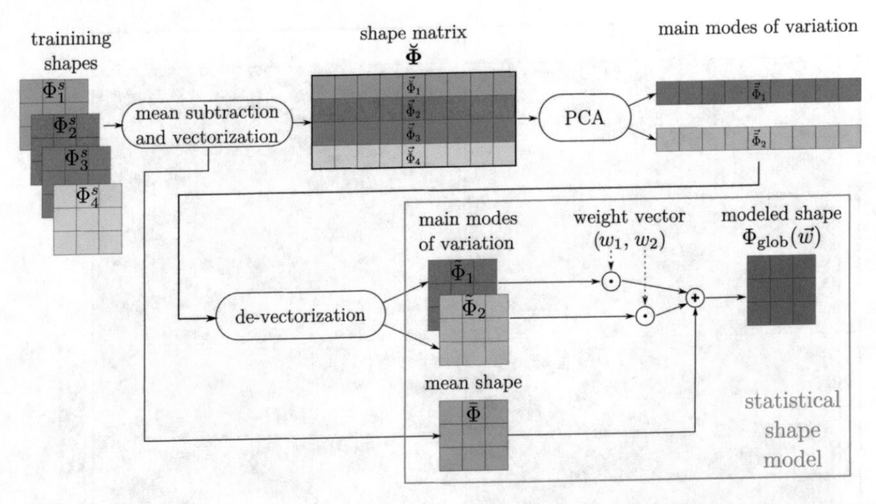

Fig. 2.8 Exemplary schematic overview of the training process of the implicit linear parametric statistical shape model from Eq. (2.34)

In Fig. 2.9, the first four main modes of variation are depicted (without addition of the mean shape). They nicely show where most of the variation occurs in the training shapes. It is clearly visible that the first mode of variation mainly encodes the variation in the frontal sinuses as these differ the most between different people (see also Fig. 4.2). The second mode of variation encodes mainly the lower expansion of the paranasal sinuses, the third allows to balance the expansion of the frontal sinuses, the fourth encodes amongst others the upper extension of the maxillary sinuses, and so on.

What can also be seen is that the maximum amplitude of the main modes of variation $\tilde{\Phi}_i$ quickly decreases with growing index i. This is due to the strong decrease of the corresponding eigenvalues σ_i^2. The scree graph from Fig. 2.10 and the cumulative variance from Fig. 2.11 confirm this finding. It can be seen that the first five main modes of variation already account for more than 90% of the total variation which is inherent in the training data and the first ten main modes of variation account for more than 97% of the total variation, respectively.

In Fig. 2.12, one can see the shape variations that occur when the mean shape $\bar{\Phi}$ is varied by plus and minus three standard deviations along the five main modes of variation $\tilde{\Phi}_1, \ldots, \tilde{\Phi}_5$. The level set functions that represent the modeled shapes are color-coded, where blue tones are assigned to negative values of the level set functions. The zero level sets are additionally shown as black curves.

The distribution of the training shapes, projected onto the first three main modes of variation, is shown in Figs. 2.13 and 2.14. It can be seen that a Gaussian distribution is a reasonable assumption to describe the distribution of the training shapes.

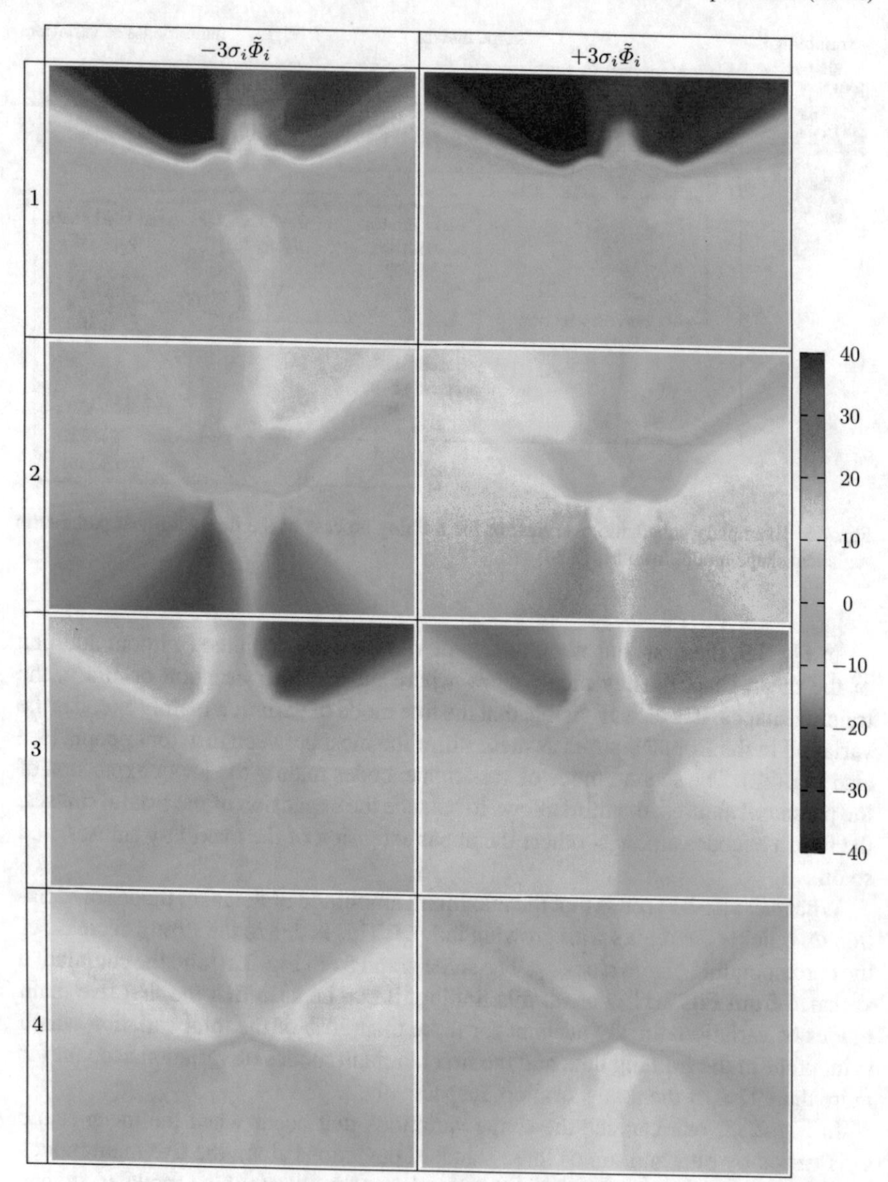

Fig. 2.9 First four main modes of variation of the implicit training shapes

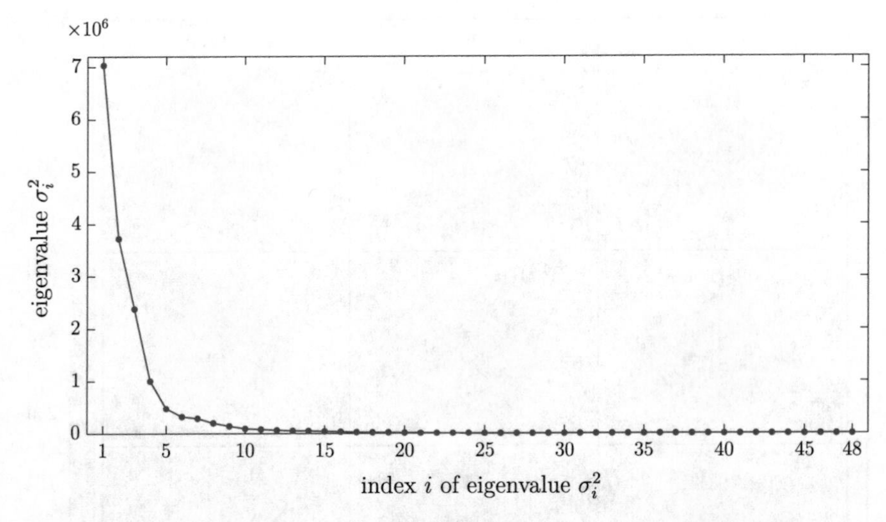

Fig. 2.10 Scree graph, showing the eigenvalues σ_i^2 in descending order of importance

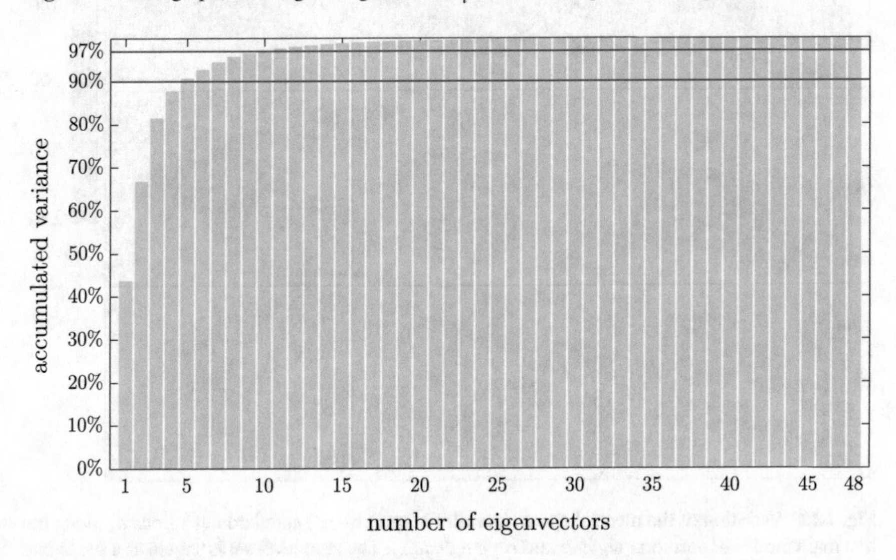

Fig. 2.11 Cumulative variance for a growing number of eigenvectors. Five eigenvectors are enough to account for 90% of the total variation and ten eigenvectors capture more than 97% of the total variation, respectively

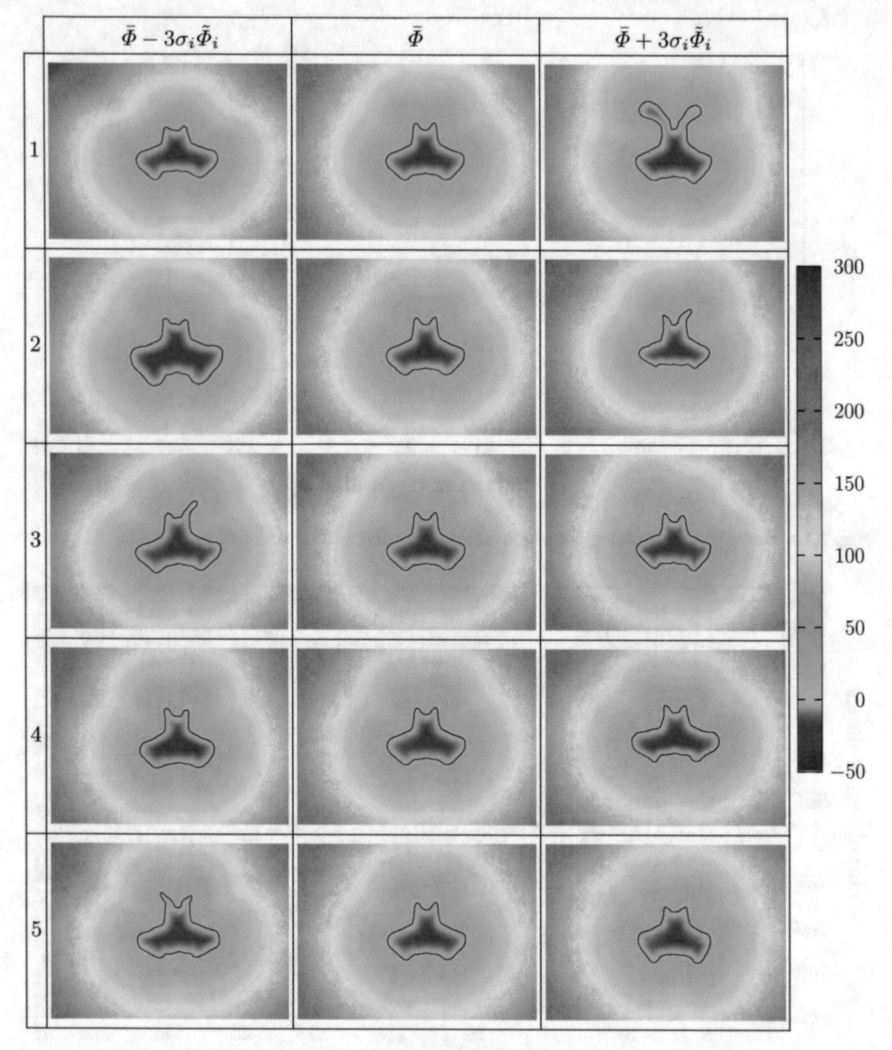

Fig. 2.12 Variation of the mean shape $\bar{\Phi}$ (*middle column*) by ± 3 standard deviations σ_i along the five main modes of variation $\tilde{\Phi}_i$ (*left* and *right columns*). The zero level set is shown as a *black line*

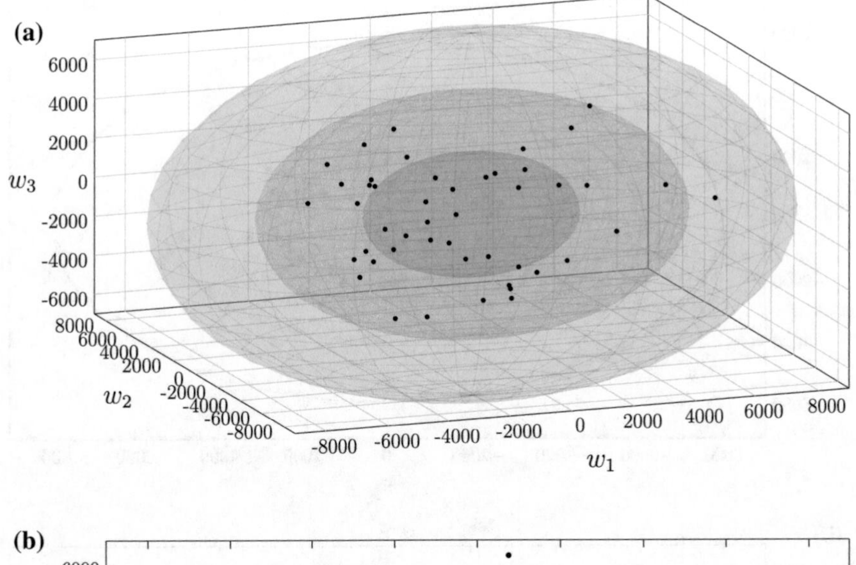

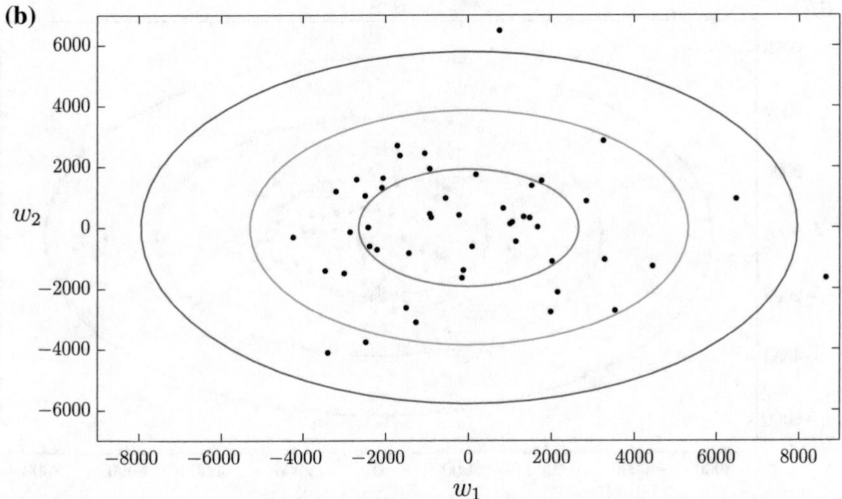

Fig. 2.13 **a** Distribution of the implicit training shapes (*black dots*), projected onto the first three main modes of variation. The *ellipses* at 1, 2, and 3 standard deviations of the estimated Gaussian distribution are shown in *red*, *green*, and *blue*, respectively. **b** Two-dimensional projection of the plot shown in (**a**) onto the first versus second main mode of variation

2.4 Rigid Shape Alignment

In the previous sections, we have demanded that the training shapes are aligned to each other with regard to rotation, translation, and scale. According to the definition in Sect. 2.1, this information does not belong to the shape information of a specific shape class.

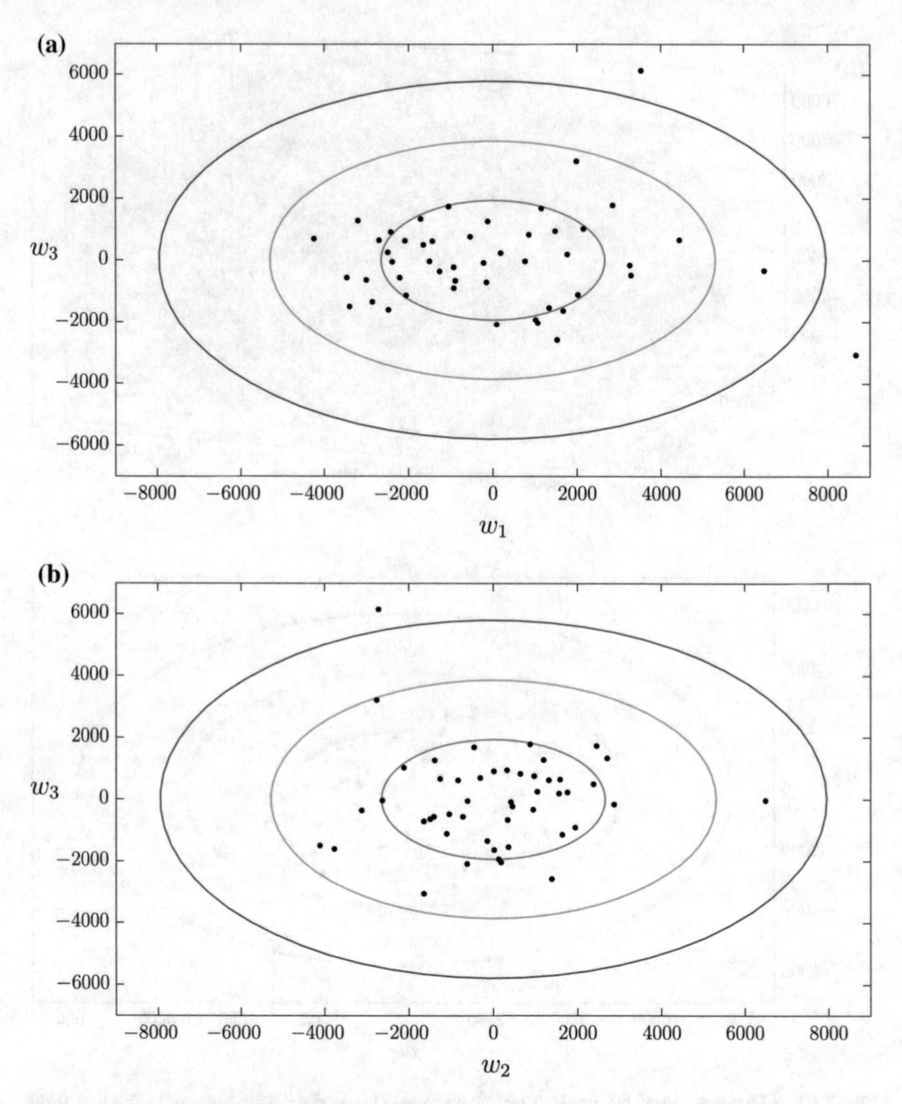

Fig. 2.14 Two-dimensional projections of the plot shown in Fig. 2.13a onto the first versus third (**a**) and second versus third (**b**) main mode of variation

So, in the following we will discuss how such a *rigid alignment* can be obtained. For this purpose, we will first define the so-called similarity transformation in Sect. 2.4.1, then we discuss the rigid alignment of two shapes in Sect. 2.4.2, and finally in Sect. 2.4.3 we give an algorithm for aligning multiple shapes.

2.4.1 *Similarity Transformation*

We remember the explicit shape representation from Eq. (2.1) or from Eq. (2.2), respectively:

$$\vec{C} = \begin{cases} (x_1, y_1, \ldots, x_r, y_r) & , \text{if } d = 2 \\ (x_1, y_1, z_1, \ldots, x_r, y_r, z_r) & , \text{if } d = 3 \,. \end{cases} \tag{2.36}$$

This explicit representation can also be denoted by an $r \times d$-dimensional matrix $\mathbf{X}$ when we write the individual points (x_i, y_i) or (x_i, y_i, z_i) row-wise one above the other:

$$\mathbf{X} = \begin{pmatrix} \vec{x}_1 \\ \vdots \\ \vec{x}_r \end{pmatrix}, \tag{2.37}$$

with $\vec{x}_i = (x_i, y_i)$ for $d = 2$ or $\vec{x}_i = (x_i, y_i, z_i)$ for $d = 3$, respectively.

The matrix $\mathbf{X}$ is also called a *configuration matrix* [64, Definition 2.1]. Now, the *Euclidean similarity transformations* of a configuration matrix $\mathbf{X}$ are defined as all the matrices $\mathbf{X}'$ that can be obtained by translating, rotating, and isotropically scaling the matrix $\mathbf{X}$ [64, Definition 3.2]:

$$\mathbf{X}' = \beta \mathbf{X} \Gamma + \vec{1}_r^T \vec{\gamma}\,, \tag{2.38}$$

where $\beta \in \mathbb{R}^+$ is a positive scale factor, $\Gamma \in SO(d)$ is an orthogonal rotation matrix, $\vec{\gamma} \in \mathbb{R}^d$ is a d-dimensional translation vector, and $\vec{1}_r$ is a r-dimensional row-vector whose entries are all equal to one.

2.4.2 *Rigid Alignment of Two Shape Configurations*

In order to rigidly align a shape configuration $\mathbf{X}^{(2)}$ to a shape configuration $\mathbf{X}^{(1)}$ with regard to rotation, translation, and scale, we need to determine the Euclidean similarity transformation that minimizes the squared Euclidean distance between both configurations:

$$\left[\hat{\beta}, \hat{\Gamma}, \hat{\vec{\gamma}}\right] = \underset{\beta, \Gamma, \vec{\gamma}}{\arg\min} \; ||\mathbf{X}^{(1)} - \beta \mathbf{X}^{(2)} \Gamma - \vec{1}_r^T \vec{\gamma}||^2 \tag{2.39}$$

1. Compute the centroid of each configuration $\mathbf{X}^{(k)}$, $k \in \{1, 2\}$, as

$$\bar{\vec{x}}^{(k)} = \frac{1}{r} \sum_{i=1}^{r} \vec{x}_i^{(k)} . \tag{2.40}$$

2. Center the configurations via

$$\begin{aligned} \tilde{\mathbf{X}}^{(k)} &= \mathbf{X}^{(k)} - \vec{1}_r^T \bar{\vec{x}}^{(k)} \\ &= \begin{pmatrix} \vec{x}_1^{(k)} \\ \vdots \\ \vec{x}_r^{(k)} \end{pmatrix} - \begin{pmatrix} \bar{\vec{x}}^{(k)} \\ \vdots \\ \bar{\vec{x}}^{(k)} \end{pmatrix} . \end{aligned} \tag{2.41}$$

3. Compute the sample covariance matrix of the mean-subtracted configurations $\tilde{\mathbf{X}}^{(1)}$ and $\tilde{\mathbf{X}}^{(2)}$ (Eq. (D.16)):

$$\begin{aligned} \hat{\Sigma} &= \frac{1}{r-1} (\tilde{\mathbf{X}}^{(1)})^T \tilde{\mathbf{X}}^{(2)} \\ &= \frac{1}{r-1} \sum_{i=1}^{r} (\vec{x}_i^{(1)})^T \vec{x}_i^{(2)} . \end{aligned} \tag{2.42}$$

4. Compute a *Singular Value Decomposition* (SVD) of the sample covariance matrix [23, Chap. 4.6.3]:

$$\hat{\Sigma} = \mathbf{U} \Lambda \mathbf{V}^T , \tag{2.43}$$

with $\mathbf{U}, \mathbf{V} \in O(d)$ being orthogonal matrices and Λ is a $d \times d$ diagonal matrix of positive elements, the so-called *singular values*.

5. Obtain the rotation estimate via

$$\hat{\Gamma} = \mathbf{V} \begin{pmatrix} 1 & \dots & 0 \\ \vdots & \ddots & \vdots \\ 0 & \dots & \det(\mathbf{UV}) \end{pmatrix} \mathbf{U}^T , \tag{2.44}$$

where $\det(\mathbf{VU}) \in \{-1, +1\}$.

6. Estimate the scale factor via

$$\hat{\beta} = \frac{\operatorname{tr}(\hat{\mathbf{\Sigma}} \hat{\Gamma})}{\frac{1}{r-1} \operatorname{tr}((\tilde{\mathbf{X}}^{(2)})^T \tilde{\mathbf{X}}^{(2)})} , \tag{2.45}$$

where $\operatorname{tr}(\mathbf{A})$ gives the trace of a square matrix $\mathbf{A}$.

7. Obtain the translation estimate via

$$\hat{\vec{\gamma}} = \bar{\vec{x}}^{(1)} - \hat{\beta} \bar{\vec{x}}^{(2)} \hat{\Gamma} . \tag{2.46}$$

Algorithm 2.3: Necessary steps to perform a full ordinary Procrustes analysis.

This means, we need to estimate a scale factor $\hat{\beta}$, a rotation matrix $\hat{\Gamma}$, and a translation vector $\hat{\vec{\gamma}}$. This problem is also known as *full ordinary Procrustes analysis* [64, Definition 7.1]. A solution to problem (2.39) can be obtained with the approach[3] in Algorithm 2.3 ([64, result 7.1] or [47, Chap. 9.4.1]).

2.4.3 Rigid Alignment of Multiple Shape Configurations

In general we have more than two shape configurations that need to be rigidly aligned to each other. So, given $n \geq 2$ shape configurations, e.g. the training shapes for the PDM from Sect. 2.2, Eq. (2.39) modifies to

$$\begin{bmatrix} \hat{\beta}_1,\ldots,\hat{\beta}_n \\ \hat{\mathbf{\Gamma}}_1,\ldots,\hat{\mathbf{\Gamma}}_n \\ \hat{\vec{\gamma}}_1,\ldots,\hat{\vec{\gamma}}_n \end{bmatrix} = \underset{\substack{\beta_1,\ldots,\beta_n \\ \mathbf{\Gamma}_1,\ldots,\mathbf{\Gamma}_n \\ \vec{\gamma}_1,\ldots,\vec{\gamma}_n}}{\arg\min}\ \frac{1}{n} \sum_{i=1}^{n} \sum_{j=i+1}^{n} \left\| \left(\beta_i \mathbf{X}^{(i)} \mathbf{\Gamma}_i + \vec{1}_r^T \vec{\gamma}_i\right) - \left(\beta_j \mathbf{X}^{(j)} \mathbf{\Gamma}_j + \vec{1}_r^T \vec{\gamma}_j\right) \right\|^2 \tag{2.47}$$

This problem is also known as *full generalized Procrustes analysis* [64, Definition 7.4]. A solution to problem (2.47) can be obtained with the simple Algorithm 2.4 [34].

Step 4 in Algorithm 2.4 is needed in order to prevent the mean from shrinking, rotating or drifting [34]. This would happen, because Eq. (2.47) defines an under-determined system of equations that has no unique solution if the mean is not constrained. Algorithm 2.4 usually converges quickly so that already one iteration provides a very good alignment result. A set of hand shapes that have been rigidly aligned with Algorithm 2.4 can be seen in Fig. 2.3b.

Algorithm 2.4 is simple and it works in most cases, however its convergence has not been formally proven. Another slightly more difficult algorithm with investigated convergence bounds can be found in [64, chap. 7.4.1]. For two-dimensional shapes there also exists a closed-form solution for the mean configuration from Eq. (2.47) so that only step 5 of Algorithm 2.4 needs to be executed once in order to obtain the rigidly aligned shapes.

---

[3] The construction in step 5 of Algorithm 2.3 ensures that $\hat{\Gamma} \in SO(d)$, i.e. $\hat{\Gamma}$ is a rotation matrix (an orthogonal matrix with determinant 1). In the case that reflections should be also allowed, i.e. $\hat{\Gamma} \in O(d)$ (an orthogonal matrix with determinant 1 or -1), Eq. (2.44) reduces to $\hat{\Gamma} = \mathbf{V}\mathbf{U}^T$ [64, Chap. 7.2.1].

The here presented approach works for the implicit shape representation from Eq. (1.32) as well when at least $r = 2$ (for $d = 2$) or $r = 3$ (for $d = 3$)[4] explicit corresponding points are additionally available on the implicit shapes. These points can e.g. be manually selected [99]. Other approaches try to align the implicit shapes by maximizing the mutual overlap of the interior regions (blue region in Fig. 1.2a) [130] or by minimizing the sum of squared differences between the signed distance functions [95] with the help of variational methods as introduced in Sect. 1.3.

1. Rigidly align each shape configuration $\mathbf{X}^{(i)}$, $2 \leq i \leq n$, to the first shape configuration $\mathbf{X}^{(1)}$ with the help of Algorithm 2.3 in order to obtain initial estimates for $\hat{\beta}_i$, $\hat{\Gamma}_i$, and $\hat{\vec{\gamma}}_i$, $2 \leq i \leq n$.
2. Set $\hat{\beta}_1 = 1$, $\hat{\Gamma}_1 = \mathbf{I}_d$, and $\hat{\vec{\gamma}}_1 = \vec{0}_d$, where $\mathbf{I}_d$ is the $d \times d$ identity matrix and $\vec{0}_d$ is a d-dimensional row-vector containing only zeros.

repeat

3. Compute the mean of the similarity transformed shape configurations

$$\bar{\mathbf{X}} = \frac{1}{n}\sum_{i=1}^{n} \mathbf{X}^{(i)\prime} = \frac{1}{n}\sum_{i=1}^{n}\left(\beta_i \mathbf{X}^{(i)}\Gamma_i + \vec{1}_r^T \vec{\gamma}_i\right). \tag{2.48}$$

4. Rigidly align the mean configuration $\bar{\mathbf{X}}$ to the first shape configuration $\mathbf{X}^{(1)}$ with the help of Algorithm 2.3 in order to obtain an adjusted mean $\bar{\mathbf{X}}'$.
5. Rigidly align each similarity transformed shape configuration $\mathbf{X}^{(i)\prime}$, $1 \leq i \leq n$, to the adjusted mean configuration $\bar{\mathbf{X}}'$ with the help of Algorithm 2.3 in order to update the estimates for $\hat{\beta}_i$, $\hat{\Gamma}_i$, and $\hat{\vec{\gamma}}_i$, $1 \leq i \leq n$.

until convergence

Algorithm 2.4: Simple procedure to perform a full generalized Procrustes analysis.

2.5 Problem: Limited Training Data

The statistical shape models from Eqs. (2.14) and (2.34) can provide an important means in the solution of segmentation problems or in the reconstruction of incomplete shapes. However, one major problem of statistical shape models is to obtain a sufficient amount of training data. Heimann and Meinzer have nicely summarized the problem of limited training shapes in [60, p. 546]:

> The power of a statistical model rises and falls with the quantity of available training data. In case of 3D SSMs, this quantity is almost always too low, ...[5]

[4]The 2D similarity transformation has 4 degrees of freedom, and the 3D similarity transformation has 7 degrees of freedom, respectively.

[5]Reprinted from [60, p. 546], © 2009, with permission from Elsevier.

The reason why there are almost always too few training shapes is that they have to be manually extracted from given image data which is a tedious and time-consuming task. For example, the manual extraction of the paranasal sinuses from CT data takes about 900 min for a dataset which consists of 98 slices, each having a resolution of 512×512 pixels [128].

The limited training data is a problem since for the above mentioned statistical shape models the assumption has been made that all plausible shapes can be described by linear combinations of the training shapes. Now, for a limited number of training shapes, this assumption is too restrictive. As can be seen in Eqs. (2.22) and 2.23 the dimension of the shape space cannot exceed the number of the training shapes minus one. The actual dimension will most likely be even smaller because the maximum dimension is only achieved for linearly independent training shapes (compare also the scree graphs from Figs. 2.6 and 2.10). So, for a small number of training shapes, the dimension of the shape space will most probably be too small to capture the full amount of variation inside a specific shape class. Much effort has therefore been spent on making statistical models more flexible.

In the following, we will discuss the most important previous contributions by other researchers which aim at obtaining more flexible SSMs. There are basically three different approaches to address the problem of limited training data. The approaches can be devided in three categories, but their boundaries are fluid [33]:

1. Artificially enlarge the shape variations which are described by the model.
2. Relax the model-constraints and locally refine the segmentation result.
3. Partition the model and describe the individual segments independently.

An overview of the approaches discussed below can be seen in Table 2.1. The approaches [22, 27, 39, 108, 109] are based on the variational level set image seg-

Table 2.1 Overview of existing approaches which aim at obtaining more flexible SSMs. Explicit approaches are given in black and implicit approaches in blue

Artificial enlargement of shape variations	Relaxation of model-constraints	Model partitioning
Cootes and Taylor [28, 31]	Cootes and Taylor [29]	Blanz and Vetter [19]
Cootes and Taylor [29]	Wang and Staib [133, 135]	de Bruijne et al [24]
Wang and Staib [134]	Shen et al. [117, 118]	Roberts et al. [104]
de Bruijne et al. [24]	Cootes and Taylor [30]	Davatzikos et al. [43]
Shen et al. [117, 118]	Weese et al. [136]	Zhao et al. [141]
Taron et al. [125]	Chen et al. [27]	Nain et al. [92]
Loog [80]	Bresson et al. [22]	Rousson and Paragios [108]
Koikkalainen et al. [72]	Shang and Dössel [116]	Knothe [71]
Lüthi et al. [83]	Rousson et al. [109]	Amberg et al. [11, 82]
Jud and Vetter [67, 68]	Cremers et al. [39]	
	Rousson and Paragios [108]	

mentation framework. We will discuss them extensively later in Sect. 5.2. The other approaches will be shortly discussed in the following.

2.5.1 Artificial Enlargement of Shape Variations

An early attempt to artificially enlarge the shape variations has been made by Cootes and Taylor [28, 31]. Instead of considering each shape as a rigid object, they assigned them additional elastic properties described by a stiffness matrix. Cootes and Taylor then performed a PCA on the stiffness matrix in order to obtain the resonant modes of vibration of each shape. This leads to a so-called *vibrational model* for each shape that is of the same form as the PDM from Eq. (2.14). However, the variation modes of these vibrational models are defined by the eigenvectors of the stiffness matrix and not by the training shapes. One can use the vibrational models in order to enlarge the number of training shapes by creating a set of vibrationally-deformed training shapes from each real training shape. The extension of the training set by these vibrationally-deformed training shapes leads to an increase of the estimated sample covariance along the resonant modes of vibration, thus enhancing the flexibility of the trained shape model in these directions.

Later, Cootes and Taylor simplified their approach by directly modifying some elements of the sample covariance matrix of the training shapes [29]. They proposed to add a constant value to the diagonal elements, representing the variance of a certain point, and to those off-diagonal elements that represent the covariance of neighboring points. The extra variance gives the individual points more freedom to move while the additional covariance also ensures that neighboring elements tend to move together. Consequently, smooth shape variations are favored.

A very similar approach has also been proposed by Wang and Staib [134]. They additionally introduced a weighting factor that tunes the influence of the smooth shape variations on the total variation. When no training samples are available their formulation relies completely on the smooth shape variations, and with an increasing number of training samples it iteratively adapts to purely trained shape-driven variations. Another very similar approach has been proposed by de Bruijne et al. [24]. In their approach the components of the artificial deformations are decoupled so that there exist no dependencies between different spatial dimensions. The decoupling has the effect that one does not need to establish the full covariance matrix, which has very large memory requirements especially for 3D shapes. The approach by de Bruijne et al. works best when the considered object class is very elongated in one spatial dimension in relation to the other dimensions.

All above-mentioned approaches make use of a synthetic covariance matrix that enhances the shape variations by additionally allowing smooth deformations. A different approach has been made by Shen et al. [117, 118]. They introduced the so-called *Active Focus Deformable Statistical Shape Model* (AFDSM). The AFDSM can *focus* on important structures by multiplying the subset of boundary points that define those structures with weighting factors greater one. By doing this, the variance of the training shapes is increased in the direction of the important structures. After PCA, this results in a modified shape space where the main modes of variation correspond to variations of the important structures. The structures can be determined manually or they can be chosen to those boundary points which correlate with strong image information (e.g. large gradient magnitudes). Subsequently, Koikkalainen et al. [72] modified this approach by smoothly deforming randomly selected local patches of the training shapes with the help of a so called *deformation sphere* [84] in order to enlarge the variations in the training set.

Other Approaches have been proposed by Taron et al. [125] and Loog [80]. Taron et al. [125] tried to include the uncertainties, that remain after establishing the point correspondences, into the model building process. This is achieved by modeling the uncertainties of the control points as a multivariate Gaussian distribution and generating new, artificial shapes by sampling from this distribution. Loog [80] proposed to fill the training shapes' covariance matrix only with those elements that correspond to the k nearest neighbors of each point and to fill the other entries with the help of a maximum entropy approach. By doing this, they model only the variations between nearby points and assume the other points as independent as possible.[6]

All so far discussed approaches rely on the PDM from Eq. (2.34). Recently, Lüthi et al. [83] proposed an extension of the approaches from [29] and [134] to those shape models where the modes of variation are given by continuous functions instead of discrete vectors, i.e. that are of the same form as the model from Eq. (2.34).[7] For this purpose, they proposed to describe the variability of the shape model and the extra variability as a combined Gaussian process, which is a multivariate Gaussian distribution of infinitely many random variables, and to use the Nyström approximation in order to obtain a low-rank approximation of this process. By doing this, they obtain a new set of continuous eigenfunctions that again represent the trained shape variations as well as the extra flexibility. Subsequently, Jud and Vetter [67, 68] extended this approach by multiplying the covariance function of the shape model's Gaussian process by a covariance function which decreases exponentially with growing distance between two points on the training shapes. This has the effect that distant shape parts are decoupled which also leads to more flexibility in the model. In fact, the basic idea behind the approach by Jud and Vetter, to decouple distant parts of the training shapes, is very similar to our idea which we are going to present in Chap. 3. However, the benefit of our approach is that globally consistent deformations can still be modeled whereas they are completely eliminated in the approach by Jud and Vetter.

[6]The maximum entropy approach is necessary to obtain a valid covariance matrix. Simply setting all remaining elements to zero, which means to assume that the corresponding points are uncorrelated, does not results in a valid covariance matrix.

[7]In the work of Lüthi et al. smooth deformation fields are modeled instead of level set functions.

2.5.2 *Relaxation of Model-Constraints*

Another approach to cope with a limited amount of training data is to use the statistical shape model only to robustly find a coarse initial solution and to relax the model-constraints afterwards. This means that also those shapes are allowed in the final solution which cannot be accurately approximated by the statistical shape model. Like for the artificially enhanced shape models, an early attempt in this direction has again been made by Cootes and Taylor. After fitting their already artificially enhanced PDM from [29], they search along lines normal to the model boundary in order to find a new set of candidate points which match a previously trained intensity distribution. They than select those points from the candidate set as final segmentation result that minimize the residual error to the best model fit. Later, they extended their approach by learning the needed local offsets based on the observed gray level differences between the optimal segmentation result and the best global model fit [30]. Similar approaches have also been proposed by Shen et al. [117, 118] and Shang and Dössel [116]. They both search for candidate points in regions with large gradient magnitudes. Wang and Staib [133, 135] proposed a physical model-based image registration approach that incorporates a statistical shape model. In their approach, they first compute the best fit of the statistical shape model and then use the boundary points of the deformed model shape as landmarks in their image registration framework. At the landmark positions, they constrain the deformation vectors of their deformation field to match the vector difference between the points on the mean shape and the deformed model shape. Similar to Cootes and Taylor, they also mentioned that using the enhanced PDM from [134] could further improve the results.

In the previously mentioned approaches, the segmentation result is still bound to the subspace of feasible shapes during the model fitting process. The common approach is to iteratively estimate a shape candidate which is then projected back into the model space. The model-constraints are relaxed only after the last iteration of the adaptation process, when the best model fit has already been found. Weese et al. [136] proposed to relax the model-constraints not only after but already during the model fitting process. For this purpose, they used an energy function which consists of a weighted combination of an external energy and an internal energy. The external energy drives the shape towards nearby gradients. Simultaneously, the distances between neighboring points on the deformed shape are compared to the distances between neighboring points on the best model approximation. These distances should be equal for a deformed shape that resembles the modeled shape. So, the internal energy penalizes the squared difference of these two distances. Weese et al. minimized their energy with the help of the conjugate gradient method in order to obtain a segmentation result which captures the image boundaries as good as possible but which resides also close to the subspace of feasible shapes. Many level set approaches with relaxed model-constraints (e.g. [22, 27, 39, 108]) are based on the same idea. They will be discussed in Sect. 5.2.

2.5.3 Model Partitioning

All the above-mentioned approaches enhance the flexibility of statistical shape models. However, their main disadvantage is that they allow variations which cannot be explained by the training shapes. A third way to address the problem of limited training data, which does not introduce artificial variations, is to partition the shapes either spatially or in the frequency domain. The assumption behind this is that the individual partitions are likely to contain less variation than the complete shape so that their deformations should be easier to model with only a few training shapes. For example, Blanz and Vetter [19] built a statistical shape model of human faces. In order to increase the expressiveness of their model, they segmented the faces into four parts, namely the mouth, nose, eyes, and the rest. This means that the space of feasible shapes is divided into four subspaces and a model fit is obtained by searching four independent parameter vectors in these subspaces. The thus obtained four shape parts have afterwards been blended together with the help of an image stitching algorithm which yields a natural-looking complete face.

A different approach has been made by de Bruijne et al. [24]. They proposed an extension of statistical shape models that is especially tailored to tubular objects. It consists of decoupling the variations in the bending of the centerline from the variations in the diameter of the object. This is achieved by building two separate models so that one describes the bending of the object and the other describes the varying diameter. Both models are subsequently combined into one model by extracting the principal modes of variation from a matrix that contains the main modes of variation from both models, the centerline model and the diameter model, respectively. Another approach that is best suited for elongated objects has been made by Roberts et al. [104]. They proposed to describe the object with a global statistical shape model that is combined with a sequence of partially overlapping sub-models. Each of the sub-models is thereby trained by using only a sub-part of all available object points. An initial solution for the global model is obtained by minimizing the mean squared error to a few user-defined landmarks. The thus obtained solution is then used as a soft constraint in the determination of the local solutions. The sub-models are fitted to the data in a user-defined sequence, and the overlapping parts of the sub-models are used as additional soft constraints in the fitting of neighboring sub-models. As a result, the segmentation clearly depends on the sequence in which the different sub-models are adapted to the data.

A straightforward further development of partitioned statistical shape models are hierarchical statistical shape models. Davatzikos et al. [43] proposed two methods to obtain those models. The basic idea of their approaches is to use one global statistical shape model that captures the global deformations of an object class and many local statistical shape models that capture the local variations, respectively. In their first approach, they partition the shapes spatially into a collection of lined-up segments, each containing a subset of all points that define the object boundary. They then compute the gravity center of each segment and build a global PDM by considering only these center points. Afterwards, one local PDM is constructed for each segment by considering only the points in the respective segment. Now, in order to approximate a shape with their hierarchical model, Davatzikos et al. first compute a fit of the global model, use this to determine the gravity centers of the sub-models, and then refine the global fit by computing local fits for all sub-models. Like in the approach by Blanz and Vetter, this approach leads to discontinuities at the segment borders so that Davatzikos et al. also compute a continuous blending in order to smoothly connect neighboring segments.

Because their first approach has been rather heuristically motivated, Davatzikos et al. additionally proposed a mathematically more sound hierarchical shape partition based on the wavelet transform. They used the discrete wavelet transform in order to divide the space-frequency domain in a number of bands that are arranged in a tree structure, i.e. the low-frequency bands have a broad spatial support whereas high-frequency bands get more and more localized. The Daubechies wavelets are used as basis functions since they have a wider support as e.g. Haar wavelets and thus provide a smoother transition between neighboring bands. The remaining spatial coupling among neighboring bands greatly reduces the discontinuities between them so that no additional blending is necessary. Now, similar to their heuristic approach, Davatzikos et al. independently model the wavelet coefficients in each frequency band by building one PDM for each band. With this model, an unknown shape can be approximated by computing the wavelet transform of the shape, projecting the wavelet coefficients into the individual PDM subspaces, and computing the inverse wavelet transform of the now shape-restricted coefficients. Nain et al. [92] extended the wavelet-based approach from Davatzikos et al. to 3D shapes by mapping the 3D shapes onto a sphere and then using spherical wavelets to divide them into space-frequency bands. In the model fitting process, they start by fitting the global PDM at the coarsest level of their wavelet tree and incrementally add the higher frequency bands when no better approximation can be obtained at the preceding level. This is supposed to increase the robustness of the fitting process against local minima, in contrary to the wavelet-based approach by Davatzikos et al. where all coefficients are used from the beginning of the fitting process.

Another hierarchical approach has been proposed by Knothe [71]. He extended the approach from [19] by using a Laplacian pyramid in order to divide the face shape variations into various frequency bands. In contrast to the wavelet-based approaches from [43] and [92], the frequency bands obtained from the Laplacian pyramid have intrinsically no local support. However, in practical applications, the variance of the high frequency components differs greatly from zero only in local regions of the shape (e.g. around the nose or the ears for face shapes) and is close to zero in the remaining regions. So, Knothe proposed to spatially decouple the high frequency components by thresholding the variance. Afterwards, a PCA is performed in each spatially decoupled frequency band and the resulting eigenvalues and eigenvectors are gathered into one combined PDM by sorting them with regard to decreasing eigenvalues. This construction leads to some global modes which control the overall shape and many modes with local support which control local face properties.

As mentioned above, shape inconsistencies can occur at the borders of the individual partitions when individual weight vectors are used for each partition and the partitions have no overlapping support. Zhao et al. [141] tried to address this problem. As in the preceding approaches, the weight vectors for each sub-part of the shape are first determined individually. Subsequently, the individual weight vectors are connected by straight line segments so that a modeled shape is now represented by a weight curve. Each training sample can likewise be presented by a weight curve. In order to prevent shape inconsistencies, Zhao et al. then determine the sample curve with closest distance to the weight curve of the modeled shape and restrict the modeled shape to be *close* to the sample curve. This is achieved by applying an affine transformation that is constructed as a trade-off between small deformations of the weight curve and minimization of the mean squared distance to the nearest sample curve. By doing this, basically only those weight combinations are allowed which also occurred for one of the training shapes. This may prevent the problem of shape inconsistencies, however, this comes at the cost of loosing the ability to adapt to different training shapes in the individual partitions, which is one of the central parts of partitioned statistical shape models.

Recently, Amberg et al. [11, 82] proposed an approach that differs from all the above-mentioned approaches. Their approach is inspired by *local linear regression* and it is able to enhance the model flexibility while avoiding to explicitly partition the model either spatially or in the frequency domain. The basic idea is to fit the statistical shape model only to a local neighborhood around a given point on the target shape instead of fitting the model to the complete target shape. The thus obtained solution than determines the modeled shape in this particular point only. So, in order to obtain a fit of the full model, the local fitting procedure has to be repeated for all points on the target shape. The size of the local neighborhood to which the model is fitted thereby determines how strictly the modeled shape adheres to the global shape properties. When the neighborhood includes all contour points, one obtains a global model fit that is strictly restricted to the subspace of feasible shapes. On the other hand, when the neighborhood contains only the actually considered point, the shape constraints are completely relaxed. The major problem with this approach is that computing a fit of the full SSM in each contour point is computationally very demanding. In order to address the computational burden, Amberg et al. proposed to compute a full model fit only for a subset of the contour points and to smoothly interpolate between the local model fits. This reduces the computation time, however, it can also be regarded as re-introducing predefined segments into the model-fitting process.

In the following, we will present a new approach to enhance the model flexibility. We call it the *Locally Deformable Statistical Shape Model* (LDSSM). Similar to [11], the main idea of our LDSSM is to allow a different model approximation in each point of the underlying data domain. However, we propose a novel framework that iteratively adapts the parameters of the local model approximations in order to approximate a given target shape. So, our approach allows to obtain a locally optimal model approximation without the necessity for any predefined segments. Additional smoothness constraints on the local parameters thereby ensure a globally consistent solution. The iterative framework removes the need to compute a full SSM fit in each point and makes our approach directly applicable also for high-resolution 3D datasets.

Chapter 3
A Locally Deformable Statistical Shape Model (LDSSM)

In Chap. 2, we have introduced the basic concepts of statistical shape models, but we have also discussed their limits, especially that a large number of training shapes is needed in order to capture the full amount of intra-class shape variation. We propose a new way to address this limitation in Sects. 3.1 and 3.2 by introducing a *Locally Deformable Statistical Shape Model* (LDSSM) that makes better use of the available training data than a global model and thus greatly reduces the number of required training shapes. Furthermore, our approach has no need for any predefined segments and it does not introduce shape variations that cannot be explained through the training data. In Sect. 3.3, we will additionally integrate our LDSSM into an iterative framework that can be used to solve image segmentation problems. The following illustrations are an extended version of the work that has been presented by us in [2, 3, 5, 7, 9].

3.1 Motivation

We begin by motivating our new approach with a simple example. Imagine that the set of training shapes consists only of two training shapes: A hand with short fingers (shown in Fig. 3.1a) and a hand with long fingers (shown in Fig. 3.1b), respectively. Despite the different lengths of the fingers, both hands are identical. Now, given these two training shapes, we want to reconstruct the partially occluded test shape from Fig. 3.1c. It is a hand where the 3 middle fingers are proportionally longer than the thumb and the little finger and where the middle finger is additionally occluded by the light gray rectangle. Hence, this hand represents a nonlinear combination of both training shapes.

With a global linear model, like the one from Eq. (2.34), one will not be able to accurately represent this test shape. The best solution that can be obtained, in a least-squares sense, is shown by the dashed black line in Fig. 3.1c. However, for a human observer it can be easily seen that a better solution can be obtained when the thumb and the little finger are approximated with the help of the first training shape

C. Last, *From Global to Local Statistical Shape Priors*, Studies in Systems, Decision and Control 98, DOI 10.1007/978-3-319-53508-1_3

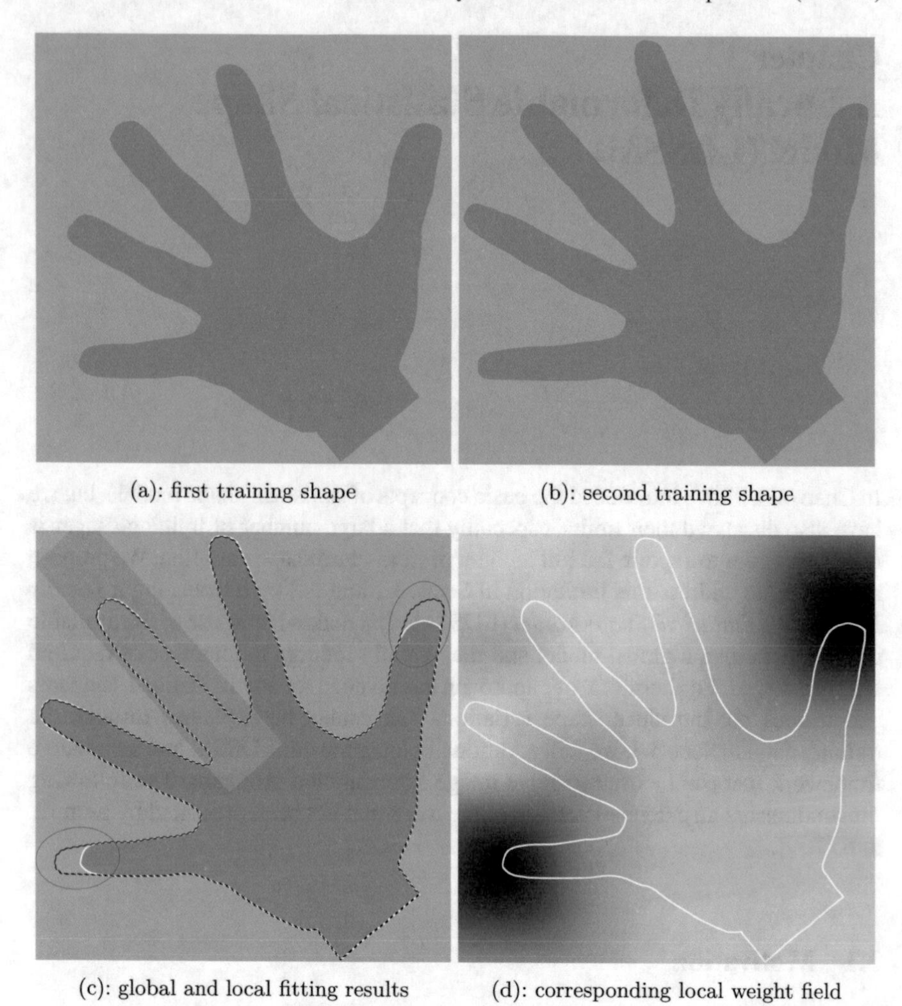

Fig. 3.1 Benefits of the local adaptation of global shape parameters: Given two training shapes (**a**) and (**b**), it is possible to reconstruct the occluded test shape (**c**) with our proposed *Locally Deformable Statistical Shape Model* (*white line*) but not with a global *Statistical Shape Model* (*black line*). The weight field (**d**), weighting the influence of the two training shapes, clearly shows the combination of both training shapes in the local solution. © [2014] IEEE. Reprinted, with permission, from [2]

and when the other fingers are approximated with the help of the second training shape, respectively. This just described approximation is shown by a white line in Fig. 3.1c.

Now, this is just a toy example. However, the same observation, that a given target shape can be approximated by local combinations of different training shapes, can also be made for many practical segmentation and shape approximation problems, especially for those that involve biological shapes. So, in the following sections, we will present what we call a *Locally Deformable Statistical Shape Model* (LDSSM). It extends the global statistical shape model from Eq. (2.34) in order to consider the above mentioned observation.

3.2 Mathematical Formulation of Our LDSSM

Before we will begin to describe our *Locally Deformable Statistical Shape Model* (LDSSM), we will shortly recapitulate the basic idea behind the spatially partitioned statistical shape models that have already been mentioned in Sect. 2.5.

In a spatially partitioned statistical shape model, the basic idea is to increase the flexibility of the global statistical shape model from Eq. (2.34) by partitioning the data domain Ω in a number of r regions Ω_u so that $\Omega_1 \cup \ldots \cup \Omega_r = \Omega$ and allowing a different model approximation in each of these partitions Ω_u. One way to achieve this is to extend the global statistical shape model from Eq. (2.34) by using a different weight vector $\vec{w}_u$ in each partition Ω_u. Each of these weight vectors $\vec{w}_u$ can be used to generate a complete shape $\Phi_{\text{glob}}(\vec{w}_u)$. However, in order to obtain a partitioned statistical shape model, one considers only that part of the generated shape $\Phi_{\text{glob}}(\vec{w}_u)$ which is located inside the partition Ω_u. This yields the following equation for a spatially partitioned statistical shape model:

$$\Phi_{\text{part}}(\vec{w}_1, \ldots, \vec{w}_r)(\vec{x}) = \bar{\Phi}(\vec{x}) + \sum_{u=1}^{r} \left[\mathbf{1}_{\Omega_u}(\vec{x}) \sum_{i=1}^{m} w_u^i \, \tilde{\Phi}_i(\vec{x}) \right] , \tag{3.1}$$

where w_u^i denotes the i-th component of the weight vector $\vec{w}_u$ and $\mathbf{1}_{\Omega_u}$ is the characteristic function of the set Ω_u. It is defined as

$$\mathbf{1}_{\Omega_u}(\vec{x}) = \begin{cases} 1 & , \text{if } \vec{x} \in \Omega_u \\ 0 & , \text{else} \end{cases} . \tag{3.2}$$

As desired, Eq. (3.1) increases the flexibility of the global statistical shape model from Eq. (2.34). However, it still has two considerable drawbacks:

1. The optimal partition into individual segments Ω_u has to be somehow determined prior to reconstructing a certain shape.
2. A smooth crossover between the generated partial shapes from adjacent partitions has to be ensured in order to obtain a continuous complete shape.

In the following, we address these drawbacks. For this purpose, we propose to further extend the partitioned model from Eq. (3.1) by allowing one weight vector in each element $\vec{x}$ of the data domain Ω. Additionally, we demand that neighboring weight vectors differ only slightly from each other. This new formulation avoids the need of predefined partitions, and the smooth transition between neighboring weights implicitly guarantees a continuous shape representation.

Mathematically speaking, our idea is to increase the flexibility of the global statistical shape model by generalizing the low-dimensional vector space of feasible shapes that is defined in Eq. (2.34). It is a vector space with basis $\{\tilde{\Phi}_1, \ldots, \tilde{\Phi}_m\}$ over the field of $\mathbb{R}$ (i.e. each weight is a real number), and we generalize it by replacing the scalar weights $w_i \in \mathbb{R}$ through smooth functions $\Psi_i : \Omega \to \mathbb{R}$. We call this approach the *Locally Deformable Statistical Shape Model* (LDSSM), and we define it as

$$\Phi_{\text{loc}}(\vec{\Psi}(\vec{x})) = \bar{\Phi}(\vec{x}) + \sum_{i=1}^{m} \Psi_i(\vec{x})\tilde{\Phi}_i(\vec{x}), \tag{3.3}$$

where $\vec{\Psi} : \Omega \to \mathbb{R}^m$ are all scalar weight functions $(\Psi_1, \ldots, \Psi_m)$ combined in one vector-valued function. Please note that it is still beneficial to perform a PCA on the set of training shapes and use only the m most significant eigenshapes in our LDSSM as the PCA identifies all the shape information that can be obtained by linear combinations of the eigenshapes. Keeping all the eigenshapes would not provide any significant additional information to our generalized shape model.

The introduced vector-valued function $\vec{\Psi}$ can be interpreted as a smooth field of weight vectors $\vec{\Psi}(\vec{x})$ or, synonymously, as a vector of smooth weight fields $\vec{\Psi} = (\Psi_1, \ldots, \Psi_m)$. Similar to the global statistical shape model from Eq. (2.34), where all shapes $C_{\text{glob}}(\vec{w})$ that can be generated with the help of the SSM are obtained by modifying the weight vector $\vec{w}$, we obtain all shapes $C_{\text{loc}}(\vec{\Psi})$ that can be generated with the help of the LDSSM by modifying the smooth field of weight-vectors $\vec{\Psi}$. The shapes $C_{\text{loc}}(\vec{\Psi})$ are given as the zero level set of the corresponding higher dimensional function $\Phi_{\text{loc}}(\vec{\Psi})$. A depiction of the differences between the global SSM from Eq. (2.34) and our new LDSSM from Eq. (3.3) can be seen in Fig. 3.2.

The restriction to smooth weight functions has two key reasons. The first reason is to ensure that the contour $C_{\text{loc}}(\vec{\Psi})$ of the modeled shape $\Phi_{\text{loc}}(\vec{\Psi})$ remains continuous: When Ψ_i and $\tilde{\Phi}_i$ are smooth functions, then also the result of the component-wise multiplication in Eq. (3.3) remains smooth, and hence the resulting zero level set $C_{\text{loc}}(\vec{\Psi})$ of the resulting local shape $\Phi_{\text{loc}}(\vec{\Psi})$ is continuous. The second reason is that the amount of extra flexibility which is additionally allowed in our LDSSM in contrast to the global SSM directly depends on the smoothness of the weight functions Ψ_i: By varying the degree of smoothness of the weight functions Ψ_i, we can control the locality of our generalized shape space. This is because the more uneven the weight functions Ψ_i, the more transitions between different local combinations of the main modes of variations are possible. In contradiction, the more smooth the weight functions Ψ_i, the less transitions between different local combinations are possible and the model becomes more global.

This becomes even more clear when we have a look at the both extremes: Constant weight functions Ψ_i and discontinuous weight functions Ψ_i. For constant weight functions Ψ_i, one obtains a linear combination of the main modes of variation $\tilde{\Phi}_i$ in Eq. (3.3) as they are multiplied by the same weights $\Psi_i(\vec{x})$ in each element $\vec{x}$ of the data domain Ω. Thus, for constant functions Ψ_i, the shapes which can be generated with the help of the LDSSM are bound to the linear SSM subspace and

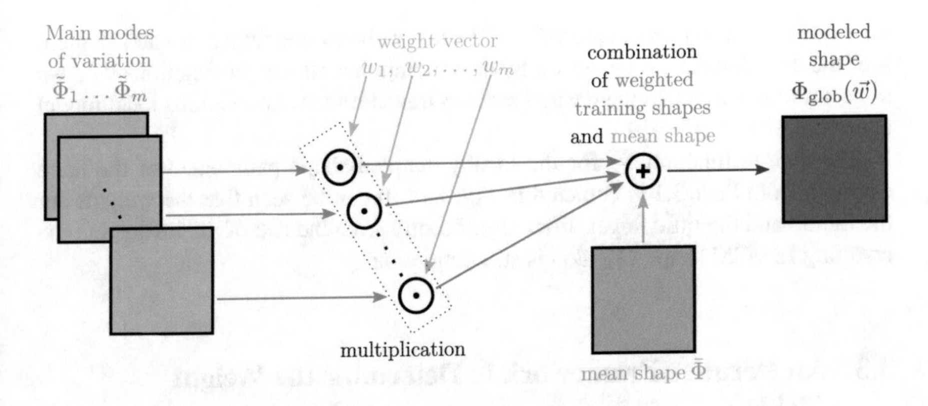

(a): global *Statistical Shape Model* (SSM)

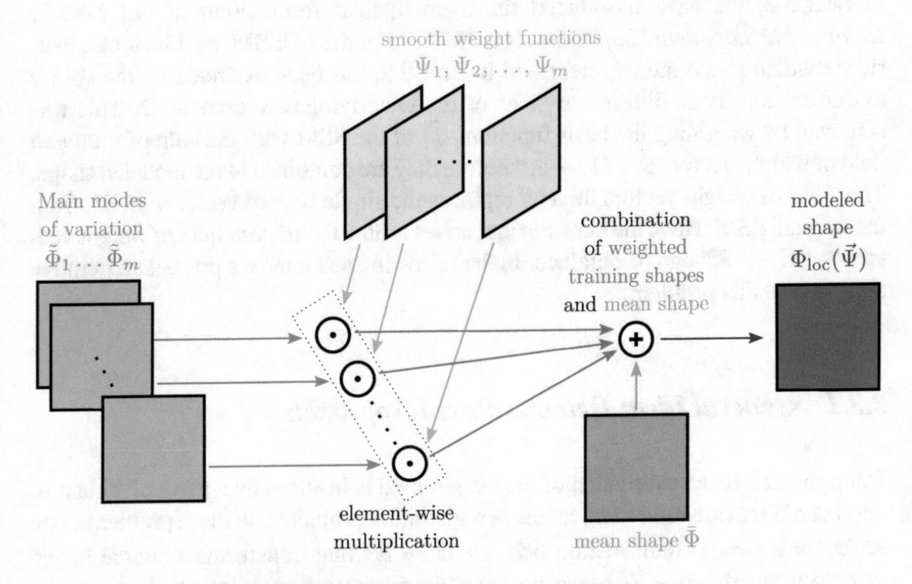

(b): *Locally Deformable Statistical Shape Model* (LDSSM)

Fig. 3.2 Schematic visualizations of the global SSM and the LDSSM. One can clearly see the difference between both models: The weight vector $\vec{w} = (w_1, \ldots, w_m)$ from the global SSM (**a**) is replaced by a vector of *smooth* weight functions $\vec{\Psi} = (\Psi_1, \ldots, \Psi_m)$ for the LDSSM (**b**)

the LDSSM becomes identical to the SSM from Eq. (2.34). The other extreme is that the smoothness constraint is completely dropped so that discontinuous weight functions Ψ_i become possible. In this case, and when we further assume that at least one basis function $\tilde{\Phi}_i$ is nonzero in each element $\vec{x}$ of the data domain Ω (which is typically the case), arbitrary shapes can be modeled with the help of the LDSSM as the value of the level set function $\Phi_{\text{loc}}(\vec{\Psi})$ can now be chosen freely for each element $\vec{x}$ of the data domain Ω. However, by demanding that all weight-functions Ψ_i have to be smooth, we obtain the desired smooth transition between various local model approximations.

The weight function Ψ_1 for the locally adapted shape parameters of the hand example from Sect. 3.1 is depicted in Fig. 3.1d. It can be seen that the weights for the thumb and the little finger differ significantly from the rest of the image, but the resulting LDSSM shape $C_{\text{loc}}(\Psi_1)$ is still continuous.

3.3 An Iterative Framework to Determine the Weight Fields

In Sect. 3.2, we have introduced the mathematical formulation of our *Locally Deformable Statistical Shape Model* (LDSSM). For the LDSSM, the linear parametric statistical shape model, presented in Sect. 2.3, has been extended by the ability to deform locally in different regions of the underlying data domain Ω. This was achieved by weighting the basis functions $\tilde{\Phi}_i$ of the SSM with the help of a *smooth* field of weight vectors $\vec{\Psi} : \Omega \rightarrow \mathbb{R}^m$ before they are combined to the modeled shape. This field of weight vectors thereby replaces the single weight vector $\vec{w} \in \mathbb{R}^m$ from the original SSM. Now, the question that arises is how the *smooth* field of weight vectors $\vec{\Psi} : \Omega \rightarrow \mathbb{R}^m$ can be obtained. In the following sections, we present an iterative approach to this problem.

3.3.1 General Idea: Demons-Based Approach

Our approach to the estimation of the weight field is inspired by a particular class of approaches that are used to solve image registration problems. In these problems, one seeks for a *smooth* deformation field $\hat{\vec{v}} : \Omega \rightarrow \mathbb{R}^2$ that transforms a source image I_1 onto a target image I_2 based on a certain minimization criterium, like e.g. the minimization of the squared intensity differences [91]:

$$\hat{\vec{v}} = \arg\min_{\vec{v}} ||I_1(\vec{v}) - I_2||^2 . \tag{3.4}$$

One particular class of algorithms to solve these kinds of problems are the so-called *demons-based approaches*. They have been presented by Thirion in 1998 [127] and have found wide acceptance in the image registration community in the following years [121].

The general idea of the demons-based approaches is to start with an initial deformation field of $\vec{v}(\vec{x}) = \vec{0}$ for all $\vec{x} \in \Omega$ and then compute an incremental update of the deformation field *individually* in each element $\vec{x}$. By doing this, the resulting

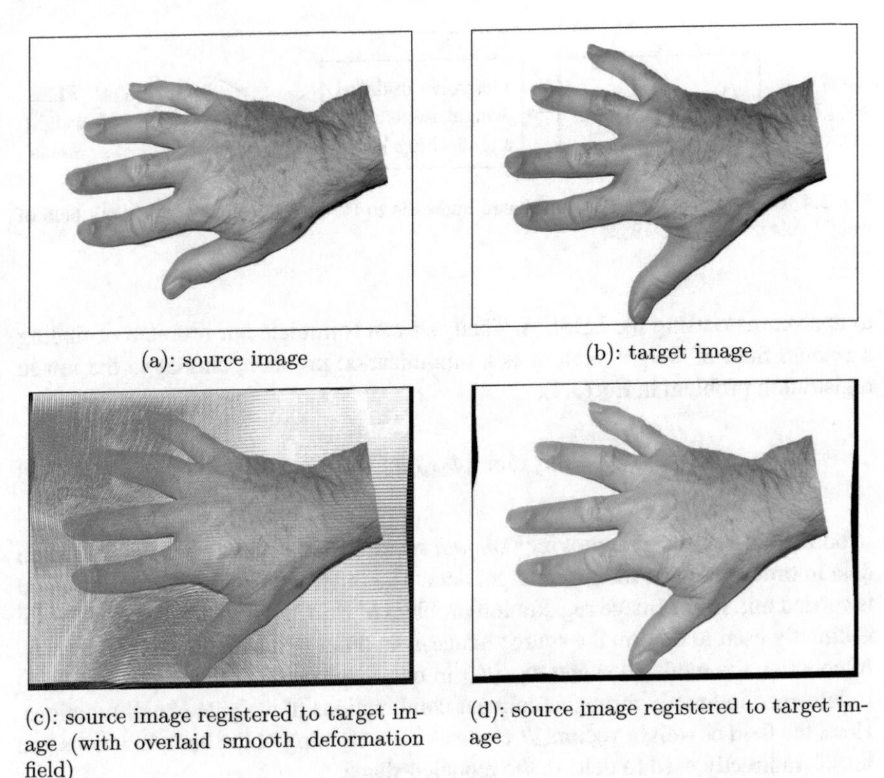

(a): source image

(b): target image

(c): source image registered to target image (with overlaid smooth deformation field)

(d): source image registered to target image

Fig. 3.3 Exemplary registration of two hands with a demons-based approach

deformation field is not required to be smooth, as no neighborhood relations are considered while estimating the deformation update. Then, after each incremental update step, the resulting weight field is convolved with a smoothing kernel K_{smooth} in order to ensure its smoothness. This process of subsequently computing element-wise updates of the deformation field and subsequently smoothing the deformation field in a local neighborhood is iterated until the obtained deformation field remains stable. An example can be seen in Fig. 3.3.

It is also possible to compute the deformation update only for a subset of all elements $\vec{x}$. The elements of this subset are called *demons*, hence the term *demons-based approaches*. By doing this, the deformation field in all the other elements, which are not part of the subset, is only modified by the convolution with the smoothing kernel K_{smooth} and not by external image forces. However, in image registration problems, it is common to compute the deformation update on all elements $\vec{x}$ of the data domain Ω [127], but we will come back to this important aspect in the following sections.

The connection between these demons-based approaches to image registration problems and our problem of finding a *smooth* field of weight vectors for the LDSSM can be seen when we assume that we have given a target shape Φ_{targ} which we want

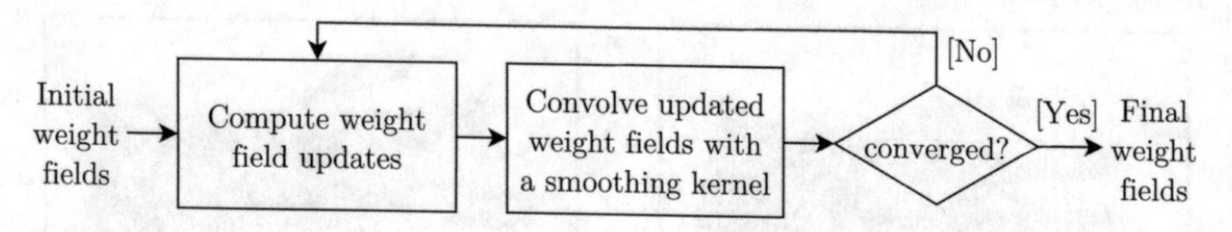

Fig. 3.4 General idea of the demons-based approach to the determination of a smooth field of weight vectors for our LDSSM

to approximate using the LDSSM. Then, we can formulate our problem of finding a *smooth* field of weight vectors as a minimization problem, similar to the image registration problem in Eq. (3.4):

$$\hat{\vec{\Psi}} = \arg\min_{\vec{\Psi}} ||\Phi_{\text{loc}}(\vec{\Psi}) - \Phi_{\text{targ}}||^2 . \tag{3.5}$$

In both problems, one searches for a *smooth* vector field that deforms the given source data in order to match the given target data. The difference is how the deformation is carried out. In the image registration problem of Eq. (3.4), the deformation field $\vec{v}$ is directly used to deform the source image I_1 in order to match the target image I_2. In contrast, the modeled shape $\Phi_{\text{loc}}(\vec{\Psi})$ in our shape reconstruction problem (3.5) is deformed indirectly through nonlinear combinations of weighted basis functions. Here, the field of weight vectors $\vec{\Psi}$ controls the weighting of the basis functions and thus is indirectly used to deform the modeled shape.

So, as both problems (3.4) and (3.5) closely resemble each other, our idea is to use the same procedure to obtain the *smooth* field of weight vectors $\hat{\vec{\Psi}} : \Omega \rightarrow \mathbb{R}^m$ that is also used in demons-based image registration to get a *smooth* deformation field $\hat{\vec{v}} : \Omega \rightarrow \mathbb{R}^2$. This means, we want to start with an initial field of weight vectors $\vec{\Psi}_{\text{init}}$ and compute element-wise updates of the field of weight vectors individually in each element $\vec{x}$ of the data domain Ω. In order to ensure its smoothness, the updated field of weight vectors is then convolved with a smoothing kernel K_{smooth} after each element has been updated. These two steps, the update and the smoothing of the field of weight vectors, shall be iterated until the obtained field of weight vectors remains stable. This general idea is also illustrated in the flow-chart of Fig. 3.4.

3.3.2 A Side Note to Optimal Weights for the SSM

In the previous section, we have presented the general idea of our approach to obtain a *smooth* field of weight vectors for the LDSSM. Before we describe this approach in detail, we first have a closer look at the minimization problem that has been stated in Eq. (3.5). As the LDSSM is an extension of the SSM, the minimization problem which we want to solve in order to obtain the *smooth* field of weight vectors should be somehow related to the SSM. So, we have a look at how to obtain an *optimal* weight vector $\hat{\vec{w}}$ for the SSM when we have given a target shape Φ_{targ}.

For this purpose, we recall the vectorial representation of the SSM, where each shape Φ is represented as a row-vector $\vec{\Phi}$ by successively appending the rows of the discretized data domain in a single vector. It has been defined in Eq. (2.33) as

$$\vec{\Phi}_{\text{glob}}(\vec{w}) = \bar{\vec{\Phi}} + \vec{w}\,\tilde{\mathbf{\Phi}}\,, \tag{3.6}$$

where $\tilde{\mathbf{\Phi}}$ is the matrix that one obtains by combining the m most important eigenvectors of the sample covariance matrix:

$$\tilde{\mathbf{\Phi}} = \begin{pmatrix} \tilde{\vec{\Phi}}_1 \\ \vdots \\ \tilde{\vec{\Phi}}_m \end{pmatrix}. \tag{3.7}$$

Making use of the fact that $\tilde{\mathbf{\Phi}}\tilde{\mathbf{\Phi}}^T = \mathbf{I}$, as the row-vectors of $\tilde{\mathbf{\Phi}}$ form an orthonormal basis of the subspace spanned by the m most important eigenvectors of the sample covariance matrix, we can dissolve Eq. (3.6) with regard to $\vec{w}$ in order to calculate the weights that correspond to a given modeled shape $\vec{\Phi}_{\text{glob}}(\vec{w})$:

$$\begin{aligned} \vec{\Phi}_{\text{glob}}(\vec{w}) &= \bar{\vec{\Phi}} + \vec{w}\,\tilde{\mathbf{\Phi}} \\ \vec{\Phi}_{\text{glob}}(\vec{w}) - \bar{\vec{\Phi}} &= \vec{w}\,\tilde{\mathbf{\Phi}} \\ \left(\vec{\Phi}_{\text{glob}}(\vec{w}) - \bar{\vec{\Phi}}\right)\tilde{\mathbf{\Phi}}^T &= \vec{w}\,\tilde{\mathbf{\Phi}}\,\tilde{\mathbf{\Phi}}^T \\ \left(\vec{\Phi}_{\text{glob}}(\vec{w}) - \bar{\vec{\Phi}}\right)\tilde{\mathbf{\Phi}}^T &= \vec{w}\,. \end{aligned} \tag{3.8}$$

Now, when we replace the modeled shape $\vec{\Phi}_{\text{glob}}(\vec{w})$ in Eq. (3.8) by our target shape $\vec{\Phi}_{\text{targ}}$ and use the same formula to calculate an *optimal* weight vector $\hat{\vec{w}}$ for this shape, we obtain

$$\hat{\vec{w}} = \left(\vec{\Phi}_{\text{targ}} - \bar{\vec{\Phi}}\right)\tilde{\mathbf{\Phi}}^T. \tag{3.9}$$

The question that arises is in which sense this weight vector $\hat{\vec{w}}$ is *optimal*. The answer to this question can be seen when we reinsert this result in the SSM from Eq. (3.7). By doing this, we get

$$\vec{\Phi}_{\text{glob}}(\hat{\vec{w}}) - \bar{\vec{\Phi}} = \left(\vec{\Phi}_{\text{targ}} - \bar{\vec{\Phi}}\right)\tilde{\mathbf{\Phi}}^T\tilde{\mathbf{\Phi}}\,. \tag{3.10}$$

This is the orthogonal projection of our target shape $\vec{\Phi}_{\text{targ}}$ into the SSM subspace [88]. It is well-known that the orthogonal projection of a point into a subspace gives a new point with minimal distance to the original point [88] (see also Sect. 2.2). Hence,

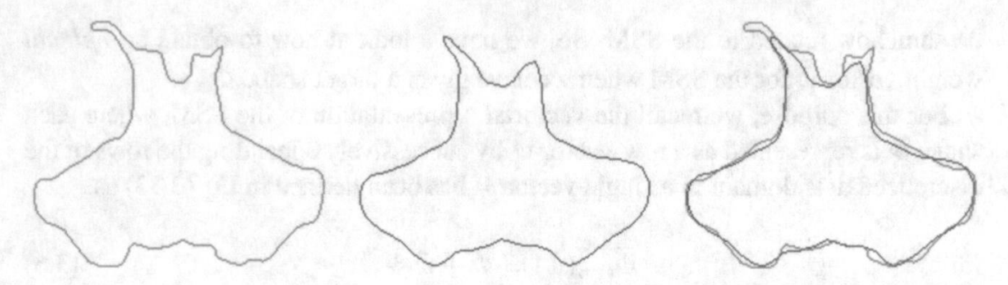

Fig. 3.5 Original shape (*green*) and best possible approximation that can be obtained with the SSM from Eq. (2.34) (*red*)

the minimization problem that is solved by Eq. (3.9) can be written as

$$\hat{\vec{w}} = \arg\min_{\vec{w}} ||\Phi_{\text{glob}}(\vec{w}) - \Phi_{\text{targ}}||^2 . \tag{3.11}$$

This means, we have shown that for the SSM one can compute a closed-form solution for the weight vector $\hat{\vec{w}}$ that belongs to the modeled shape $\Phi_{\text{glob}}(\hat{\vec{w}})$ which minimizes the distance to a given target shape Φ_{targ}. An example for this can be seen in Fig. 3.5. So, it is straightforward to use the same error function also for the determination of the *smooth* field of weight vectors from the LDSSM.

3.3.3 Determination of the Weights for a Known Target Shape

So far, we have argued in Sect. 3.3.2 which energy functional shall be minimized in order to obtain our *smooth* weight fields, and we have presented the general idea of how to approach this minimization problem in Sect. 3.3.1. In the following, we will now have a closer look at the individual steps of the proposed approach.

As stated in Eq. (3.5), the goal is to minimize the distance between the modeled shape $\Phi_{\text{loc}}(\hat{\vec{\Psi}})$ and the target shape Φ_{targ} under the constraint that the resulting weight fields $\hat{\vec{\Psi}}$ have to be smooth. Under the assumption that the weight fields as well as the modeled shape and the target shape are given as sampled functions, the distance which is defined in Eq. (3.5) can be written as

$$d(\vec{\Psi}) = \sum_{i=1}^{t} \left(\Phi^i_{\text{loc}}(\vec{\Psi}^i) - \Phi^i_{\text{targ}} \right)^2 , \tag{3.12}$$

where $t = jk$ if $d = 2$ and $t = jkl$ if $d = 3$, respectively (see also Eqs. (2.1) and (2.2)). In Eq. (3.12), Φ^i_{loc} and Φ^i_{targ} denote the i-th entry of the t-dimensional shape vectors $\vec{\Phi}_{\text{loc}}$ and $\vec{\Phi}_{\text{targ}}$, and $\vec{\Psi}^i$ denotes the corresponding i-th element from the field of weight vectors. So, the total distance is given as the sum of the element-wise squared differences $d^i(\vec{\Psi}^i)$ between the two shapes:

$$d(\vec{\Psi}) = \sum_{i=1}^{t} d^i(\vec{\Psi}^i), \tag{3.13}$$

where each element-wise distance $d^i(\vec{\Psi}^i)$ is given as

$$d^i(\vec{\Psi}^i) = \left(\Phi^i_{\text{loc}}(\vec{\Psi}^i) - \Phi^i_{\text{targ}}\right)^2. \tag{3.14}$$

In contrary to the optimal weight vector $\hat{\vec{w}}$ of the SSM, it is in general not possible to compute a closed-form solution for the optimal weight fields $\hat{\vec{\Psi}}$ of the LDSSM.[1] Instead, Eq. (3.12) has to be solved numerically. This is were the demons-based approach from Sect. 3.3.1 comes into play. We recall that the key idea is to calculate a weight update *independently* for each element of the data domain and use the following smoothing step in order to consider dependencies between adjacent elements. For our approach, this means that we assume independence between the element-wise weight vectors $\vec{\Psi}^i$ in the weight update step. As a result, we can simply calculate the gradient $\nabla d^i(\vec{\Psi}^i)$ of each element-wise error term in Eq. (3.14) and perform gradient descent individually for each element of the data domain. With the help of the LDSSM from Eq. (3.3), the element-wise distance from Eq. (3.14) can be rewritten as

$$d^i(\vec{\Psi}^i) = \left(\bar{\Phi}^i + \langle \vec{\Psi}^i, \vec{\tilde{\Phi}}^i \rangle - \Phi^i_{\text{targ}}\right)^2, \tag{3.15}$$

where $\vec{\tilde{\Phi}}^i$ denotes the i-th column of the Matrix $\tilde{\mathbf{\Phi}}$ which defines the basis of the SSM subspace. It contains the i-th element of each of the m base vectors:

$$\vec{\tilde{\Phi}}^i = \left(\tilde{\Phi}^i_1, \ldots, \tilde{\Phi}^i_m\right). \tag{3.16}$$

When we differentiate this with regard to $\vec{\Psi}^i$, the gradient of Eq. (3.14) is given by

$$\nabla d^i(\vec{\Psi}^i) = 2\left(\bar{\Phi}^i + \langle \vec{\Psi}^i, \vec{\tilde{\Phi}}^i \rangle - \Phi^i_{\text{targ}}\right)\vec{\tilde{\Phi}}^i. \tag{3.17}$$

Now, one can compute an update of the field of weight vectors by performing one or more gradient descent steps independently for each element of the data domain Ω:

[1] Equation (3.8) does not hold in this case because there is no scalar product defined in the generalized vector space of the LDSSM.

$$\vec{\Psi}_{k+1}^{i} = \vec{\Psi}_{k}^{i} - \alpha^{i}(\vec{\Psi}_{k}^{i})\, \nabla d^{i}(\vec{\Psi}_{k}^{i}), \tag{3.18}$$

where the index k denotes the k-th iteration and $\alpha^{i}(\vec{\Psi}_{k}^{i})$ denotes the element-wise step size of the gradient descent in each iteration. Our experiments show that is is sufficient to perform one gradient descent step using Cauchy's step size in order to compute a weight field update. Cauchy's step size is the step size which gives the most reduction of the function value along the current negative gradient direction [23, p. 931, Eq. (18.76)]. For the minimization problem in Eq. (3.14), Cauchy's step size is given by [23, p. 931, Eq. (18.77b)]

$$\alpha^{i}(\vec{\Psi}_{k}^{i}) = \frac{\langle \nabla d^{i}(\vec{\Psi}_{k}^{i}), \nabla d^{i}(\vec{\Psi}_{k}^{i}) \rangle}{2\, \langle \nabla d^{i}(\vec{\Psi}_{k}^{i})\, (\vec{\vec{\Phi}}^{i} \otimes \vec{\vec{\Phi}}^{i}), \nabla d^{i}(\vec{\Psi}_{k}^{i}) \rangle}. \tag{3.19}$$

After this update step, the updated weight field is convolved with a smoothing kernel K_{smooth} in order to obtain the desired *smooth* field of weight vectors.

At this point, we have derived a method to obtain a field of weight vectors $\hat{\vec{\Psi}}$ that minimizes the distance between the model approximation $\Phi_{\text{loc}}(\vec{\Psi})$ and the target shape Φ_{targ}, under the constraint that the weight fields have to be smooth. However, we have not yet considered the trained shape distribution so that the described approach might result into shapes which are very unlikely according to the training data. In order to prevent this, we constrain our weight field to the trained shape distribution.

As explained in Sect. 2.2, this can either be achieved by projecting each weight vector which lies outside the hyper-ellipse that describes the three-sigma boundary of the trained Gaussian distribution back onto this hyper-ellipse, or by truncating each weight vector's elements individually to lie within the three-sigma range. We decided to use the computationally much less expensive element-wise truncation of the weight vectors as an additionally regularization step before the smoothing step. Now, the modified flow-chart that incorporates the trained shape distribution can be seen in Fig. 3.6, and the detailed algorithm of our demons-based weight field update approach is given in Algorithm 3.1.

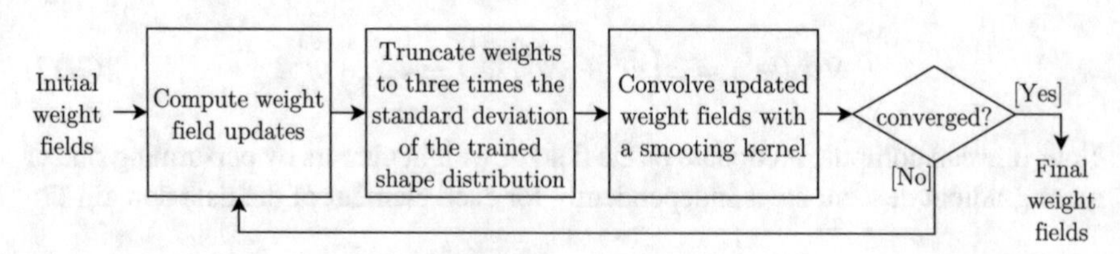

Fig. 3.6 Demons-based approach to the determination of a smooth field of weight vectors for our LDSSM that is constrained by the trained shape distribution

```
1: procedure LOCALSHAPEFIT(Ψ⃗_init, Φ_targ)
2:     Ψ⃗ ← Ψ⃗_init
3:     repeat
4:         for all weight vectors Ψ⃗^i in the weight field Ψ⃗ do
5:             if Φ^i_targ is given then
6:                 Ψ⃗^i ← Ψ⃗^i − α^i(Ψ⃗^i) ∇d^i(Ψ⃗^i)
7:                 Ψ⃗^i ← min(max(Ψ⃗^i, −3σ⃗), 3σ⃗)      ▷ comp.-wise min, max
8:             end if
9:         end for
10:        Ψ⃗ ← Ψ⃗ ∗ K_smooth      ▷ component-wise convolution
11:    until stopping criterion fulfilled
12:    return Ψ⃗
13: end procedure
```

Algorithm 3.1: Iterative approximation of a given (partial) target shape with the LDSSM.

As already mentioned in Sect. 3.3.1, the here presented iterative weight update approach also works for target shapes Φ_{targ} which are only partially known. This is the reason why we check for given shape information in line 5 of Algorithm 3.1. If no shape information is given for any element i, we skip the weight update and truncation step. So, the weight vectors $\vec{\Psi}^i$ for those elements i of the data domain Ω where no information about the target shape is present are only altered by the smoothing step in line 10 but not by the weight update loop in lines 4 to 9. As a consequence, the weight updates from the elements i with known target shape information Φ^i_{targ} are propagated to the remaining elements by the convolution with the smoothing kernel K_{smooth}.

A possible application for Algorithm 3.1 is the reconstruction of incomplete range scans with the help of the LDSSM. This will be investigated in Sect. 4.3. For image and volume segmentation problems, we have to further extend the presented approach, as in these problems the target shape is usually not given, not even partially. This problem will be addressed in the following subsections.

3.3.4 *Determination of the Weights for Segmentation Problems*

The iterative weight field update approach, presented in the previous section, can be applied when a target shape is at least partially known. However, in image segmentation problems, the target shape is not given explicitly. Instead, the goal of the segmentation process is to locate an unknown target shape within the image. So, in order to use the LDSSM for image segmentation problems, the weight field update approach from Sect. 3.3.3 has to be extended for an unknown target shape.

A common assumption in image segmentation problems is that large parts of the desired object boundary are located at edges in the image [46, Sect. 1.6.2]. So, we first present an approach to apply our iterative weight field update Algorithm 3.1 in order to extract the object of interest from binary edge images. This approach is then further extended to work also on gradient-magnitude images. Finally, we present a general formulation of our approach in order to make it applicable to a wide range of image segmentation problems, where the desired object boundary is no longer required to be located at the edges of the input image.

3.3.4.1 Weight Field Estimation for Binary Edge Images

In this subsection, we will derive a method that allows us to determine the weight fields that belong to an unknown target shape $\hat{\Phi}_{\text{targ}}$ using binary edge data I_{bin} as input. The image edges have been extracted by a suitable preprocessing approach from given image data I. In order to derive such a method, we have to deal with two major problems:

1. Only a subset of the image edges belongs to the boundary of our object of interest (e.g. the bird in Fig. 3.7a). The other edges may belong to other structures in the image or they may correspond to noise from the image formation process. So, we have to estimate which edges belong to the object of interest before we can compute a weight field estimate.
2. In our LDSSM, the interior of an object is represented by negative distance values and the exterior of the object is represented by positive distance values, respectively. However, for an edge in the image it is not a trivial task to decide on which side of the edge is the interior of the object and on which side is the exterior. In general, we can only compute the unsigned distance from the image edges (c.f. Fig. 3.7c).

(a) I (b) I_{bin} (c) I_{dist}

Fig. 3.7 Original image I, edge image I_{bin}, and unsigned distance image I_{dist}

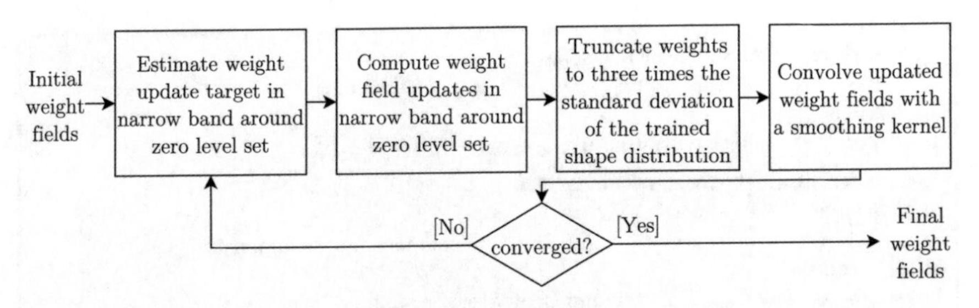

Fig. 3.8 Weight field update loop for an unknown target shape

Our solution to these problems is to assume that the evolving contour of our LDSSM is already close to the desired object boundaries. Consequently, we consider only those edges as relevant which lie inside the narrow band $\text{NB}(C_{\text{loc}})$ around the zero level contour C_{loc} of our LDSSM Φ_{loc} (c.f. Sect. 1.4.3). We then use only the relevant edges together with the evolving shape Φ_{loc} in order to estimate a *weight update target* $\hat{\Phi}_{\text{targ}}^{\text{weight}}$ in the narrow band around the evolving contour. This weight update target can then be used as an input for Algorithm 3.1 to update the weight vectors which reside inside the narrow band. As discussed in the previous section, the update of the weight vectors is then propagated to the rest of the image by the smoothing step of our iterative weight field update approach.

As the contour of our local model moves when we modify the field of weight vectors, the weight update target has to be recomputed in each iteration of our iterative weight field update approach. Consequently, Algorithm 3.1 has to be modified in order to consider the estimation of a weight update target in each iteration. Additionally, we replace the known target shape in the distance computation from Eq. (3.14) (line 6 of Algorithm 3.2) by the estimated weight update target:

$$d^i(\vec{\Psi}^i) = \left(\Phi_{\text{loc}}^i(\vec{\Psi}^i) - \Phi_{\text{targ}}^{i,\text{weight}}\right)^2. \tag{3.20}$$

The modified Algorithm 3.2 can now be used to extract an object from given image data I with the help of the LDSSM. The corresponding flowchart of our modified weight field update for an unknown target shape can be seen in Fig. 3.8.

Next, we answer the question how the estimated weight update target $\hat{\Phi}_{\text{targ}}^{\text{weight}}$ in the narrow band is obtained by the procedure EstimateWeightUpdateTarget in line 4 of Algorithm 3.2. Our basic idea to address this problem is that the contour $\hat{C}_{\text{targ}}^{\text{weight}}$ of the weight update target shall be located closer to the image edges than the contour C_{loc} of the currently modeled shape $\Phi_{\text{loc}}(\vec{\Psi})$. As a consequence, the gradient descent from line 6 will also drive the contour C_{loc} of the modeled shape $\Phi_{\text{loc}}(\vec{\Psi})$ closer to the image edges when we try to approximate the weight update target $\hat{\Phi}_{\text{targ}}^{\text{weight}}$ with the help of the LDSSM.

Two one-dimensional examples that illustrate the procedure are shown in Figs. 3.9 and 3.10, respectively. They depict cross-sections through the image plane Ω along

```
1: procedure LocalSegmentation(Ψ_init, I)
2:     Ψ ← Ψ_init
3:     repeat
4:         Φ̂_targ^weight ← EstimateWeightUpdateTarget(I, Φ_loc(Ψ))
5:         for all weight vectors Ψ^i in the narrow band NB(C_loc) do
6:             Ψ^i ← Ψ^i − α^i(Ψ^i) ∇d^i(Ψ^i)
7:             Ψ^i ← min(max(Ψ^i, −3σ), 3σ)    ▷ component-wise min, max
8:         end for
9:         Ψ ← Ψ ∗ K_smooth    ▷ component-wise convolution
10:    until stopping criterion fulfilled
11:    return Ψ
12: end procedure
```

Algorithm 3.2: Iterative segmentation of given image data I with the LDSSM.

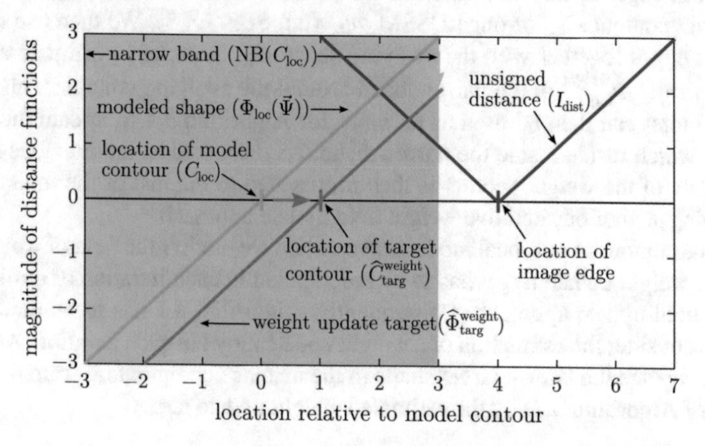

Fig. 3.9 Weight update target estimation for an image edge outside the modeled shape

the normal of the model contour C_{loc}. Figure 3.9 shows the case where an image edge is located outside the modeled shape, and Fig. 3.10 depicts the opposite case where an image edge is located inside the modeled shape, respectively. It can be assumed that the image edges are parallel to the model contour C_{loc} in the depicted cross-sections.

As said before, in both cases, the goal is to estimate a weight update target $\hat{\Phi}_{\text{targ}}^{\text{weight}}$ in the narrow band NB(C_{loc}) around the zero level set of our modeled shape $\Phi_{\text{loc}}(\vec{\Psi})$ so that the target contour $\hat{C}_{\text{targ}}^{\text{weight}}$ is located closer to the image edge than the model contour C_{loc}. This is depicted by a magenta arrow.

Now, it can be seen in Fig. 3.9 that when we subtract some value (cyan arrows) from the modeled shape $\Phi_{\text{loc}}(\vec{\Psi})$ (green line) inside the narrow band NB(C_{loc}) around the model contour, we obtain a weight update target estimate $\hat{\Phi}_{\text{targ}}^{\text{weight}}$ (red line) which fulfills our requirement. So, we can use the currently modeled shape minus some

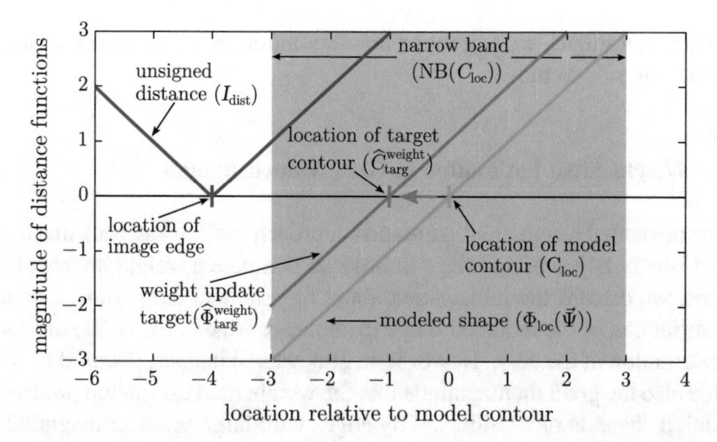

Fig. 3.10 Weight update target estimation for an image edge inside the modeled shape

value as a weight update target in those image regions where we want the model contour to move outwards. Likewise, when we want the contour to move inwards towards an image edge, we have to add some value to the modeled shape. This is shown in Fig. 3.10.

The problem that remains open is how to determine whether the image edge is located inside or outside the modeled shape. In order to figure this out, we consider the unsigned distance transformation I_{dist} of the edge image I_{bin} (blue line in Figs. 3.9 and 3.10). When we compare this unsigned distance representation with the modeled shape inside the narrow band, we can see that the gradient ∇I_{dist} of the signed distance image and the gradient $\nabla \Phi_{\text{loc}}(\vec{\Psi})$ of the modeled shape are oriented in opposite directions when the image edge is located outside the modeled shape, and they are oriented in the same direction when the image edge is located inside the modeled shape, respectively. So, we can compute the dot product between the gradient of the modeled shape and the gradient of the unsigned distance edge representation in order to estimate the correct sign of the offset from the modeled shape.

Consequently, we propose the following equation in order to estimate a weight update target for binary edge images:

$$\hat{\Phi}^{\text{weight}}_{\text{targ}}(\vec{x}) = \Phi_{\text{loc}}(\vec{\Psi}(\vec{x})) + \langle \nabla I_{\text{dist}}(\vec{x}), \nabla \Phi_{\text{loc}}(\vec{\Psi}(\vec{x})) \rangle . \tag{3.21}$$

As both gradients have unit length (under the assumption that the modeled shape is represented through a signed distance function), the magnitude of the scalar product will always be in the range $[-1, 1]$. The maximal offsets of $+1$ and -1 are thereby obtained for edges that are parallel to the model contour and which are located inside or outside the modeled shape, respectively. The minimal offset of zero is obtained for edges that are perpendicular to the model contour. This means that the modeled shape is mostly attracted by edges parallel to the current contour and it is not attracted by edges perpendicular to the current contour at all. This is not surprising, as in the case of perpendicular edges, we can make no assumption in which direction the model contour is supposed to move.

3.3.4.2 Weight Field Estimation in Gray Valued Images

So far, the presented weight field estimation approach has the drawback that it requires to extract binary edges of an image in order to compute a weight update target. By doing this, we discard the information about the gradient magnitude. The result is that the attraction of the modeled shape to an image edge in Eq. (3.21) depends only on the orientation of the edge. However, in gray valued images, it would be desirable to include also the gradient magnitude into the weight field estimation process so that the modeled shape is more attracted by edges with large gradient magnitudes than by edges with small gradient magnitudes. So, we extend the weight field estimation approach from the previous section in order to work on the gradient magnitude $|\nabla I|$ of the input image I instead of the binarized edge image I_{bin}.

The gradient magnitude $|\nabla I|$ has the property that it quickly decreases to zero with increasing distance from an image edge. However, for our weight field estimation approach it would be favorable to have non-zero values for the gradient magnitude in a larger area around the image edge in order to be able to identify the location of the nearest image edge. To achieve this, we convolve the gradient magnitude image $|\nabla I|$ with a Gaussian smoothing kernel K_{gauss} in order to spread the influence of the image gradient over a larger area. The thus obtained smoothed gradient magnitude image G is given as

$$G = K_{\text{gauss}} * |\nabla I| \,. \tag{3.22}$$

Thereby, the standard deviation of the Gaussian distribution determines the degree of the spreading. A one-dimensional example for the function curve of the smoothed gradient magnitude around an image edge is shown in Fig. 3.11 (blue line). It can be seen that in the narrow band $\text{NB}(C_{\text{loc}})$ this function curve is approximately proportional to the negative distance from the image edge. Consequently, we propose to replace the gradient of the distance image ∇I_{dist} in Eq. (3.21) by the negative gradient of the smoothed gradient magnitude image $-\nabla G$ in order to estimate a weight update target for gray valued images:

$$\hat{\Phi}^{\text{weight}}_{\text{targ}}(\vec{x}) = \Phi_{\text{loc}}(\vec{\Psi}(\vec{x})) - \langle \nabla G(\vec{x}), \nabla \Phi_{\text{loc}}(\vec{\Psi}(\vec{x})) \rangle \,. \tag{3.23}$$

The problem which arises is that now the magnitude of the dot product between the negative gradient of the smoothed gradient magnitude image $-\nabla G$ and the gradient of the modeled shape $\nabla \Phi(\vec{\Psi})$ may result in arbitrary large values. This is because the steepness of the smoothed gradient magnitude curve G is determined by the gradient magnitude $|\nabla I|$ and the standard deviation σ_{gauss} of the Gaussian smoothing kernel K_{gauss}, respectively.

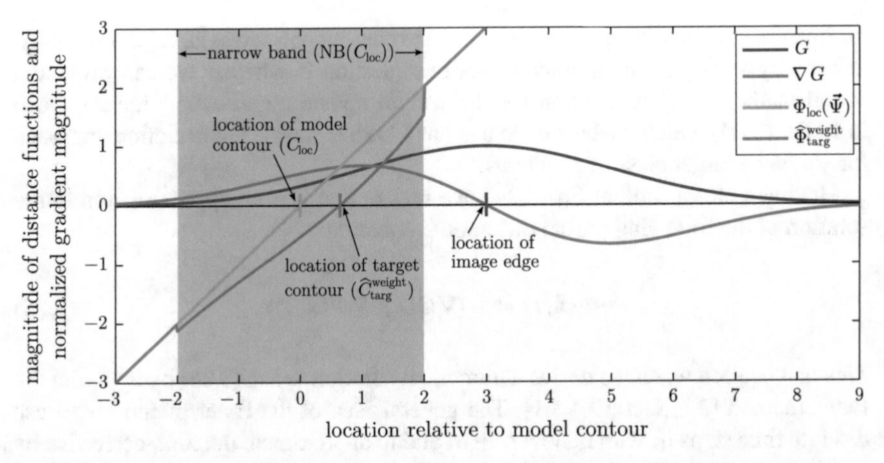

Fig. 3.11 Weight update target estimation based on the smoothed gradient magnitude

Now, a simple normalization of the gradient ∇G would eliminate the edge strength information which we wanted to included into our weight update target estimate. Instead, we heuristically determined that it is a good approach to limit the magnitude of the dot product from Eq. (3.23) to the range [0, 1] in order to stay in the same range of values as the results of the dot product from Eq. (3.21). For this purpose, we normalize the dot product by the maximal magnitude that arises in the narrow band $\mathrm{NB}(C_{\mathrm{loc}})$ around the model contour. So, our weight update target estimate for gray valued images can be finally formulated as

$$\hat{\Phi}^{\mathrm{weight}}_{\mathrm{targ}}(\vec{x}) = \Phi_{\mathrm{loc}}(\vec{\Psi}(\vec{x})) - \frac{\langle \nabla G(\vec{x}), \nabla \Phi_{\mathrm{loc}}(\vec{\Psi}(\vec{x})) \rangle}{\max_{\vec{x} \in \mathrm{NB}(C)} \left| \langle \nabla G(\vec{x}), \nabla \Phi_{\mathrm{loc}}(\vec{\Psi}(\vec{x})) \rangle \right|}. \tag{3.24}$$

An example for the estimated weight update target is depicted in Fig. 3.11. Similar to Figs. 3.9 and 3.10, a one-dimensional cross-section of the image plane Ω along the normal of the model contour C_{loc} is shown. The nearest image edge is located to the right of the model contour. It can be assumed that the image edge and the model contour C_{loc} are parallel to each other in this cross-section. The weight update target estimation is performed in the narrow band $\mathrm{NB}(C_{\mathrm{loc}})$ around the current contour by the approach from Eq. (3.24). It can be seen that, as desired, the target contour $\hat{C}_{\mathrm{targ}}$ is located closer to the nearest edge then the model contour C_{loc}.

3.3.4.3 Connection to Level Set Segmentation Methods

At this point, we have extended our weight field estimation approach in order to work on gradient magnitude images instead of binary edge images. However, one drawback which remains is that the method is heuristically motivated and works only for image segmentation approaches where the contour of interest is, at least partially, defined by strong image gradients. So, the question is whether we can provide a sound mathematical motivation for the weight update target estimation approach from Eq. (3.24) which enables us to use our LDSSM image segmentation approach for solving a larger class of problems.

Having a closer look at Eq. (3.24), we realize that it is an approximation to the solution of the following partial differential equation:

$$\frac{\partial}{\partial t}\Phi(\vec{x},t) = -\langle \nabla G(\vec{x}), \nabla \Phi(\vec{x},t)\rangle \,. \tag{3.25}$$

This can be seen when we derive a numerical solution to Eq. (3.25) by applying the Euler method [23, Sect. 19.4.1.1]. The general idea of the Euler method is to use discrete time steps t_k, with $t_{k+1} = t_k + \alpha$, and to approximate the time-derivative by a difference quotient:

$$\frac{\Phi(\vec{x},t_{k+1}) - \Phi(\vec{x},t_k)}{\alpha} = -\langle \nabla G(\vec{x}), \nabla \Phi(\vec{x},t_k)\rangle \,. \tag{3.26}$$

By doing this, one obtains a first-order time explicit scheme to calculate the new function value at time step t_{k+1} from the previous function value at time t_k:

$$\Phi(\vec{x},t_{k+1}) = \Phi(\vec{x},t_k) - \alpha\langle \nabla G(\vec{x}), \nabla \Phi(\vec{x},t_k)\rangle \,. \tag{3.27}$$

Now, when we assume that we have given the initial value $\Phi(\vec{x},t_0) = \Phi_{\mathrm{loc}}(\vec{\Psi}(\vec{x}))$ at time step t_0, Eq. (3.24) can be interpreted as the forward Euler approximation of Eq. (3.25) with step size

$$\alpha = \left[\max_{\vec{x}\in \mathrm{NB}(C_k)} \left|\langle \nabla G(\vec{x}), \nabla \Phi_{\mathrm{loc}}(\vec{\Psi}(\vec{x}))\rangle\right|\right]^{-1} , \tag{3.28}$$

where the solution $\Phi(\vec{x},t_1) = \hat{\Phi}_{\mathrm{targ}}^{\mathrm{weight}}(\vec{x})$ at time step t_1 yields our weight update target estimate.

When we further compare Eqs. (3.25) to (1.39), it can be seen that it describes a transport problem where the velocity field is given by ∇G. As outlined in Sect. 1.4.2, the corresponding level set formulation is then specified by Eq. (1.43) with scalar velocity

$$F(\vec{x}) = |\nabla G(\vec{x})| \cos(\nabla G(\vec{x}), \nabla \Phi(\vec{x},t)) \,. \tag{3.29}$$

This is the orthogonal projection of ∇G onto the outward normal of the level set $\nabla \Phi$. As ∇G determines the direction of increasing gradient magnitude values, the sign of the velocity F in a point $\vec{x}$ is positive when the outward normal $\nabla \Phi$ of the level set is oriented in the direction of increasing gradient magnitude values, and the sign is negative when the outward normal is oriented in the opposite direction, respectively. As a consequence, the direction of the level set evolution for a point $\vec{x}$

is always in the direction of increasing gradient magnitudes, no matter whether they are located inside or outside the evolving level set. The magnitude of the velocity F in a point $\vec{x}$ reaches its maximal amplitude when the direction of increasing gradient magnitudes is parallel to the normal of the level set in this point, and the velocity magnitude decreases to zero when the direction of increasing gradient magnitudes is perpendicular to the normal of the level set, respectively. This means that the zero level set is mostly attracted by parallel edges and it remains unchanged for perpendicular edges.

So, we have shown that our heuristically motivated weight update target estimation approach from Eq. (3.24) is mathematically well justified by the transport problem from Eq. (3.25). Also, it is particularly noteworthy that with our heuristically determined step size α from Eq. (3.28) already one Euler step is sufficient in order to obtain a good estimation of the weight update target $\hat{\Phi}_{\text{targ}}^{\text{weight}}$.

What we have achieved by this mathematical justification of our heuristic approach is that we are not limited to the particular velocity field ∇G anymore. Instead, we can now make use of all speed functions that have been published in the level set literature [115]. This will be addressed in more detail in Chap. 5.

Chapter 4
Evaluation of the Locally Deformable Statistical Shape Model

In Sect. 3.3, we presented an iterative framework that allows us to obtain a *smooth* field of weight vectors so that our *Locally Deformable Statistical Shape Model* (LDSSM), which we introduced in Sect. 3.2, can be used to approximate a (partially) known target shape as well as to solve image segmentation problems. In image segmentation problems, the target shape is initially unknown and has to be extracted from the image data.

In the following sections, we will evaluate the performance of our new LDSSM-based segmentation approach by segmenting the outer bony border of the nasal cavities and the paranasal sinuses (Sect. 4.1) and by segmenting the lower end of the femur (thighbone) as well as the upper end of the tibia (shinbone) in the vicinity of the right human knee joint (Sect. 4.2), respectively. The approximation of a partially known target shape will be subsequently evaluated by fitting the LDSSM to incomplete range scans of faces (Sect. 4.3).

In all cases, we will compare the performance of our new LDSSM-based shape fitting/segmentation approach to approaches that make use of global model information. The following evaluations are an extended version of the work that we have published in [5, 7, 9, 10].

4.1 Segmenting the Nasal Cavity and the Paranasal Sinuses

In the following subsections, we will demonstrate the potential of our LDSSM by extracting the combined outer bony boundary of the nasal cavity and the paranasal sinuses from *Computed Tomography* (CT) data. The results and approaches from the following subsections are an extended version of the work that we have presented in [9, 10].

The knowledge of the outer nasal cavity and paranasal sinuses boundary is useful in many clinical applications. It can be used e.g. for diagnosis of sinus pathologies,

C. Last, *From Global to Local Statistical Shape Priors*, Studies in Systems, Decision and Control 98, DOI 10.1007/978-3-319-53508-1_4

simulation of endonasal surgeries, surgical planning, and, in particular, robot assisted surgery. In the last few years, *Functional Endoscopic Sinus Surgery* (FESS) has been established as the state of the art technique for the treatment of endonasal pathologies. One important disadvantage of this approach is that the surgeon has to keep the endoscope in his hand during the whole surgery and consequently has only one hand left for the other surgical instruments. *Robot Assisted FESS* (RAFESS) may help to overcome this problem by passing the tedious job of endoscope guidance to a robot (see e.g. [1, 102, 103]). To exactly define the workspace of the robot, it is crucial to know where the outer nasal cavity and paranasal sinuses boundary is located. In order to extract this information from CT data, a segmentation of the structures of interest is unavoidable. However, the purely manual segmentation of the paranasal sinuses takes about 900 min [128] what is infeasible for the daily surgical workflow. Consequently, automatic segmentation approaches are required.

To the best of our knowledge, no fully-automatic approach to the segmentation of the paranasal sinuses exists so far. This is because of the great anatomical complexity and the high inter-patient variability of the endonasal structures (see Fig. 4.1). In 2004, Apelt et al. [12] published a semi-automatic framework for the segmentation of the inner and outer bony paranasal sinus boundaries. The user has to define an accurate *Volume of Interest* (VOI) around every object that should be segmented: The object boundaries have to be labeled manually in every second to tenth slice of the CT dataset. Coarse object boundaries in the remaining slices are then obtained via interpolation. Afterwards, an interactive watershed transform is applied inside the so-obtained user-defined VOIs in order to refine the object boundaries. Apelt et al. reported segmentation times of about one hour for a complete segmentation of the paranasal sinus boundaries where most time is spent for the manual generation of the different VOIs. This amount of manual interaction is still not feasible for the daily surgical workflow.

Other semi-automatic approaches that do not require per-slice user interaction have been proposed by Salah et al. [111], Seo et al. [114], and us [6]. All these approaches have in common that they are based on a 3D region-growing method that is used to segment the air-filled parts of the nasal cavity and the paranasal sinuses. Some manual pre- and post-processing is required in order to define the seed-points for the region-growing method and to prevent the region-growing method to leak into unwanted parts like e.g. the throat. With these semi-automatic approaches, it is possible to achieve segmentation times of about five to ten minutes [111] and, more importantly, to reduce also the manual interaction time to only a few minutes.

However, this speedup is mostly due to the fact that these approaches concentrate on extracting the air-filled parts of the paranasal sinuses. As reported in [12, 99], much more user action is required when the goal is to extract the outer bony border of the nasal cavity and the paranasal sinuses so that the segmented region includes also the mucosa. This is especially necessary in case of pathological sinuses with modified mucosa which occupies large portions of the paranasal sinuses. An example to clarify this assumption can be seen in Fig. 4.1a.

In summary, it is apparent that the segmentation of the paranasal sinuses is a complicated task. Even medical experts interpret CT data in different ways [129]. For

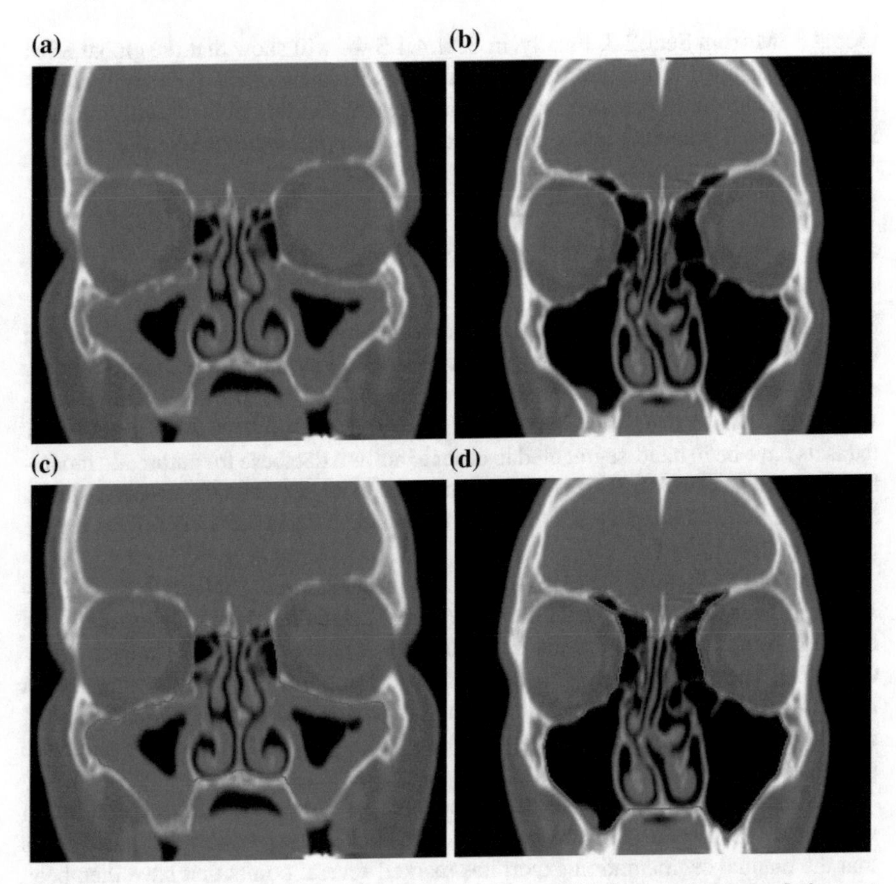

Fig. 4.1 Exemplary slices in frontal view from two different CT datasets, showing **a**, **b** the great inter-patient variability and **c**, **d** the overlaid desired boundaries. © Springer-Verlag Berlin Heidelberg 2010. Reprinted from [10] with permission of Springer

a non-expert it is practically impossible to produce a correct segmentation because detailed anatomical knowledge is required to identify the boundaries of the paranasal sinuses. However, it is not possible in the daily surgical workflow to keep an medical expert occupied for a couple of hours with manually or semi-automatically segmenting the paranasal sinuses of one patient. So, a fully-automatic segmentation approach is needed [99]. Due to the high anatomical complexity, we believe that such a fully-automatic segmentation is only possible with approaches that include anatomical model information of the endonasal structures.

The following subsections are structured as follows: In Sect. 4.1.1, we introduce the paranasal sinuses database on which we conduct our experiments. Afterwards, in Sect. 4.1.2, we present a fully-automatic approach that tackles the above-mentioned difficult segmentation problem by restricting potential solutions to those shapes that lie inside the low-dimensional subspace of feasible shapes that is spanned by the global SSM from Sect. 2.3. Finally, in Sect. 4.1.3 we will show that the global SSM is too constrained to capture the full intra-class variance of the paranasal sinuses and that much more accurate segmentation results can be obtained with our new LDSSM-based segmentation approach that has been presented in Sect. 3.3.4.

4.1.1 Description of the Paranasal Sinuses Database

The database for our experiment has been created by a manual segmentation expert with detailed anatomical knowledge of the nasal cavity and the paranasal sinuses at the *Klinik und Poliklinik für Hals-Nasen-Ohrenheilkunde/Chirurgie, Universitätsklinikum Bonn* in the period between December 2003 and May 2006 [99]. 49 CT datasets have been hand-segmented in order to act as a database for automatic model-based segmentation.[1] A 3D reconstruction of the hand-segmented datasets is shown in Fig. 4.2. One can clearly see the great inter-patient variability, especially for the frontal sinuses.

All used CT datasets have been acquired at the *Radiologische Klinik, Universitätsklinikum Bonn* by a spiral CT from Philips. The datasets have a slice thickness between 1 and 2 mm and the pixel resolution lies between 0.3×0.3 mm to 0.6×0.6 mm. These high-quality CT datasets ensure high-quality hand-segmentation results. A balanced male/female-ratio has been chosen (25 male patients and 24 female patients) with an age range from 16 to 78 years. The average age of the patients in the database is 39 years.

The manual segmentation has been conducted by the manual segmentation expert with a self-developed software tool that provides line segmentation. This means that the manual segmentation expert has marked several points that have then been connected via straight line segments. The segmentation has been performed layer by layer in each frontal view plane. So, the manual segmentation expert started the segmentation procedure in front of the patients face and then moved gradually towards the back of the head, marking the structures of interest in each of the 150–200 CT layers per dataset. All paranasal sinuses, including the nasal cavity, have been marked on the inner edge of their bony boundary so that the segmented regions contain the air-filled parts as well as the inner mucosa surface (c.f. Fig. 4.1). Neither the individual ethmoid cells nor the turbinates have been marked. With this procedure, the manual segmentation took about 8–10 h for each CT dataset.

In addition to the manual segmentation, 24 anatomical landmarks have been marked in each CT dataset. Anatomical landmarks are biologically-meaningful points with a reproducible, corresponding location along all datasets of the database. They can be used for example to rigidly align the individual datasets, i.e. to align them with regard to rotation, translation, and uniform scale. The landmarks which are best suited for a registration of the individual CT datasets are the anterior

[1] Please note that in [99] the database consisted of 50 datasets. However, one dataset had to be excluded because of the lack of anatomical landmarks.

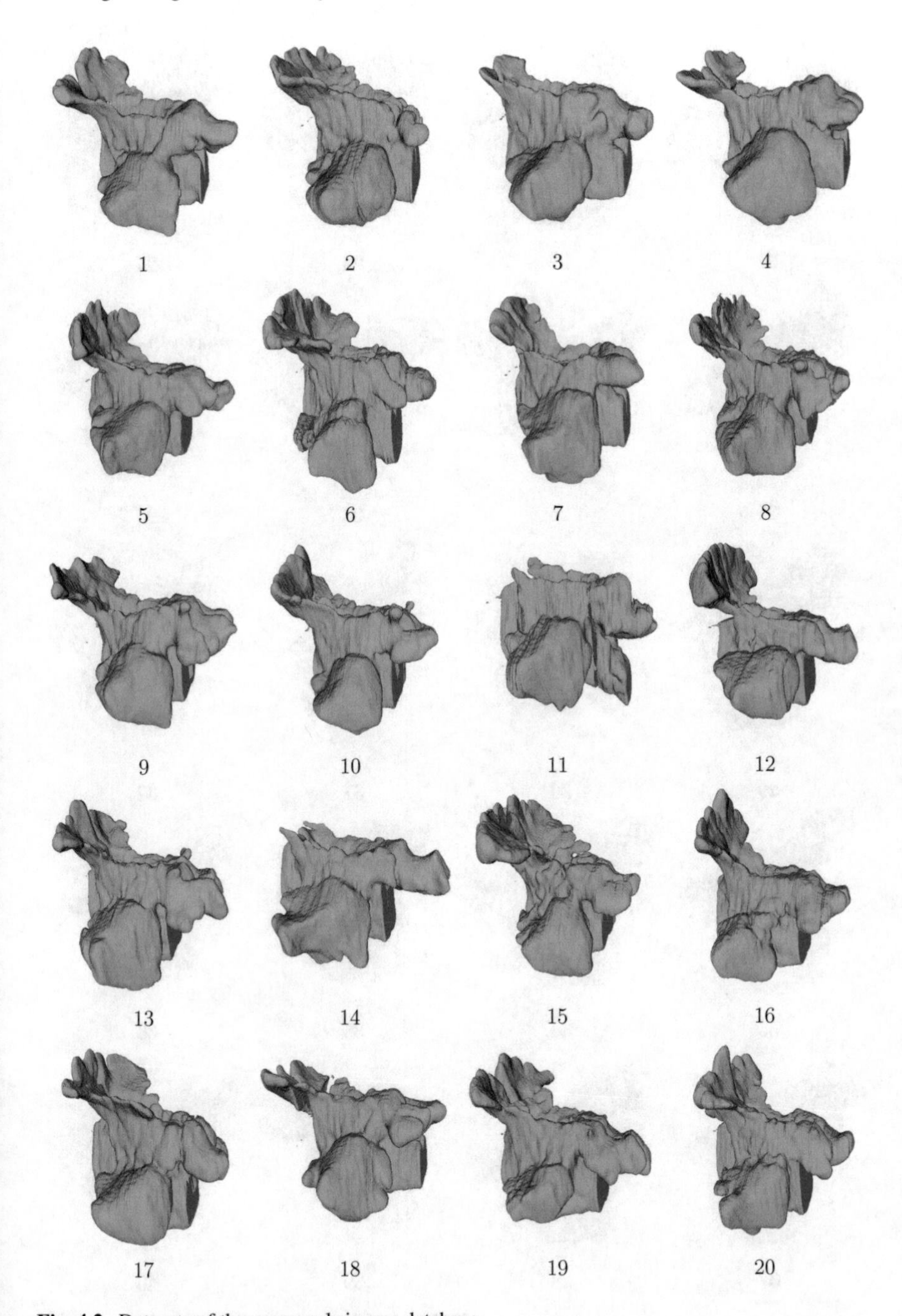

Fig. 4.2 Datasets of the paranasal sinuses database

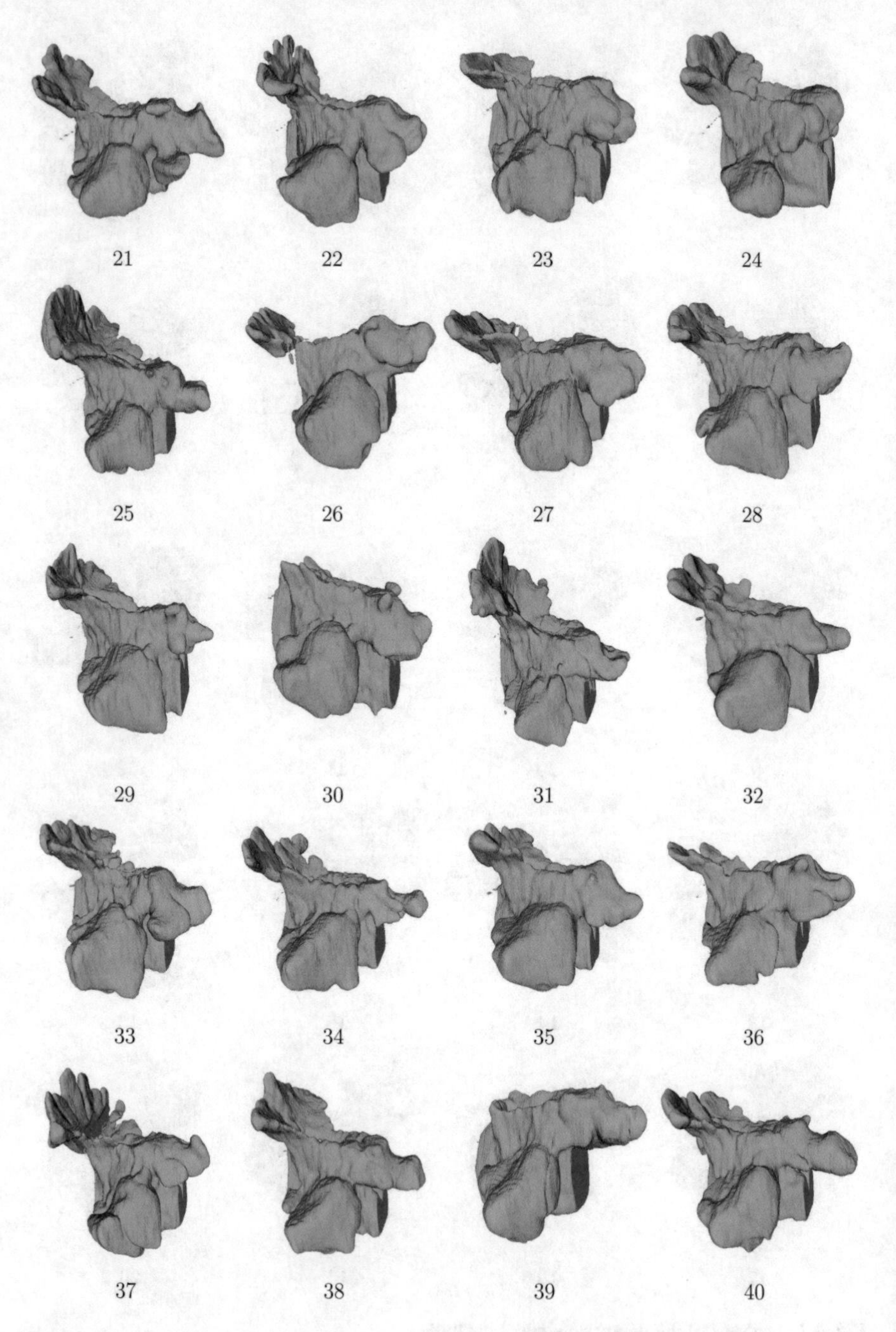

Fig. 4.2 (continued)

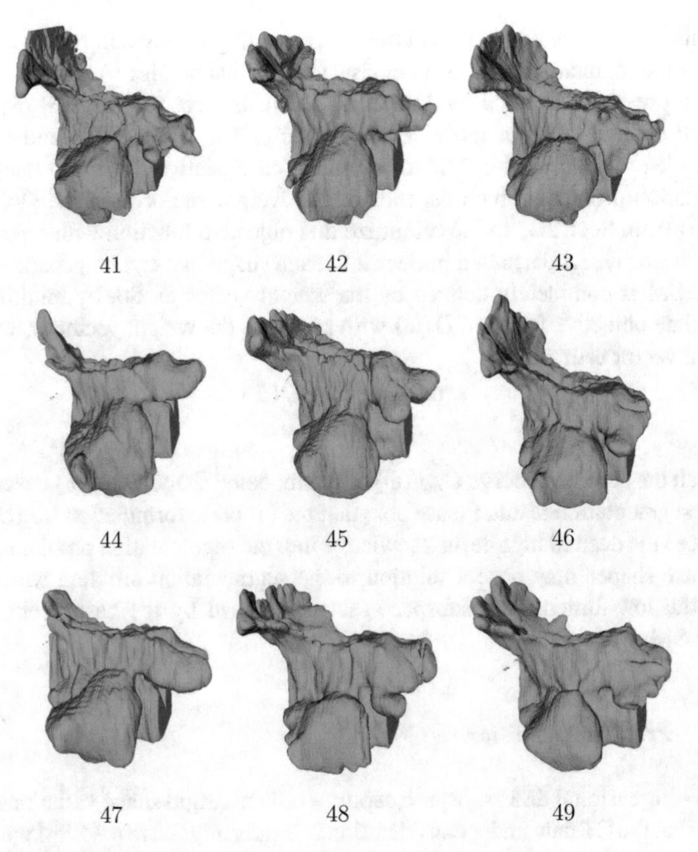

Fig. 4.2 (continued)

nasal spine, the posterior nasal spine, the upper tip of the crista galli, and the left and right styloid process [99]. These are 5 small-sized bony markers which can be located with a high accuracy between the individual datasets. Additional landmarks that have been marked are amongst others the infraorbital foramina and the orbit's centroids.

4.1.2 Segmentation with Global Model Information

As mentioned in the introduction of Sect. 4.1, so far it exists no fully-automatic method to extract the outer bony border of the nasal cavity and the paranasal sinuses from CT data. However, in order to show the potential of our LDSSM, we need a reference segmentation that automatically extracts the desired boundary with the help of statistical shape information provided by the global SSM from Sect. 2.3.

In this section, we address this issue by presenting a fully-automatic processing chain for the segmentation of the paranasal sinuses that builds up on the work which has been presented in 2003 by Tsai et al. [130]. In Sect. 5.3, we will discuss the approach of Tsai et al. in more detail. For now, it is sufficient to know that the general idea of their approach is to formulate an objective function $O(\vec{w})$ for the segmentation problem, which depends on the weight vector $\vec{w}$ of the global SSM $\Phi_{\text{glob}}(\vec{w})$ from Sect. 2.3, and to minimize this objective function with regard to the weight vector $\vec{w}$. As discussed in Sect. 2.2, each shape that can be generated by the global SSM is completely defined by the weight vector $\vec{w}$. So, by minimizing an appropriate objective function $O(\vec{w})$ with regard to the weight vector $\vec{w}$, we obtain a weight vector estimate

$$\hat{\vec{w}} = \arg\min_{\vec{w}} O(\vec{w}) \tag{4.1}$$

for which the zero level curve $C_{\text{glob}}(\hat{\vec{w}})$ of the modeled shape $\Phi_{\text{glob}}(\hat{\vec{w}})$ describes the desired segmentation result. Please note that the problem formulation from Eq. (4.1) introduces the desired high-level knowledge into the segmentation problem, because only those shapes may pose a solution to the segmentation problem which reside within the low-dimensional subspace that is spanned by the base vectors of the global SSM.

4.1.2.1 Problem Definition

The low-dimensional SSM subspace captures only nonrigid shape variations. So, we assume that the CT data under consideration is rigidly aligned to the SSM with regard to rotation, translation, and scale. If this requirement is met, we can define an objective function $O(\vec{w})$ that describes the segmentation problem. For this purpose, we remind ourselves that we have already become acquainted with some energy functionals that are used to describe segmentation problems in the context of variational image segmentation (c.f. Sect. 1.5), namely the region-based energy functional by Chan and Vese [26] which has been defined in Eq. (1.58) and the edge-based *Geodesic Active Contours* approach that is given in Eq. (1.53). These energy functionals can be modified to serve as an objective function for the above-mentioned approach by replacing the level set function Φ through the global model $\Phi_{\text{glob}}(\vec{w})$. In fact, the thus obtained modified region-based function by Chan and Vese has been considered as a possible objective function in the original work by Tsai et al. (c.f. Sect. 5.3.2).

The energy functional of Chan and Vese is designed to segment the data into two regions with different mean intensities. However, this does not reflect the properties of our segmentation problem. As mentioned above, the goal here is to extract the combined outer border of the nasal cavity and paranasal sinuses. Two exemplary CT slices and the corresponding hand-segmented reference segmentations can be seen in Fig. 4.1. It is clearly visible that the region inside the desired boundary consists of materials with very different intensities – bone (white), mucosa (grey), and air (black) – and that the region outside the desired boundary is made up of very similar intensities. One way to address this problem is to transform the data under consideration by applying some kind of mapping function that takes into account local intensity correlations with the aim that the transformed data can then be segmented into two regions with different mean values. These local intensity correlations are commonly referred to as *image texture*. For further information we refer the reader to [40]. The most common mapping function, which has also been used by Tsai et al. in [130], is the gradient magnitude.

An example for the gradient magnitude can be seen in Fig. 4.5a. However, in our case there exist similar local intensity correlations on the boundary of the paranasal cavities as well as inside and outside the paranasal cavities. So, we think that a region-based objective function, like the one by Chan and Vese, is not an optimal choice for our problem.

Now, what we know from the hand-segmented training data, which has been described in Sect. 4.1.1, is that always the edges of the bony structures have been marked. This means, we expect high gradient magnitude values on the segmentation border. So, we propose to use the following edge-based objective function to describe the segmentation problem:

$$O(\vec{w}) = -\frac{1}{|C_{\text{glob}}(\vec{w})|} \sum_{\vec{x} \in C_{\text{glob}}(\vec{w})} \left[K_{\text{gauss}} * |\nabla I| \right] (\vec{x}) , \tag{4.2}$$

where K_{gauss} defines a Gaussian smoothing kernel, $|\nabla I|$ is the gradient magnitude of the input data I, $C_{\text{glob}}(\vec{w})$ is the zero level curve of the modeled shape $\Phi_{\text{glob}}(\vec{w})$, and $|C_{\text{glob}}(\vec{w})|$ is the length of the zero level curve.

Equation (4.2) is a slightly modified version of the *Geodesic Active Contours* energy functional from Eq. (1.53). Consequently, it reaches its minimum when large parts of the zero level curve of the modeled shape are located on the edges of the input data I. The Gaussian smoothing kernel K_{gauss} is used here to reduce the likelihood of local minima of the objective function, caused by noise or weak gradients, by spreading the influence of strong gradients over neighboring elements. After the convolution with the smoothing kernel, one obtains a smoother course of the gradient magnitude. In the original GAC approach, shorter contours are additionally preferred over long contours in order to make the segmentation result more robust against noise (c.f. Sect. 1.5.1). However, as we consider each shape in the SSM subspace equally likely, regardless of the length of its zero level set, we normalize the objective function by the contour length.

When we take a look at Fig. 4.5a, we see that there exist many edges with high gradient magnitudes so that it is not possible to extract the desired boundary solely based on this low-level image information. This is why Eq. (4.2) contains additional high-level knowledge in form of the global SSM from Sect. 2.3. So, many of the edges from Fig. 4.5a should be irrelevant for the determination of the segmentation result as possible results of Eq. (4.2) are restricted to the subspace of feasible shapes that is spanned by the SSM.

4.1.2.2 Processing Chain

Now that we have defined an objective function $O(\vec{w})$ for our segmentation problem in Eq. (4.2), we have to minimize this objective function in order to obtain the desired segmentation result. For this purpose, we propose to use a standard minimization method, namely the well established *Nelder–Mead simplex method* [74]. The *Nelder–Mead simplex method* is one of the most widely used methods for unconstrained nonlinear optimization problems. This is because it is known to produce significant improvements of the objective function's value already in the first few iterations and it often requires substantially fewer function evaluations than alternative methods [74]. It is a direct search method, which means that it tries to find the minimum of the objective function without using any derivatives. For an n-dimensional optimization problem, the method is initialized with $n+1$ starting points that form a simplex. A simplex is a geometric figure which is obtained as the convex hull of its vertices (e.g. a triangle for two-dimensional optimization problems). In each iteration, the method evaluates the objective function at the simplex vertices and updates one or more vertices based on a set of rules in order to reduce the value of the objective function. This is continued until no further improvement of the objective function can be achieved. The optimization result is then obtained as the simplex vertex with the smallest function value. An example of the algorithm applied to Himmelblau's function [61], a well-known two-dimensional test function for optimization algorithms, can be seen in Fig. 4.3.

Initial attempts to minimize equation (4.2) with the *Nelder–Mead simplex method* showed that, despite the high-level knowledge, the objective function $O(\vec{w})$ seems to be highly non-convex. This means that Eq. (4.2) contains many local minima so that the solution tends to be very sensitive to the initial values of the weights $\vec{w}$ that are

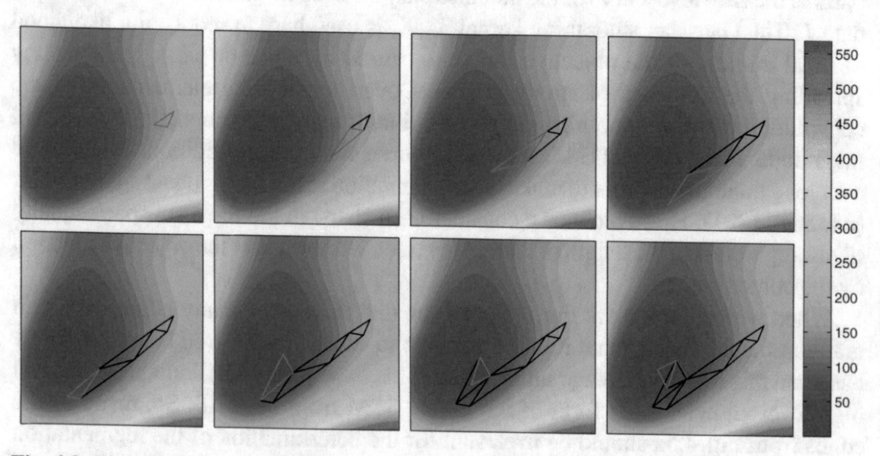

Fig. 4.3 Eight iterations of the *Nelder–Mead simplex method*, searching for a minimum of Himmelblau's two-dimensional test function in the range $[-5, 0] \times [-5, 0]$. The current simplex is shown in *red* and past simplices are shown in *black*

presented to the minimization algorithm. In general, the resulting segmentations were thus unsatisfactory. To circumvent this drawback, we need to find good initial values for the minimization of $O(\vec{w})$. This can be achieved for example by minimizing a more robust, but also more inaccurate, error function prior to the minimization of equation (4.2) and to use the thus obtained result as the initial solution for the minimization of equation (4.2).

The reason for the non-convexity of equation (4.2) is that the CT data I contains many edges, only a few of which are part of the desired boundary (c.f. Fig. 4.5a), and that the gradient magnitude falls off rapidly with increasing distance from an edge. So, Eq. (4.2) may yield large values even if the contour $C_{\text{glob}}(\vec{w})$ is located only a few elements away from the desired boundary and it may easily be trapped in local minima that are caused by other edges which are not part of the desired boundary. As mentioned above, the risk to get caught in a local minimum caused by noise or weak gradients can be reduced by convolving the gradient magnitude $|\nabla I|$ with a Gaussian smoothing kernel K_{gauss}. Besides noise reduction, this has also the benefit that the edges are smeared so that the influence of strong gradients is spread over a greater area.

As a result, the contour gets more attracted by nearby edges with strong gradient magnitudes. This is because the objective function now yields small values even if the contour is not perfectly centered with the edges. The rate of decay of the gradient magnitude, and hence the *capture range* of strong edges, can thereby be adjusted up to a certain degree by varying the standard deviation of the Gaussian smoothing kernel K_{gauss}. However, the larger the standard deviation, the more neighboring edges get blurred together so that the original position of an edge gets lost. This can be seen in Fig. 4.4 and is the reason why in practice the standard deviation of the Gaussian kernel is limited.

So, in order to obtain a more robust error function, we need to find another way to reduce the influence of irrelevant edges and to spread the *capture range* of relevant edges over a large area without corrupting their positions. For this purpose, we make once more use of the fact that we search only for bony structures. In CT data, bony structures are located in the range from 300 to 1000 Hounsfield units [14]. Gradients which are caused by edges that lie in this intensity range should thus get a higher weight than gradients which are caused by edges from other intensity ranges. In order to achieve this, we can define a one-dimensional Gaussian distribution with a mean value of $\mu_e = 650$, i.e. $\mu_e = \frac{1000-300}{2} + 300$, and an empirically determined standard deviation of $\sigma_e = 250$, and we can use this intensity distribution to weight the squared gradient magnitude data $|\nabla I|^2$ [131].

By doing this, we obtain the intensity-weighted gradient magnitude data E with

$$E(\vec{x}) = |\nabla I(\vec{x})|^2 \frac{1}{\sqrt{2\pi}\,\sigma_e} \exp\left(-\frac{1}{2}\left(\frac{I(\vec{x}) - \mu_e}{\sigma_e}\right)^2\right) \tag{4.3}$$

for all elements $\vec{x}$. It can be seen in Fig. 4.5b.

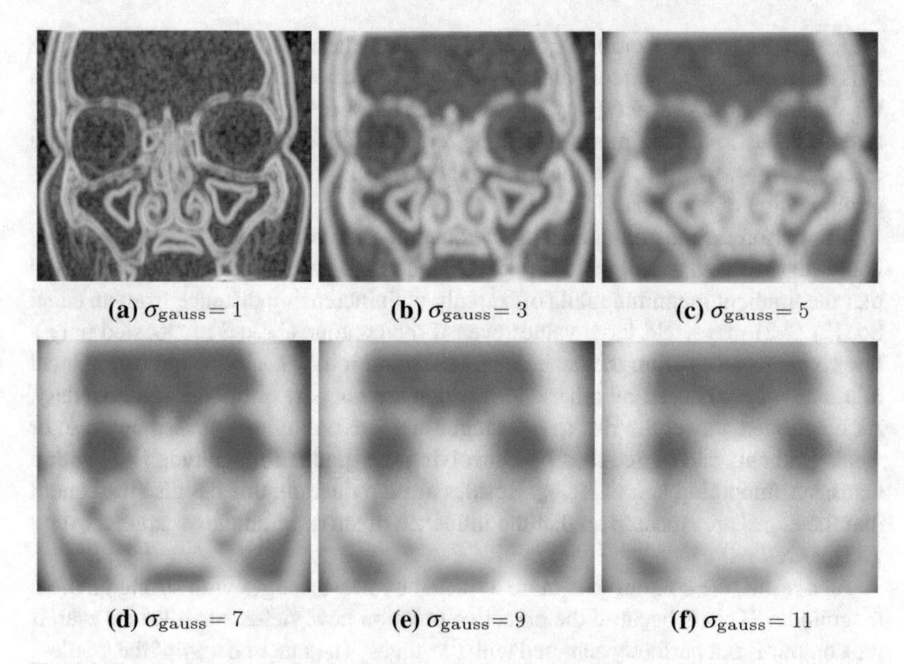

Fig. 4.4 Gradient magnitude image $|\nabla I|$, convolved with Gaussian smoothing kernels K_{gauss} that differ in the standard deviation σ_{gauss} (logarithmic representation)

As mentioned above, the intensity-weighted gradient magnitude data is then convolved with a Gaussian smoothing kernel in order to emphasize nearby strong gradients and to remove weak gradients (Fig. 4.5c). Afterwards, we extract binary edge data E_{bin} that contains only the strongest edges by performing non-maxima suppression [73] followed by thresholding with a threshold t (Fig. 4.5d). On this binary data, a distance transform [52] is applied that yields the distance-transformed data E_{dist} which contains the distance to the nearest edge in each element $\vec{x}$ (Fig. 4.5e). The metric used therein is the euclidean distance.

This distance-transformed edge data E_{dist} can now be used to define a new objective function $O_{\mathrm{dist}}(\vec{w})$ which contains less local minima than the original objective function $O(\vec{w})$ and has a large *capture range* because it contains the distance to the nearest strong edge in each element:

$$O_{\mathrm{dist}}(\vec{w}) = \frac{1}{|C_{\mathrm{glob}}(\vec{w})|} \sum_{\vec{x} \in C_{\mathrm{glob}}(\vec{w})} E_{\mathrm{dist}}(\vec{x}) \,. \tag{4.4}$$

As for the minimization of equation (4.2), we also use the *Nelder–Mead simplex method* to minimize equation (4.4). However, this time the objective function is robust enough to converge to the desired result when we start the minimization from the mean shape $\bar{C}$. This means that we use $\vec{w}_{\mathrm{init}} = (0, \ldots, 0)$ as initial weight vector for the minimization of $O_{\mathrm{dist}}(\vec{w})$. We denote the thus obtained minimization result as

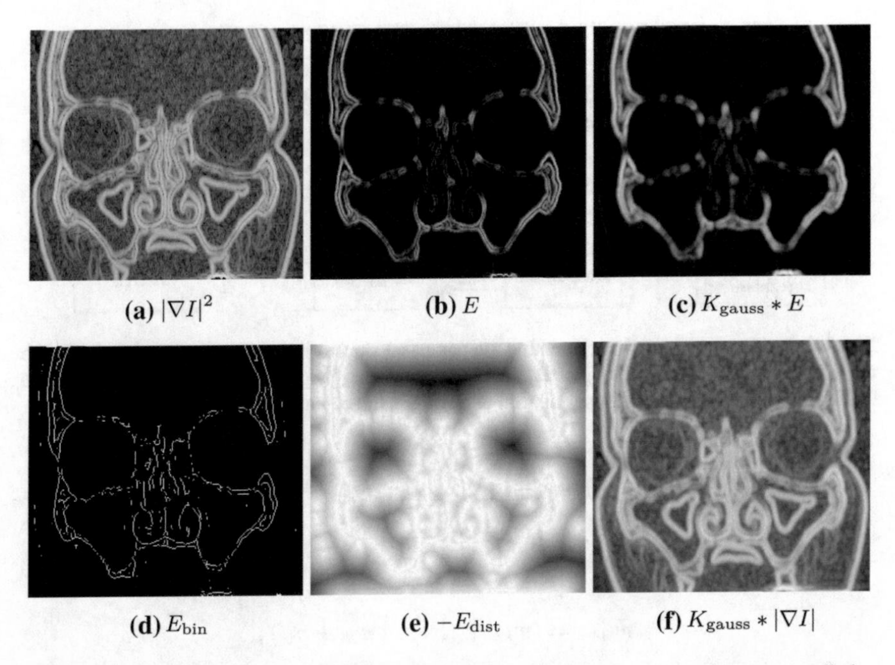

Fig. 4.5 Intermediate steps of our processing chain for segmenting the paranasal sinuses, applied to the CT data in Fig. 4.1a (logarithmic representation). Subfigures **(a)**–**(d)**: © Springer-Verlag Berlin Heidelberg 2010. Reprinted from [10] with permission of Springer

$$\vec{w}_{\text{dist}} = \arg\min_{\vec{w}} O_{\text{dist}}(\vec{w}) , \tag{4.5}$$

and we use it as the initial value for the minimization of our primary objective function $O(\vec{w})$ from Eq. (4.2). With this initial value, the minimization of equation (4.2) now yields correct results that improve the initial solution.

The complete processing chain for extracting the outer bony border of the nasal cavity and the paranasal sinuses is depicted in Fig. 4.6. The magnitude of the gradient $|\nabla I|$ is approximated from the input data I by using central differences. Please note that we use the squared gradient magnitude in the process of defining the input data for the intermediate objective function $O_{\text{dist}}(\vec{w})$, because this allows a more robust choice of the binarization-threshold t. For the final objective function $O(\vec{w})$, we use the gradient magnitude without squaring it (see Fig. 4.5f).

What has not been mentioned so far is how the $m + 1$ vertices from the initial simplex of the *Nelder–Mead simplex method* are defined. Given an initial weight vector $\vec{w} = (w_1, \ldots, w_m)$ (i.e. $\vec{w} = \vec{w}_{\text{init}}$ or $\vec{w} = \vec{w}_{\text{dist}}$ for our processing chain), the vertices of the initial simplex are initialized as

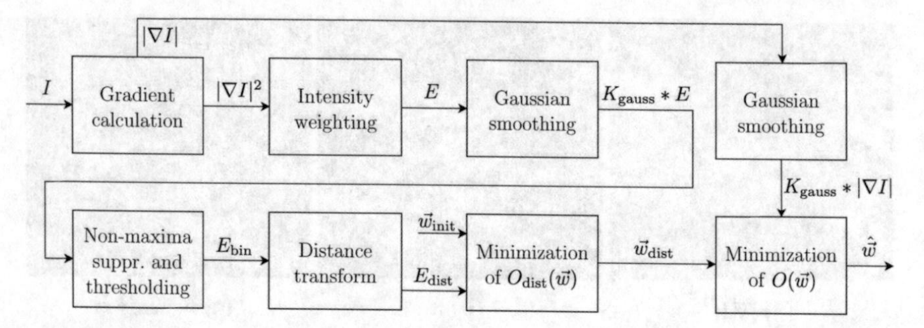

Fig. 4.6 Processing chain for segmenting the paranasal sinuses by consecutively minimizing two objective functions that contain global model information. ©Springer-Verlag Berlin Heidelberg 2010. Reprinted from [10] with permission of Springer and extended according to [9]

$$\begin{aligned}
\vec{w}_1 &= (w_1 \qquad , \ \ldots \ , w_m \qquad) \\
\vec{w}_2 &= (w_1 + \sigma_1, \ \ldots \ , w_m \qquad) \\
&\vdots \\
\vec{w}_{m+1} &= (w_1 \qquad , \ \ldots \ , w_m + \sigma_m) \, ,
\end{aligned} \tag{4.6}$$

where $\sigma_1, \ldots, \sigma_m$ are the standard deviations of the Gaussian distribution from Eq. (2.9).

4.1.3 Experimental Setup and Results

Now that we have described how the global approach by Tsai et al. can be adapted to the problem of segmenting the outer nasal cavity and paranasal sinus boundary, we can compare it to our new LDSSM-based segmentation approach that has been presented in Sect. 3.3.4. Both approaches have been explained in a general form that is applicable in two, three, or even more dimensions. However, for the moment we concentrate on segmenting the outer nasal cavity and paranasal sinuses boundary from a single two-dimensional slice of each complete CT dataset only in order to point out the advantages of our local approach over a global approach.

An application of our new LDSSM-based approach to three-dimensional data will be presented in Sect. 4.3 and the three-dimensional segmentation of the paranasal sinuses will be evaluated later in Chap. 6.

4.1.3.1 Error Measures

We use four error measures, two contour-based error measures and two volume-based error measures, to compare the automatic segmentation results with the hand-

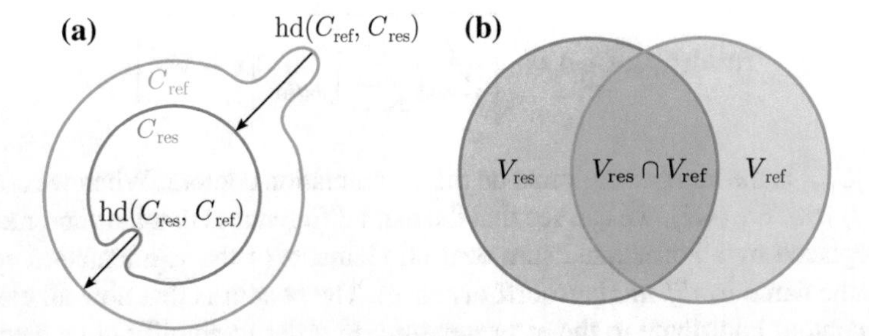

Fig. 4.7 **a** *Directed* Hausdorff distances from the segmentation contour C_{res} to the reference contour C_{ref} and vice versa. **b** Venn diagram showing the overlap of the segmented region V_{res} and the reference region V_{ref}

segmented reference. Contour-based error measure means that the contour of the segmentation result is compared to the hand-segmented reference contour and volume-based error measure means that the volumes (or areas in 2D) which are enclosed by the automatic and hand-segmented segmentations, respectively, are compared.

The first one of the two contour-based error measures is the Hausdorff distance [69]. The Hausdorff distance has been proposed by Felix Hausdorff in order to compare two point sets. The *directed* Hausdorff distance that measures the distance from the segmentation contour C_{res} to the reference contour C_{ref} is defined as

$$\mathrm{hd}(C_{res}, C_{ref}) = \sup_{\vec{x} \in C_{res}} \left[\inf_{\vec{y} \in C_{ref}} \|\vec{x} - \vec{y}\| \right] , \tag{4.7}$$

where “sup” denotes the supremum, “inf” the infimum, and $\|\cdot\|$ is the the euclidean distance. So, the *directed* Hausdorff distance $\mathrm{hd}(C_{res}, C_{ref})$ is obtained as the maximum euclidean distance from any point $\vec{x}$ on the segmentation contour to the closest point $\vec{y}$ on the reference contour (see Fig. 4.7a). The *directed* Hausdorff distance is not a metric since it is not symmetric. This means that $\mathrm{hd}(C_{res}, C_{ref})$ delivers usually not the same result as $\mathrm{hd}(C_{ref}, C_{res})$. In order to obtain a symmetric error measure, the Hausdorff distance is defined as the maximum of the two *directed* Hausdorff distances:

$$\mathrm{HD} = \max \{\mathrm{hd}(C_{res}, C_{ref}), \mathrm{hd}(C_{ref}, C_{res})\} . \tag{4.8}$$

The Hausdorff distance is well-suited to detect strong deviations between the segmentation and the reference. However, it is also error-prone to noise as a single outlier can severely change the result.

Another well-known error measure to compare two data series is the root-mean-square deviation. In the context of object matching and segmentation, it is also known as *modified* Hausdorff distance [48]. We define the *directed* root-mean-square deviation as

$$\mathrm{rmsd}(C_{\mathrm{res}}, C_{\mathrm{ref}}) = \sqrt{\frac{1}{|C_{\mathrm{res}}|} \sum_{\vec{x} \in C_{\mathrm{res}}} \left[\inf_{\vec{y} \in C_{\mathrm{ref}}} \|\vec{x} - \vec{y}\|^2 \right]}, \tag{4.9}$$

where $|C_{\mathrm{res}}|$ is the number of points on the segmentation contour. When we compare Eq. (4.9) with Eq. (4.7), we can see that the main difference is that the supremum has been replaced by a normalized sum over all elements of the segmentation contour (hence the name *modified* Hausdorff distance). The benefit is that now all elements of the contour contribute to the error measure. In order to amplify large deviations from the reference contour, we additionally use the squared euclidean distance in the computation of the error. Like the *directed* Hausdorff distance, the *directed* root-mean-square deviation is not symmetric and we obtain the root-mean-square deviation as the maximum of both *directed* root-mean-square deviations:

$$\mathrm{RMSD} = \max\{\mathrm{rmsd}(C_{\mathrm{res}}, C_{\mathrm{ref}}), \mathrm{rmsd}(C_{\mathrm{ref}}, C_{\mathrm{res}})\}. \tag{4.10}$$

As our segmentation results always have closed contours, we can compare also the regions V_{res} and V_{ref} that are enclosed by the segmentation contour C_{res} and the reference contour C_{ref}, respectively. The most commonly used measures for spatial overlap are the Jaccard coefficient j and the Dice coefficient d [69]. They are computed as

$$j = \frac{|V_{\mathrm{res}} \cap V_{\mathrm{ref}}|}{|V_{\mathrm{res}} \cup V_{\mathrm{ref}}|} \quad \text{and} \quad d = \frac{2|V_{\mathrm{res}} \cap V_{\mathrm{ref}}|}{|V_{\mathrm{res}}| + |V_{\mathrm{ref}}|}, \tag{4.11}$$

where $|V_{\mathrm{res}}|$ and $|V_{\mathrm{ref}}|$ denote the volume (or area) of the segmented region and the reference region, respectively, $|V_{\mathrm{res}} \cap V_{\mathrm{ref}}|$ indicates the volume of the intersection of both regions, and $|V_{\mathrm{res}} \cup V_{\mathrm{ref}}|$ is the volume of the union. Both coefficients measure how much the segmented region V_{res} and the reference region V_{ref} overlap (see Fig. 4.7b).

The values for both coefficients are located in the range from zero to one, where zero means no overlap and one denotes a perfect match. Additionally, both coefficients can be related to one another via

$$j = \frac{d}{2-d} \quad \text{and} \quad d = \frac{2j}{1+j} \tag{4.12}$$

so that it is in general sufficient to use only one of them. However, for the sake of comparability with other approaches, we compute them both.

The two coefficients from Eq. (4.11) measure the similarity of the regions V_{res} and V_{ref}. Now, one can define the corresponding Jaccard distance J and the corresponding Dice distance D that measure the dissimilarity of both regions as

$$J = 1 - j = 1 - \frac{|V_{\mathrm{res}} \cap V_{\mathrm{ref}}|}{|V_{\mathrm{res}} \cup V_{\mathrm{ref}}|} \tag{4.13}$$

and

$$D = 1 - d = 1 - \frac{2|V_{\mathrm{res}} \cap V_{\mathrm{ref}}|}{|V_{\mathrm{res}}| + |V_{\mathrm{ref}}|} \,. \tag{4.14}$$

These distances are our volume-based error measures. Their values are also located in the range from zero to one, where zero now denotes a perfect segmentation and one means that a completely wrong region has been segmented. Similar to the corresponding indices, both distances can be related via

$$J = \frac{2D}{1+D} \quad \text{and} \quad D = \frac{J}{2-J} \,. \tag{4.15}$$

4.1.3.2 Experimental Setup

As mentioned above, we are interested in extracting the combined outer bony boundary of the nasal cavity and the paranasal sinuses. Outer boundary means that we are only interested in the bony structures that separate the nasal cavity and the paranasal sinuses from other anatomical structures, but not in the structures that separate the individual sinuses from each other or from the nasal cavity.

We use the database that has been introduced in Sect. 4.1.1 in order to train the global SSM as well as our LDSSM. As both models capture only nonrigid shape variations, all CT datasets of the database are rigidly aligned with regard to rotation, translation, and scale prior to the model generation process. This is achieved by keeping the first CT dataset fixed and computing for each of the following CT datasets a similarity transformation (c.f. Sect. 2.4.1) that aligns them to the first dataset. However, in order to be able to identify a similarity transformation between two CT datasets, we need corresponding points in both datasets. So, we make use of five anatomical landmarks (right styloid process, left styloid process, crista galli, anterior nasal spine, and posterior nasal spine) which are available in each dataset of the above mentioned database (c.f. Sect. 4.1.1). These landmarks also can be easily and robustly identified in a new CT dataset of the paranasal sinuses with just a few mouse clicks. After the similarity alignment, a corresponding frontal view slice is extracted from each aligned dataset via trilinear interpolation. The extracted slices are shown in Appendix A. Each extracted CT slice I has a resolution of 512×366 pixels and a pixel spacing of 0.46 mm.

As already mentioned, we compare our new LDSSM-based segmentation approach from Sect. 3.3.4 to the global approach from Sect. 4.1.2 in two dimensions. So, only the extracted slices are used for further processing. For the comparison we use a leave-one-out approach. This means that from our 49 hand-segmented CT slices we incorporate $n = 48$ slices in the model generation process. The remaining CT slice is used to compare the results of both automatic approaches to the hand-segmented reference. Leaving out each CT slice in turn, we thus obtain automatic segmentation results for each CT slice where the statistical shape models are based solely on the data of the remaining CT slices and not on the CT slice that is currently evaluated.

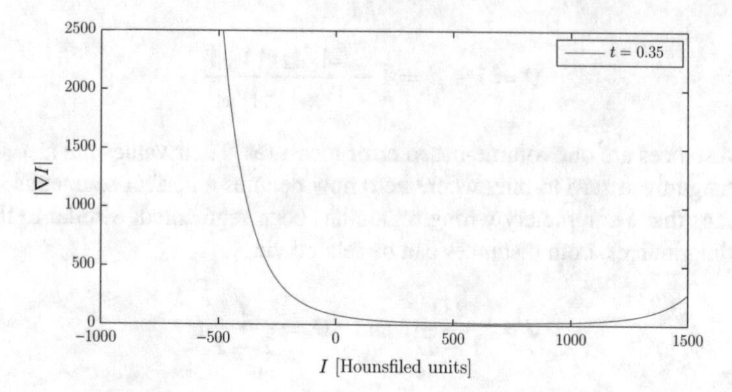

Fig. 4.8 Intensity-dependent binarization threshold for the original, unweighted gradient magnitude $|\nabla I|$ when a fixed threshold $t = 0.35$ is chosen for the intensity weighted gradient magnitude E

In order to train the global implicit SSM, the hand-segmented reference contours are converted to an implicit signed-distance representation by the approach presented in [52], where negative distances are assigned to the elements inside the contour and positive distances are assigned to the elements outside the contour, respectively. The training of the global SSM is then carried out as explained in Sect. 2.3. After the PCA, we choose the $m = 10$ main modes of variation in order to define the SSM subspace. These main modes of variation are then also used by our LDSSM from Chap. 3.

The global solution is obtained as explained in Sect. 4.1.2 and depicted in the processing chain of Fig. 4.6. Additionally, we limit the maximum intensity values of the CT slice I to 1500 Hounsfield units, as larger Hounsfield values are caused by metallic artifacts like tooth crowns. Some parameters of the approach have already been discussed in Sect. 4.1.2. Furthermore, we choose a standard deviation of $\sigma_{\text{gauss}} = 1.0$ for the Gaussian smoothing kernel K_{gauss} (c.f. Fig. 4.4a) and the binarization threshold t for the intensity weighted gradient magnitude E is chosen to 0.35. The resulting intensity-dependent binarization threshold for the original, unweighted gradient magnitude image $|\nabla I|$ is depicted in Fig. 4.8. At the limits of the bony intensity range, i.e. at 300 and 1000 Hounsfield units, the binarization threshold for the unweighted gradient magnitude is 24.9. Inside the bony intensity range, the binarization threshold remains almost constant with a minimum of 14.83 at $I = 650$. Outside this intensity range, we can see an exponential increase of the binarization threshold. The termination condition for the *Nelder–Mead simplex method* is chosen according to [100] as

$$\frac{2\,|f_{\text{best}} - f_{\text{worst}}|}{|f_{\text{best}}| + |f_{\text{worst}}|} < 1e^{-10}, \tag{4.16}$$

where f_{best} denotes the lowest function value of all simplex vertices, and f_{worst} denotes the highest value, respectively. So, the *Nelder–Mead simplex method* terminates when the absolute difference of the normalized function values of the objective function is less than $1e^{-10}$ between all simplex vertices.

Starting from the global solution, we obtain the local solution as explained in Sect. 3.3.4. This means, we use Algorithm 3.2 in order to approximate a smooth field of weight vectors $\hat{\vec{\Psi}}$ so that the zero level curve $C_{\text{loc}}(\hat{\vec{\Psi}})$ of the modeled shape $\Phi_{\text{loc}}(\hat{\vec{\Psi}})$ describes the desired segmentation result. The full processing chain is depicted in Fig. 3.8. The initial weight field $\vec{\Psi}_{\text{init}}$ is thereby obtained by setting each local weight vector $\vec{\Psi}^{i}_{\text{init}}$ of the initial weight field to the global solution $\hat{\vec{w}}$ that has been explained above. The weight update target $\hat{\Phi}^{\text{weight}}_{\text{targ}}$ in each iteration of Algorithm 3.2 is approximated as detailed in Sect. 3.3.4.2 (Eq. (3.24)) based on the smoothed gradient magnitude G from Eq. (3.22). Like for the global approach, the Gaussian smoothing kernel in Eq. (3.22) is chosen to have a standard deviation of $\sigma_{\text{gauss}} = 1.0$. By doing this, the objective function $O(\vec{w})$ from our global approach (Eq. (4.2)) uses the same input information as the target function proposed in Eq. (3.24) and a fair comparison of the global and local fitting result is possible.

However, informal tests showed that the local approach takes many iterations to reach the desired result based on the above-mentioned target function. So, in order to speed up the segmentation process, we slightly modify the estimation of the weight update target $\hat{\Phi}^{\text{weight}}_{\text{targ}}$ and use the modified weight update target $\hat{\Phi}^{\text{weight}}_{\text{targ, mod}}$ for the weight field update. The first modification is that we always force the model contour outwards when it is located in an air-filled region, because it is very likely that an air-filled region belongs to the paranasal sinuses. Additionally, we always force the model contour inwards when it is located in a bony region.

These modifications are realized by adding an offset $\gamma(\vec{x})$ to Eq. 3.24 so that

$$\hat{\Phi}^{\text{weight}}_{\text{targ, mod}}(\vec{x}) = \hat{\Phi}^{\text{weight}}_{\text{targ}}(\vec{x}) + \gamma(\vec{x}), \tag{4.17}$$

where

$$\gamma(\vec{x}) = \begin{cases} -0.1 & \text{, if } I(\vec{x}) < -200 \text{ and } \vec{x} \in \text{NB}(C_{\text{loc}}) \\ 0.1 & \text{, if } I(\vec{x}) > 200 \text{ and } \vec{x} \in \text{NB}(C_{\text{loc}}) \\ 0.0 & \text{, else.} \end{cases} \tag{4.18}$$

The narrow band $\text{NB}(C_{\text{loc}})$ around the zero level curve C_{loc} in which the weight update target is estimated (cf. Fig. 3.11) is chosen to have a diameter of 15.56 pixel and the smoothing kernel K_{smooth} that is used to smooth the weight fields in each iteration of Algorithm 3.2 (line 9) is chosen to be an 11×11 average filter. We tried different stopping criteria for Algorithm 3.2 like the absolute error between the weight fields from consecutive iterations or the absolute error between the corresponding level set functions. However, none of them has been entirely satisfactory. The reason is that the error between the weight fields from consecutive iterations, and hence the error between the corresponding level set functions, can be very small for a couple of iterations before it rises again.

This occurs when parts of the model contour have approached the relevant image structures and other parts of the model contour are located in image regions with diffuse or no relevant information. In this case, the weight fields are no longer (or rather only a little bit) changed by the weight update loop in lines 5–8 of Algorithm 3.2. However, the weight fields can still change due to the smoothing step in line 9 of Algorithm 3.2. So, the smoothing step can cause parts of the model contour to move from image regions with diffuse or no relevant information into image regions with more relevant information. This in turn raises the influence of the weight update loop on these parts of the model contour. As the changes in the weight fields due to the weight update loop are usually much larger than the changes due to the smoothing step, the error between consecutive weight fields, and hence the error between the corresponding level sets, rises when parts of the contour enter an image region with distinctive image information.

So, it is hard to define a stopping criterion based on the error between consecutive weight fields, or on the error between the corresponding level sets, as one risks to get stuck in a suboptimal solution if one chooses the minimal allowed error to high. However, choosing the minimal allowed error to low can lead in contrast to the fact that the algorithm performs many unnecessary iterations without any visible change in the position of the model contour. To circumvent this problem, we use a fixed number of 500 iterations as stopping criterion. The number of iterations has been deliberately chosen very high to make sure that our method converges for each dataset. Because of the modified target function, in many datasets of the database, a smaller number of iterations would have also been enough.

4.1.3.3 Results

Some exemplary results that show the advantage of our local approach over our global approach are depicted in Fig. 4.9. The resulting errors for all datasets are shown in Figs. 4.10 and 4.11, and the corresponding segmentation results for all datasets are depicted in Appendix A.

For the global approach, the average root-mean-square deviation for all 49 datasets is 3.39 mm with a standard deviation of 2.15 mm, the average Hausdorff distance is 11.33 mm with a standard deviation of 6.55 mm, the average Jaccard distance is 0.17 with a standard deviation of 0.06, and the average Dice distance is 0.09 with a standard deviation of 0.04. It can be clearly seen that the overall shape of the paranasal sinuses is well approximated in many of the datasets. However, it is also visible that even for the best global results there is still a small difference between the hand-segmented reference and the global result in local regions of the data under consideration (see e.g. dataset 34 in Fig. 4.9). This is because the global approach comes to its limits when the task is to adapt to local deviations that are not represented by the global shape model which is especially the case for the frontal sinuses as they are the parts that vary the most between the individual datasets [99] (see e.g. dataset 10, 15, 32, or 36 in Appendix A). This assertion is supported by the fact that there are only 5 datasets where the Hausdorff distance exceeds the mean value by more than one standard

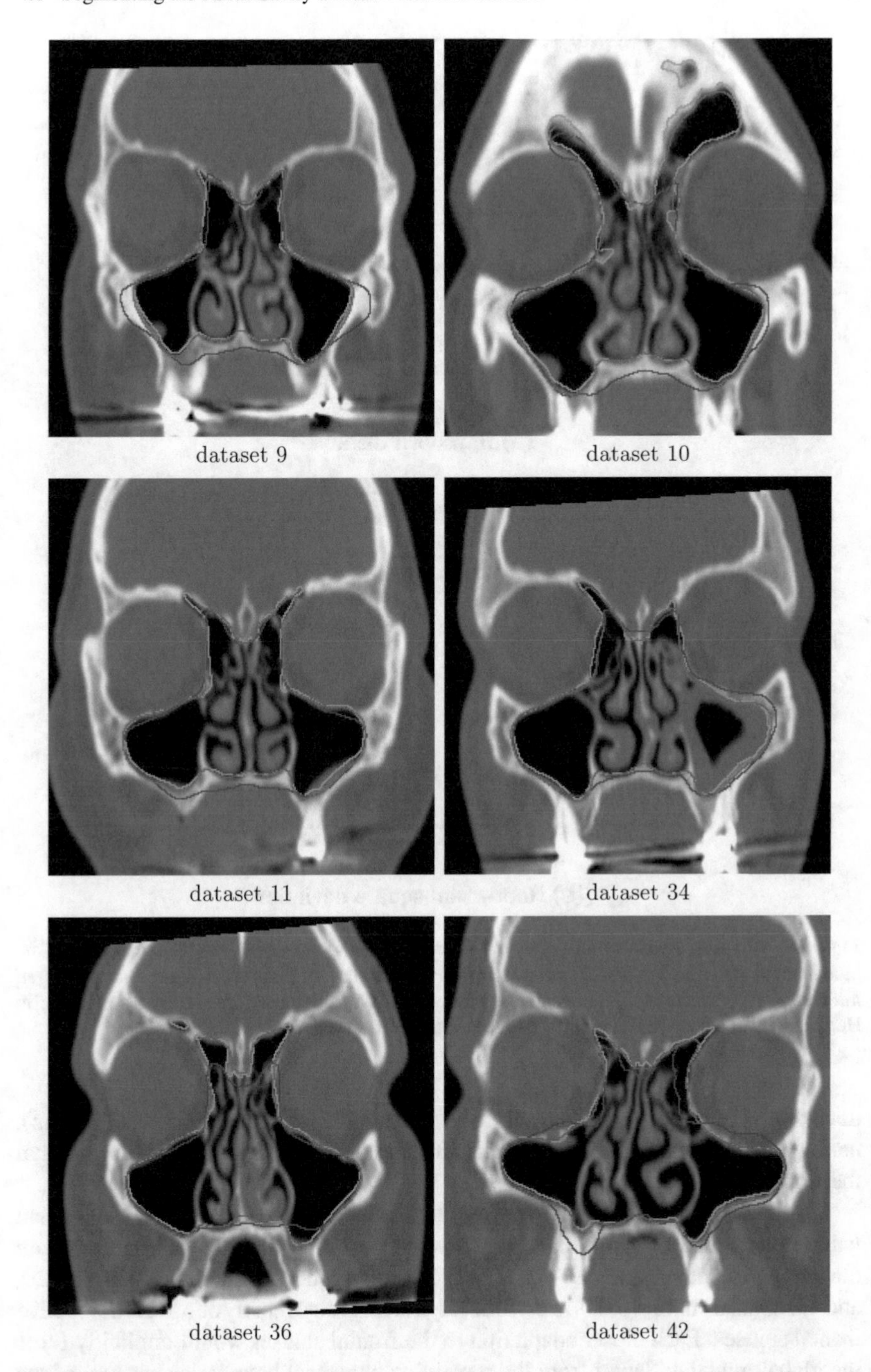

Fig. 4.9 Some exemplary results that show the advantage of our local approach (*red line*) over our global approach (*blue line*). The hand-segmented reference contour is shown as a *green line*.

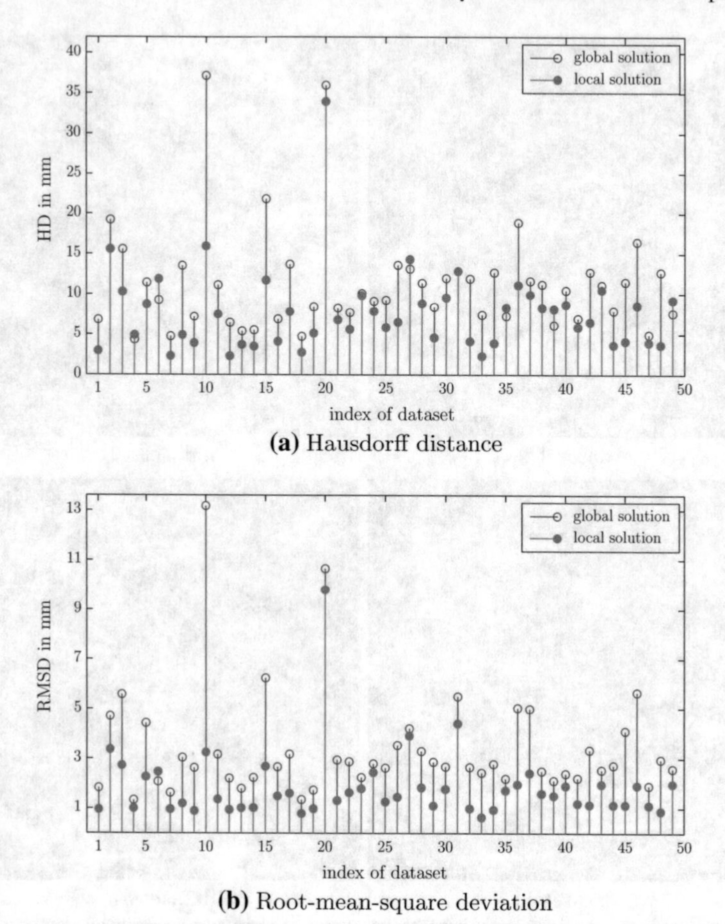

Fig. 4.10 Resulting Hausdorff distances (**a**) and root-mean-square deviations (**b**) obtained with the global approach from Sect. 4.1.2 (*blue empty circles*) and the local approach from Sect. 3.3.4 (*red filled circles*), respectively. The results are plotted against each dataset. © Springer-Verlag Berlin Heidelberg 2011. Reprinted from [9] with permission of Springer

deviation (i.e. the Hausdorff distance is greater than $11.33 + 6.55 = 17.88$ mm), namely the datasets 2, 10, 15, 20, and 36. These datasets all have in common that the frontal sinuses are very distinctive.

For the root-mean-square deviation there exist also 5 datasets where the mean value is exceeded by more than one standard deviation (i.e. the root-mean-square deviation is greater than $3.39 + 2.15 = 5.54$ mm), namely the datasets 3, 10, 15, 20, and 46. In datasets 10, 15, 20, and 46 the large error is also mostly due to the distinctive frontal sinuses. Each closer adaptation to the frontal sinuses would implicitly force the global model to depart from the remaining paranasal boundaries because of the strong coupling of the various paranasal sinuses in the global model. In dataset 3 there are no distinctive frontal sinuses. Instead, the overall shape of dataset 3 seems

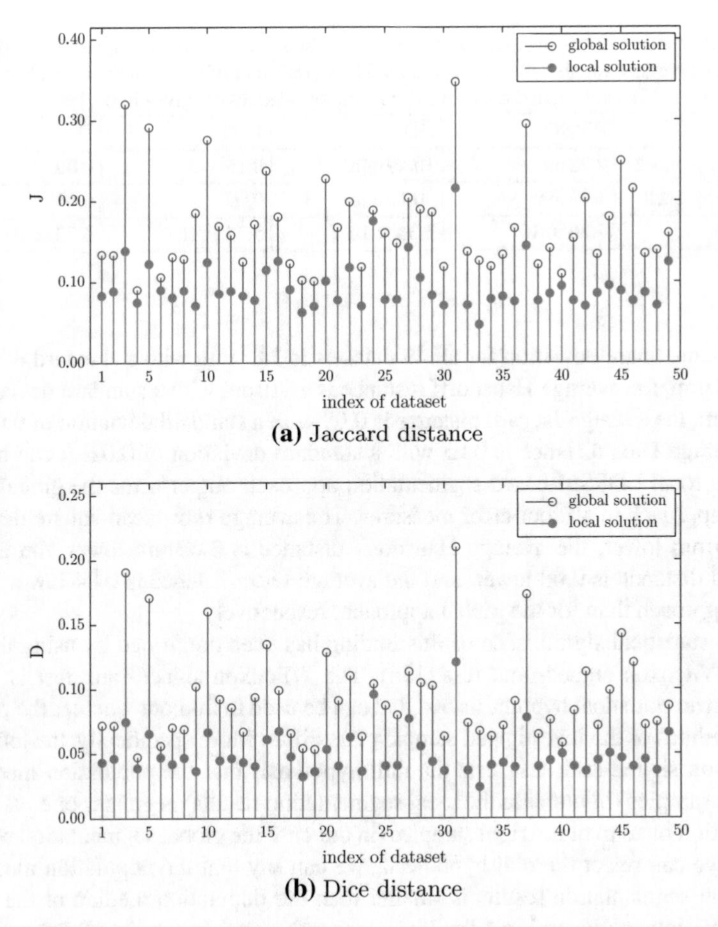

Fig. 4.11 Resulting Jaccard distances (**a**) and Dice distances (**b**) obtained with the global approach from Sect. 4.1.2 (*blue empty circles*) and the local approach from Sect. 3.3.4 (*red filled circles*), respectively. The results are plotted against each dataset. ©Springer-Verlag Berlin Heidelberg 2011. Reprinted from [9] with permission of Springer

to greatly differ from the other datasets so that no satisfactory solution could be found with the global model. Nevertheless, the results obtained with the global approach in general represent a good initial solution that, in most cases, can be refined with our proposed LDSSM-based segmentation approach.

As can be seen in Fig. 4.9, our proposed LDSSM-based segmentation approach is able to deliver more accurate segmentation results compared to the global SSM-based approach. Especially for distinctive frontal sinuses the LDSSM is clearly better suited (see e.g. dataset 10 in Fig. 4.9). The better performance is supported by the segmentation errors that are shown in Figs. 4.10 and 4.11.

Additionally, the median values of all four error measures are shown in Table 4.1 for the global and the local approach, respectively. For the local approach, the average

Table 4.1 Median errors over all 49 datasets. It can be seen that the local approach outperforms the global approach for all four error measures. The significance of this finding was evaluated using the left-sided Wilcoxon signed-rank test. The resulting ρ-values are given in the last row

	RMSD	HD	J	D
Global approach	2.72 mm	10.49 mm	0.16	0.09
Local approach	1.49 mm	6.79 mm	0.09	0.04
ρ-value	7.33×10^{-10}	3.33×10^{-8}	5.73×10^{-10}	5.73×10^{-10}

root-mean-square deviation for all 49 datasets is 1.83 mm with a standard deviation of 1.42 mm, the average Hausdorff distance is 7.70 mm with a standard deviation of 5.20 mm, the average Jaccard distance is 0.09 with a standard deviation of 0.03, and the average Dice distance is 0.05 with a standard deviation of 0.02. It can be seen that the local LDSSM-based segmentation approach outperforms the global SSM-based approach in all four error measures. The average root-mean-square deviation is 1.56 mm lower, the average Hausdorff distance is 3.62 mm lower, the average Jaccard distance is 0.08 lower, and the average Dice distance is 0.04 lower for the local approach than for the global approach, respectively.

The statistical significance of this finding has been confirmed by using the left-sided Wilcoxon signed-rank test [138]. The Wilcoxon signed-rank test is a non-parametric statistical hypothesis test that can be used to find out whether the population medians of the investigated sample pairs differ. More specifically, the left-sided Wilcoxon signed-rank test tests the null hypothesis that the population median of the left samples (in our case the local segmentation results) is greater or equal to the population median of the right samples (in our case the global segmentation results). So, if we can reject the null hypothesis, we can say that the population median of the local segmentation results is smaller than the population median of the global segmentation results and that the local approach outperforms the global approach. The resulting ρ-values for the left-sided Wilcoxon signed-rank test are given in the last row of Table 4.1. It can be seen that the null hypothesis can be rejected for all four error measures at the commonly used significance level of 0.05.

For the Jaccard distance and the Dice distance, it can also be seen that the local approach yields a smaller distance than the global approach for all 49 datasets (c.f. Fig. 4.11). However, despite the generally better performance of the local approach compared to the global approach, there exist six datasets where the Hausdorff distance of the global fit is smaller than the corresponding local result (datasets 4, 6, 27, 35, 39, and 49).

For dataset 6, also the root-mean-square deviation of the global solution is 0.39 mm better than the root-mean-square deviation of the local solution. The reason is a conflict between our definition of the segmentation problem and the location of the hand-segmented reference. For example in the lower right part of dataset 6, the hand-segmented reference does not correspond to any edge of the input data. However, there exists a linear combination of the training shapes so that the global solution has been able to approach the image edges in this region. Starting from this wrong

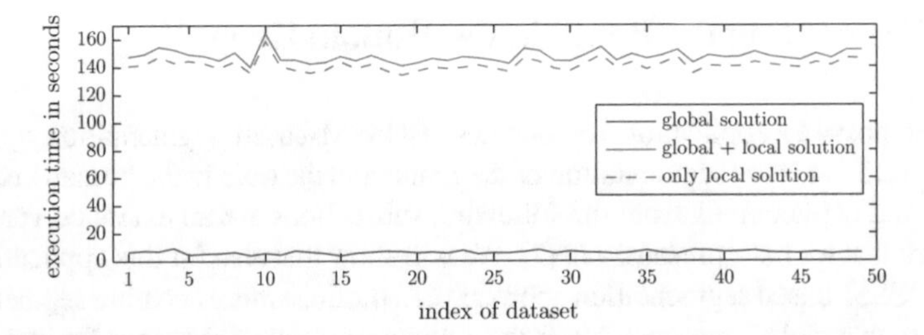

Fig. 4.12 Execution times of the global approach (*blue line*) and the local approach (*red line*) for each dataset on an Intel Core2Quad CPU with 2.83 GHz

initial solution, the LDSSM consequently adapts even more to the nearby edges that can be expressed through local combinations of the training shapes. The same is true for the lower part of dataset 27 and the lower left part of dataset 35, resulting in a worse Hausdorff distance for the local approach than for the global approach. In dataset 49, the problem is that global initial solution runs through the pharynx which is filled with air. This contradicts our assumption for the local approach that only the paranasal sinuses are filled with air and prevents the local contour to move away from the pharynx. So, a better global initial solution would also improve the local segmentation result in most cases.

What has not been mentioned so far is that the increase in segmentation accuracy comes at the price of an increased execution time. The above-mentioned approaches have been implemented in C++ (single-threaded program) and the evaluation has been performed on an Intel Core2Quad CPU with 2.83 GHz clock speed. The individual execution times for all datasets are depicted in Fig. 4.12. On this processor, the average segmentation time per dataset for the global approach is 6.20 s with a standard deviation of 1.05 s (solid blue line in Fig. 4.12) and the average segmentation time per dataset for the global approach followed by the local approach is 147.37 s with a standard deviation of 4.24 s (solid red line in Fig. 4.12). This means that the additional time cost for the local adaptation is on average 141.17 s with a standard deviation of 4.20 s (dashed red line in Fig. 4.12).

So, the runtime of the local approach is approximately one order of magnitude (roughly factor 23) larger than the runtime of the global approach. However, the runtime of the local approach can be reduced significantly when the local weight update loop in lines 5–8 of Algorithm 3.2 would be parallelized as the local weight updates are first of all independent of one another until they are connected through the smoothing step in line 9 of Algorithm 3.2. Furthermore, preoperative CT scans are usually acquired one day before the surgery so that a greater accuracy is more important than a fast execution time as the automatic segmentation algorithm has all night available in order to produce an accurate segmentation result.

4.2 Segmenting the Bones in the Human Knee

Another possible application for our new LDSSM-based segmentation approach from Sect. 3.3.4 is the segmentation of the femur and the tibia in the human knee. The results and explanations from the following subsections are an extended version of the work that we have presented in [7]. We will show that also for this application our new LDSSM-based segmentation approach can produce more accurate segmentation results than a global approach when the amount of training shapes is limited.

4.2.1 Motivation

Osteoarthritis is one of the most common health problems among the aging population of developed countries. It is a degenerative disease which results in loss of articular cartilage within the joints of the human body. Especially the knee joint is affected by osteoarthritis due to the permanent stress caused by the upper body weight. The medical treatment of knee joints affected by osteoarthritis often involves knee replacement surgery or high tibia osteotomy. Nowadays, the planning for these surgeries is typically done purely geometrical. A patient-specific knee model could improve the plan for such surgeries by also considering biomechanical aspects like joint pressure distribution [120] (see Fig. 4.13). Also it enables new technologies like robot-assisted displacement osteotomy [137]. Such a patient-specific knee model can be obtained by segmenting preoperatively acquired CT- or MRI-images.

4.2.2 Experimental Setup

The database for our experiment consists of 6 hand-segmented CT datasets of the lower end of the femur (thighbone) and the upper end of the tibia (shinbone) in the vicinity of the right human knee joint. All CT datasets have been acquired at the *Institut für Radiologie, Medizinische Hochschule Hannover* in the period between April 2006 and June 2010. The datasets all have a slice thickness of 0.4 mm and the pixel resolution lies between 0.31×0.31 mm to 0.70×0.70 mm. They have been hand-segmented at the *Institut für Robotik und Prozessinformatik, Technische Universität Braunschweig* by a medical image segmentation expert with tools provided by the Amira software package [50]. Both bones have been marked at the outer edge of their boundary. The segmentation took about 80 min for each dataset. A 3D reconstruction of the hand-segmented datasets is shown in Fig. 4.14.

Like for the segmentation of the paranasal sinuses, we demonstrate the advantages of our local approach over a global approach by segmenting the femur and the tibia from a single two-dimensional slice of each complete CT dataset only. The extracted slices can be seen in Fig. 4.15. They all have a pixel resolution of 0.4×0.4 mm. Prior

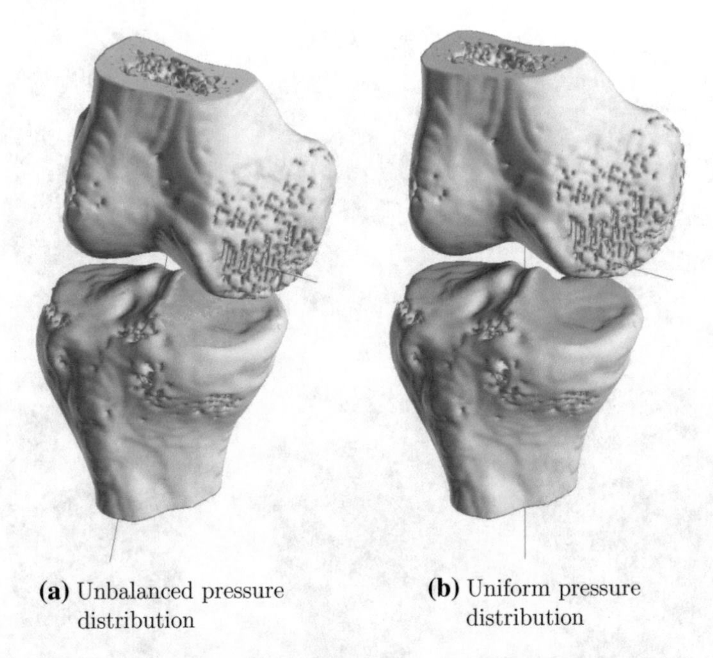

(a) Unbalanced pressure distribution

(b) Uniform pressure distribution

Fig. 4.13 Simulation of the pressure distribution in the human knee joint with a rigid body spring model (segmentations obtained via thresholding). Simulations by courtesy of A. Sommerkorn (c.f. [120])

to the extraction of the slices, the CT datasets have been rigidly aligned to the first dataset with regard to translation, rotation, and scale by using the affine alignment algorithm of the Amira software package. The error metric has been chosen in such a way that the overlap of the segmented femurs is maximized. Hence, the trained shape models contain information about the nonrigid deformations of the tibia and the femur as well as the rigid transformations between these adjacent bones.

In order to make maximum use of the available training data, we use a leave-one-out approach for the evaluation, i.e. we trained the shape models using $n = 5$ datasets in order to segment the remaining dataset, and we use all $m = 4$ available eigenshapes in the models. Prior to the model generation, the hand-segmented contours have been converted to an implicit signed-distance representation by the approach presented in [52]. Again, the eigenshapes and resulting eigenvalues that result from the training of the global SSM are also used by our LDSSM. Now, to obtain the global solution, we use the same approach as for the segmentation of the paranasal sinuses. It has been explained in Sect. 4.1.2.

This means, we search for a weight vector $\hat{\vec{w}}$ for which the zero level curve $C_{\text{glob}}(\hat{\vec{w}})$ of the modeled shape $\Phi_{\text{glob}}(\hat{\vec{w}})$ describes the desired segmentation result. For this purpose, we start from the initial weights $\vec{w}_{\text{init}} = (0, 0, 0, 0)$ and use the *Nelder–Mead simplex method* in order to minimize equation (4.4). The thus obtained result is then used as an initial value for the minimization of equation (4.2). One can see the corresponding processing chain in Fig. 4.6. The only difference to the

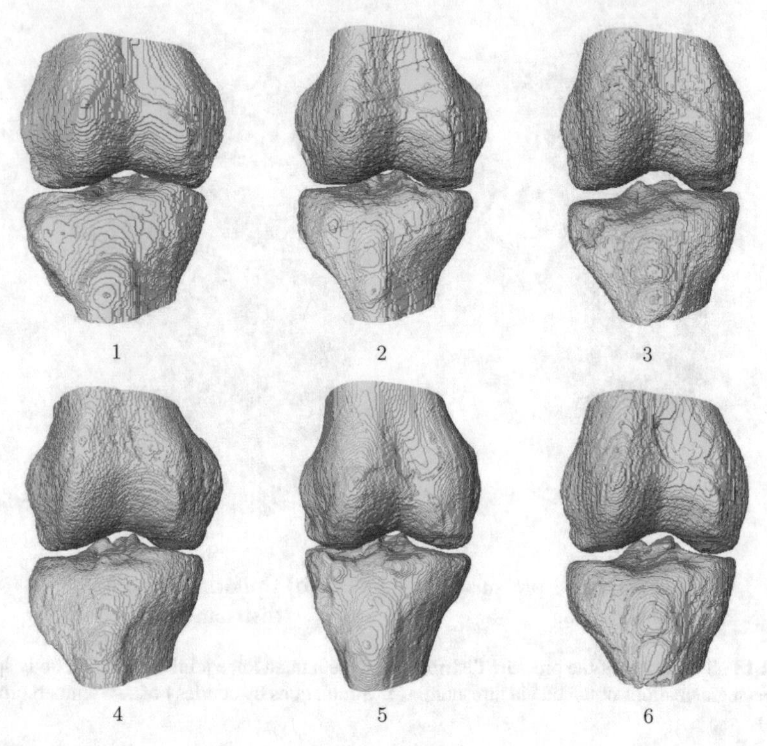

Fig. 4.14 Hand-segmentations of the *lower* end of the femur (*thighbone*) and the *upper* end of the tibia (shinbone) in the *right* human knee. ©CARS 2012. Reprinted from [7] with permission of Springer

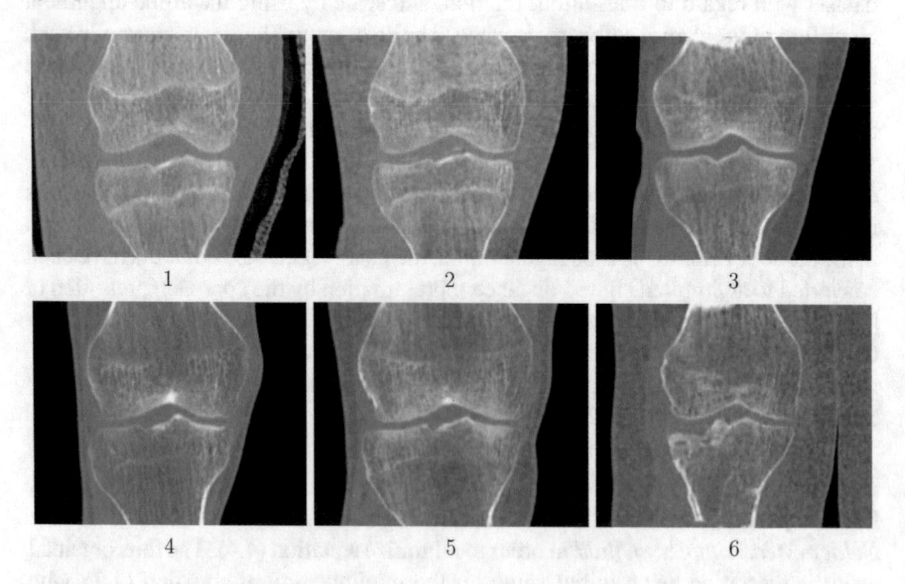

Fig. 4.15 Extracted CT slices of the human knee that are used for the evaluation. Subfigures **2 + 5**: ©CARS 2012. Reprinted from [7] with permission of Springer

paranasal sinus segmentation is that we waive the intensity weighting of the gradient values so that Eq. (4.3) modifies to

$$E(\vec{x}) = |\nabla I(\vec{x})|^2 \,. \tag{4.19}$$

This is because the intensities at the gradients caused by the bone borders have similar values than the intensities at the gradients caused by inner structures of the bones so that an intensity weighting gives no advantage here (c.f. Fig. 4.15).

The binarization threshold t for the squared gradient magnitude E is chosen to 5000, the standard deviation of the Gaussian smoothing kernel K_{gauss}, which is used to smooth the gradient magnitude, is chosen to $\sigma_{\text{gauss}} = 1.0$ (similar to Sect. 4.1.3.2), and the termination condition for the *Nelder–Mead simplex method* is chosen as in Eq. 4.16 (i.e. the absolute difference of the normalized function values of the objective function has to be less than $1e^{-10}$ between all simplex vertices).

Similar to the segmentation of the paranasal sinuses, the local solution is obtained as explained in Sect. 3.3.4: Algorithm 3.2 is used to approximate a smooth field of weight vectors $\hat{\vec{\Psi}}$ so that the zero level curve $C_{\text{loc}}(\hat{\vec{\Psi}})$ of the modeled shape $\Phi_{\text{loc}}(\hat{\vec{\Psi}})$ describes the desired segmentation result. The processing chain of the local approach is depicted in Fig. 3.8. Again, the global solution is used as initial solution for each local weight vector. Like in Sect. 4.1.3.2, we use the modified weight update target from Eq. (4.17). However, we slightly modify the offset $\gamma(\vec{x})$ that has been defined in Eq. (4.18). We heuristically determined that one can achieve a faster convergence when the local contour is slightly forced outwards when it is located in a bony image region. So, the new offset is defined as

$$\gamma(\vec{x}) = \begin{cases} -0.03 & \text{, if } I(\vec{x}) > 120 \text{ and } \vec{x} \in \text{NB}(C_{\text{loc}}) \\ 0.00 & \text{, else} \,. \end{cases} \tag{4.20}$$

Identical to the global approach, the standard deviation of the Gaussian smoothing kernel K_{gauss} (c.f. Eq. (3.22)) is chosen to $\sigma_{\text{gauss}} = 1.0$.

We choose a fixed number of 4000 iterations as the termination criterion for Algorithm 3.2. The smoothing kernel K_{smooth} in line 9 of Algorithm 3.2 is selected to be an average filter, and we heuristically determined that one obtains better segmentation results when we reduce the size of the filter kernel during the segmentation process. So, we start with a filter kernel having a size of 15×15 pixels, and we reduce the filter size in each dimension by 4 every 1000 iterations so that we end up with a 3×3 filter kernel. This is summarized in Table 4.2.

The diameter of the narrow band $\text{NB}(C_{\text{loc}})$ around the modeled contour C_{loc} in which the weight update target is estimated is chosen to match the diameter of the smoothing kernel. So, it also changes according to Table 4.2. The small filter size of 3×3 pixels at the end of the local segmentation process is necessary in order to obtain good segmentation results for this problem because of the very limited amount of training data.

Table 4.2 Size of the averaging filter in line 9 of Algorithm 3.2 and diameter of the narrow band $\mathrm{NB}(C_{\mathrm{loc}})$ in different iterations

Iterations	Filter size (in pixels)	Diameter of narrow band (in pixel)
0 … 999	15 × 15	21.21
1000 …1999	11 × 11	15.56
2000 …2999	7 × 7	9.90
3000 …3999	3 × 3	4.24

The reduction in the filter size has the added benefit that the large filter size at the beginning of the local segmentation process prevents the local model contour to adapt to the wrong image edges that are located near the poor global initial solution. Through the reduction of the filter size, the solution gets more and more local instead of switching directly from a global solution to a local solution. This will be treated in more detail in Sect. 5.5.

4.2.3 Results

The segmentation results obtained with the global and the local approach are shown in Fig. 4.16 and the corresponding errors are given in Fig. 4.17. We have calculated

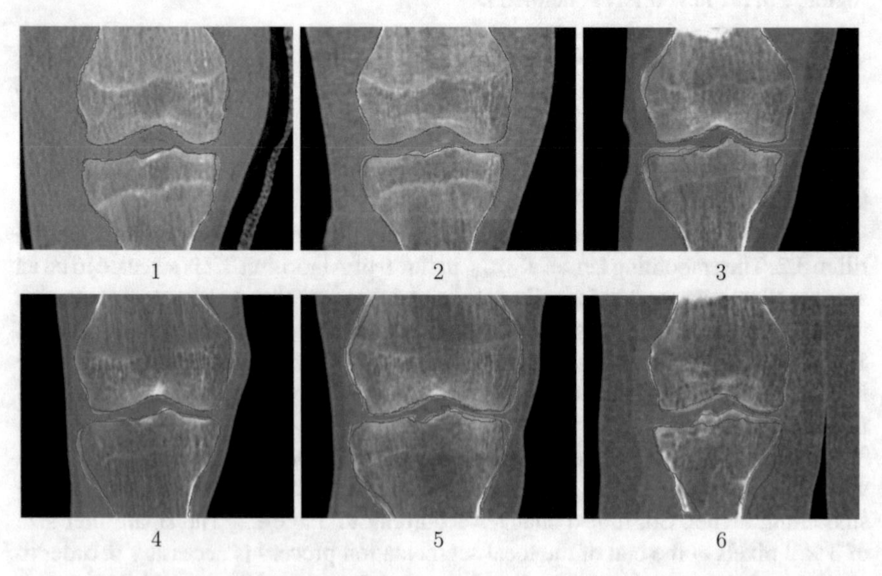

Fig. 4.16 Segmentation results: Hand-segmented reference (*green line*), results of the global approach (*blue line*), and results of our new local approach (*red line*). Subfigures **2 + 5**: © CARS 2012. Reprinted from [7] with permission of Springer

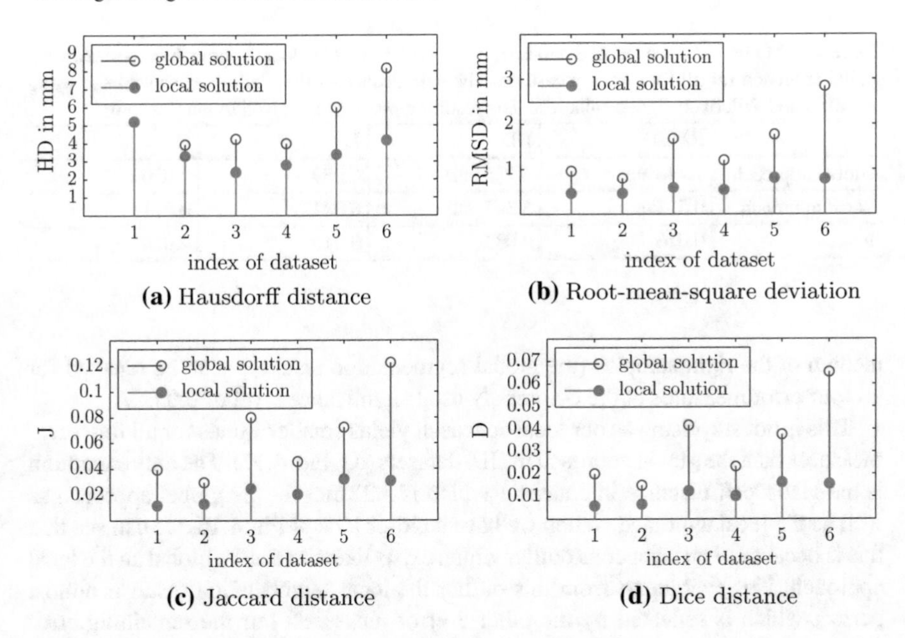

Fig. 4.17 Resulting errors for each dataset obtained with the global approach (*blue empty circles*) and the local approach (*red filled circles*), respectively

the same four error measures as for the paranasal sinus segmentation in order to compare the automatic segmentation results to the hand-segmented reference. They have been defined in Sect. 4.1.3.1. It can be seen that our LDSSM-based approach is able to deliver very accurate segmentation results even with very limited training data at hand (in this case 5 training shapes). This is supported by the quantitative evaluation.

For our local approach, the average root-mean-square deviation, averaged over all datasets, is 0.608 mm with a standard deviation of 0.157 mm, the average Hausdorff distance is 3.548 mm with a standard deviation of 1.016 mm, the average Jaccard distance is 0.022 with a standard deviation of 0.009, and the average Dice distance is 0.011 with a standard deviation of 0.004. For comparison, the average root-mean-square deviation for the global approach is 1.542 mm with a standard deviation of 0.743 mm, the average Hausdorff distance is 5.262 mm with a standard deviation of 1.632 mm, the average Jaccard distance is 0.065 with a standard deviation of 0.034, and the average Dice distance is 0.034 with a standard deviation of 0.019.

So, our local approach clearly outperforms the global approach: The average root-mean-square deviation is 0.93 mm lower, the average Hausdorff distance is 1.71 mm lower, the average Jaccard distance is 0.04 lower, and the average Dice distance is 0.02 lower. As in Sect. 4.1.3, the statistical significance of this finding has been confirmed by a left-sided Wilcoxon signed-rank test [138]. Again, it can be seen from the ρ-values in the last row of Table 4.3 that the null hypothesis – the population median of the left samples (the local segmentation results) is greater or equal to the population

Table 4.3 Median errors over all 6 datasets. It can be seen that the local approach outperforms the global approach for all four error measures. The significance of this finding was evaluated using the left-sided Wilcoxon signed-rank test. The resulting ρ-values are given in the last row

	RMSD	HD	J	D
Global approach	1.440 mm	4.735 mm	0.059	0.030
Local approach	0.565 mm	3.345 mm	0.022	0.011
ρ-value	0.016	0.031	0.016	0.016

median of the right samples (the global segmentation results) – can be rejected for all four error measures at the commonly used significance level of 0.05.

This is not surprising as our local approach yields smaller values for all four error measures than the global approach in all 6 datasets (c.f. Fig. 4.17). The only exception is the Hausdorff distance in dataset 1 which is 5.22 mm for the global approach as well as the local approach. When we have a closer look at Fig. 4.16, we can see that this is because of an erroneous outlier which exists likewise for the global an the local approach. However, apart from this outlier the local segmentation result is almost perfect which is reflected by the other 3 error measures: For the remaining error measures, the local approach reaches the lowest value of all datasets in dataset 1.

The largest values of these 3 error measures are obtained in dataset 5. This is mostly due to the local segmentation result on the right side of dataset 5, at the transition from the femur to the tibia (c.f Fig. 4.16). There, one can see that the local segmentation result deviates from the hand-segmented reference. However, when we have a look at the original data in Fig. 4.15, we can see that even for a human observer it is very hard to determine the correct boundary (especially of the femur) without additional information from adjacent CT slices. So, because of the missing local image information, the local segmentation result stays close to the global segmentation result.

In summary, we have shown the great potential of our local LDSSM-based segmentation approach in a knee CT image segmentation task. Even with only 5 training datasets at hand, we were able to get very promising segmentation results.

4.3 Fitting the LDSSM to Range Scans of Faces

In the following subsections, we will evaluate our LDSSM on another class of data, namely range scans of faces. The results and approaches from the following subsections are an extended version of the work that we published in [5]. The data employed in our experiments will be described in Sect. 4.3.1, and we use this data in Sect. 4.3.2 to show the advantages of our LDSSM over a global SSM by evaluating the best possible model approximation of a given face scan that can be obtained, depending on the number of training shapes that have been used to build the model.

Subsequently, in Sect. 4.3.3 we will present another possible application for the LDSSM: The reconstruction of missing regions in incomplete 3D face scans. For this purpose, we extend the iterative determination of the weight fields for a partially-known target shape from Sect. 3.3.3 by embedding it into a fully-automatic framework that automatically identifies:

1. the rigid transformation between the face scan and the mean shape of our LDSSM
2. an initial solution for the field of weight vectors
3. a reconstruction of the complete face scan with the help of the LDSSM

The experiments will show that the LDSSM is able to represent a natural-looking approximation of the complete face scan with only a very small amount of training shapes at hand.

4.3.1 The Basel Face Model: A Meta-Database for 3D Faces

The data used in the following sections has been provided by Prof. Dr. T. Vetter, *Department of Computer Science, University of Basel*, Switzerland [98]. He developed a statistical shape model of 3D faces, called the *Basel Face Model* (BFM), and made it publicly available [57] in order to emphasize the use of statistical shape models in various research areas. The BFM is an explicit statistical shape model of the form presented in Sect. 2.2. It has been built using face scans of 100 male and 100 female persons, mostly Europeans, for which dense point correspondences have been established. Each of the face scans is represented by a 2-dimensional triangulated surface mesh in the 3-dimensional Euclidean space. The mesh consists of 53490 vertices $(x_i, y_i, z_i) \in \mathbb{R}^3, i = 1 \ldots 53490$, which are in correspondence throughout the training data.

Instead of making the training data publicly available, Vetter et al. chose to publish the statistical shape model that has been trained using this training data, i.e. the mean shape, the base vectors that define the subspace of feasible shapes, and the standard deviations along those base vectors (c.f. Sect. 2.2).

We cannot use the BFM directly in our experiments, as the BFM uses an explicit shape representation, where each shape is represented as a triangulated surface mesh, and we use an implicit shape representation throughout this thesis, where each shape is represented as the zero level set of a signed distance function. However, the BFM can be regarded as a *meta-database*, because it can be used to generate new, synthetic faces with the help of Eq. (2.14) [98]. So, we can generate a set of synthetic training shapes, convert these shapes from an explicit to an implicit shape representation, and use this new set of implicit training shapes to train the implicit models that have been addressed in Sects. 2.3 and 3.2. The weight vectors that are used to generate the synthetic training shapes are thereby drawn from the Gaussian distribution that is given in Eq. (2.9). A set of 30 random synthetic training shapes can be seen in Fig. 4.18.

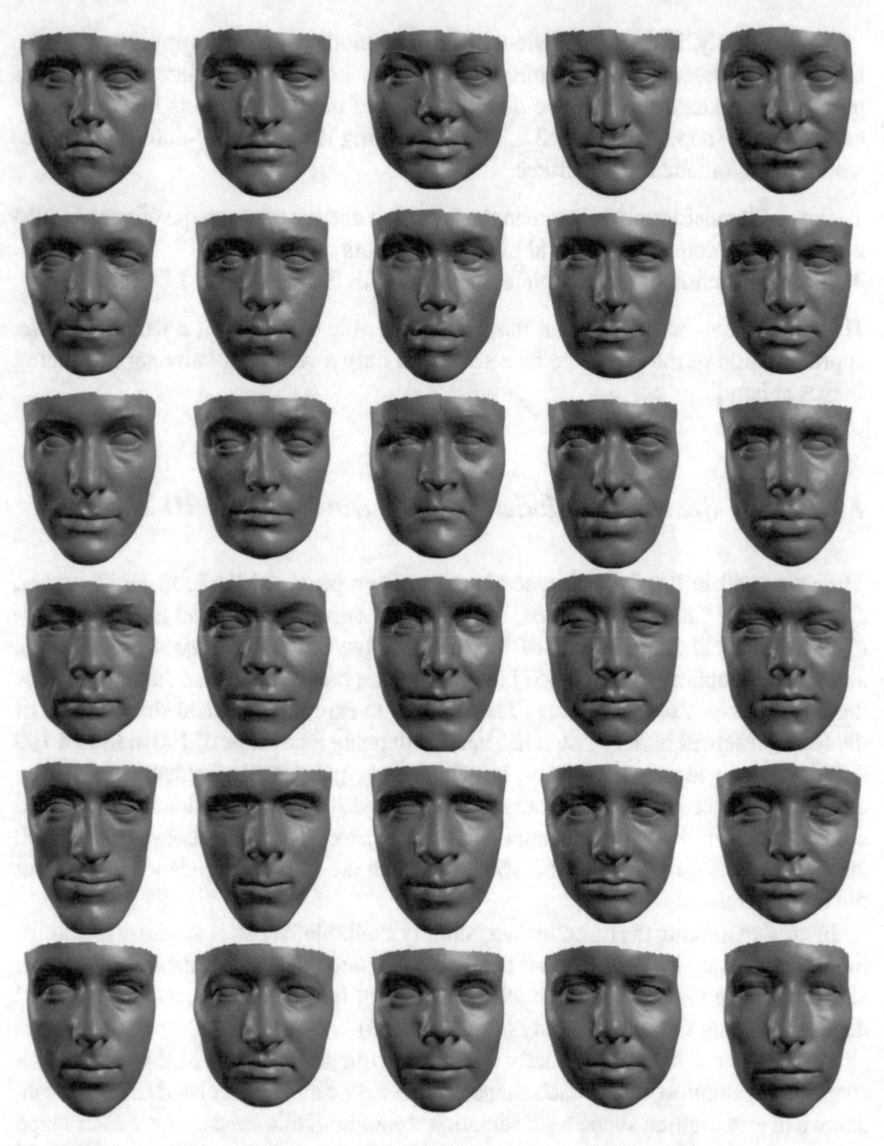

Fig. 4.18 A set of 30 synthetic training faces that have been generated by the BFM

Together with the model data, Vetter et al. also made ten exemplary face scans available that have not been used in the model generation process but are in correspondence to the modeled shapes. They can be seen in the leftmost column of Fig. 4.22.

4.3.2 *Evaluating the Shape Approximation for Known Target Shapes*

In Sect. 2.5, it has been argued that a large amount of training shapes is needed in order to capture the full amount of intra-class shape variation inside a given shape class. Below, we want to substantiate these arguments by evaluating the best possible approximation of a given target shape that can be obtained with the help of the global implicit SSM from Sect. 2.3 and our LDSSM from Sect. 3.2, respectively, depending on the number of training shapes that have been used to train the models. For comparison, we additionally compute the best possible approximation that can be obtained with the help of the global explicit SSM from Sect. 2.2. This comparison is possible, as dense point correspondences are available for the above-mentioned dataset.

4.3.2.1 Methods

In order to obtain the training data for our experiment, we use the BFM to generate nine sets of synthetic training faces with a different amount of faces in each training set: Starting with 10 training faces, the amount of faces in each training set is increased by 10 until a training set size of 90 is reached.

As these synthetic training faces are given in an explicit mesh-based representation, we have to convert them to an implicit signed distance representation. For this purpose, we use the approach that has been presented in [42]. This approach approximates a volumetric signed distance representation in a small hull around the surface mesh.

This signed distance hull is then propagated to the rest of the volume using the approach described in [52] in order to obtain the complete signed distance representation of the synthetic training shapes. The size of the volume that stores the signed distance values is chosen to be $320 \times 340 \times 200$ voxels with a uniform voxel resolution of 0.5 mm in each dimension. The thus obtained implicit shape representations are then used to train the global implicit SSM from Sect. 2.3. All available eigenvectors are used to define the SSM subspace. This means, no dimension-reduction is performed after the PCA in order to make use of all available training shapes. Again, these eigenvectors are also used by our LDSSM from Sect. 3.2.

The synthetic training faces are also used directly to train the global explicit SSM from Sect. 2.2 without converting them to a signed distance representation. The mean shape and the mean shape plus/minus 3 standard deviations of the first, fifth, and tenth principal component are depicted in Fig. 4.19 for the global implicit SSM and in Fig. 4.20 for the global explicit SSM, respectively. Both models have been trained with the shapes that are depicted in Fig. 4.18.

The target shapes for our experiment are given by the ten exemplary face scans from the BFM dataset that are shown in the leftmost columns of Fig. 4.22. As the statistical shape models capture only the nonrigid variations, each of the target shapes

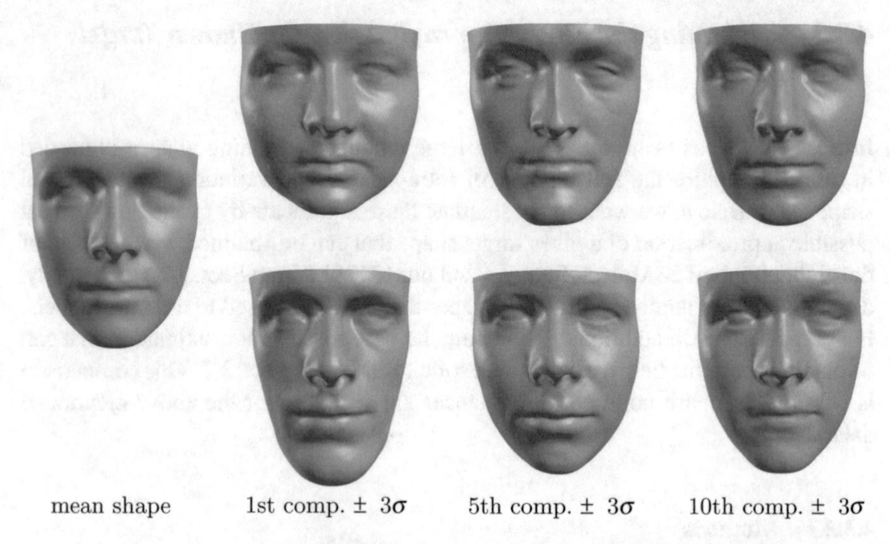

Fig. 4.19 Mean shape and mean shape plus/minus 3 standard deviations for the 1st, 5th, and 10th component of the global implicit SSM from Sect. 2.3

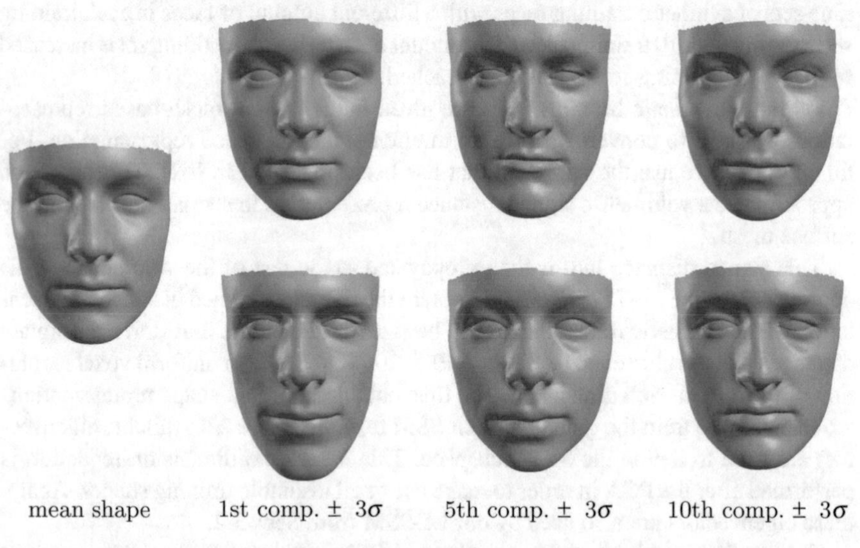

Fig. 4.20 Mean shape and mean shape plus/minus 3 standard deviations for the 1st, 5th, and 10th component of the global explicit SSM from Sect. 2.2

is rigidly aligned to the mean shape of the BFM with regard to rotation, translation, and scale by using a similarity transformation as explained in Sect. 2.4.2. In order to being able to approximate the target shapes with the global implicit SSM and the LDSSM, all target shapes are additionally converted to a signed distance representation by the approach presented above. For the global explicit SSM, the aligned shapes are used directly to compute the shape approximation without converting them to a signed distance representation. To compare the results of both implicit models with the result of the explicit SSM, the resulting shape approximations of both implicit models are transformed back to an explicit mesh-based representation using the *Marching Cubes* algorithm [81].

Now that we have described the data for our experiment, we have to take a look at how to approximate the target shapes with the three shape models mentioned above. For the global implicit SSM from Sect. 2.3, we have presented a closed-form solution in Sect. 3.3.2 that minimizes the mean-square error between the model approximation and the signed distance representation of the target shape $\vec{\Phi}_{\text{targ}}$. So, we use Eq. (3.9) in order to compute the weights $\vec{w}_{\text{impl.}}$ that describe the approximation of the target shape with the global implicit SSM:

$$\vec{w}_{\text{impl.}} = \left(\vec{\Phi}_{\text{targ}} - \bar{\vec{\Phi}}\right) \tilde{\mathbf{\Phi}}^{T} . \tag{4.21}$$

For the global explicit SSM from Sect. 2.2, the same argumentation as in Sect. 3.3.2 applies when we replace the implicit level set representations by the explicit representations from Eq. (2.2).

By doing this, the optimal weights $\vec{w}_{\text{expl.}}$ that minimize the mean-square error between the model approximation and the target shape $\vec{C}_{\text{targ}}$ can be computed as

$$\vec{w}_{\text{expl.}} = \left(\vec{C}_{\text{targ}} - \bar{\vec{C}}\right) \tilde{\mathbf{C}}^{T} . \tag{4.22}$$

Please note that this solution is only possible as dense point-correspondences are available between the target shape and the explicit SSM. For the implicit SSM, the only requirement is that the target shape has to be rigidly aligned to the model.

How to approximate a known target shape Φ_{targ} with the help of our LDSSM has been addressed in Sect. 3.3.3. So, we use Algorithm 3.1 in order to obtain a smooth field of weight vectors $\vec{\Psi}_{\text{loc.}}$ that describes the local shape approximation. The initial field of weight vectors, which is needed as an input to the algorithm, is obtained by setting each local weight vector to the global implicit shape approximation result $\vec{w}_{\text{impl.}}$ from Eq. (4.21). The smoothing kernel is chosen to be a cubic averaging kernel. We initially choose the dimension of the filter kernel to $64 \times 64 \times 64$ mm (i.e. $(2^k) \times (2^k) \times (2^k)$ mm, $k = 6$) and perform 10 iterations of the local fitting loop in lines 3–11 of Algorithm 3.1. Afterwards, the parameter k is reduced by one so that the dimension of the filter kernel reduces to $32 \times 32 \times 32$ mm. With this new filter kernel, another 10 iterations of the local fitting loop are executed. This process is repeated until we reach a final filter size of $2 \times 2 \times 2$ mm so that in total 60 iterations of the local fitting loop in Algorithm 3.1 are performed.

The second input to Algorithm 3.1, next to the initial field of weight vectors, is the signed distance representation of the target shape Φ_{targ}. The problem is that computing the local weight update in lines 6 and 7 of Algorithm 3.1 for all 21,760,000 local weight vectors is computationally very expensive (see below). So, in order to reduce the computation time, we do not consider the complete signed distance representation of the target shape Φ_{targ}, but we use only a signed distance hull with a diameter of 20 voxels around the zero level set of the target shape as a partial target shape for the Algorithm 3.1. As a result, the local weight update in lines 6 and 7 has to be computed only for a small subset of all local weight vectors. The remaining weight vectors are affected only by the smoothing step in line 10 of the algorithm. This is similar to the narrow band approach for an unknown target shape that has been depicted in Fig. 3.8, with the difference that the target shape in the narrow band does not have to be estimated, but is already known.

4.3.2.2 Results

We evaluate the quality of the different shape model approximations by calculating the root-mean-square error between each model approximation and the target shape $\vec{C}_{\text{targ}}$:

$$e_{\text{rms}} = \sqrt{\frac{1}{v} \sum_{i=1}^{v} \left\| d\left((x_i, y_i, z_i), \vec{C}_{\text{targ}} \right) \right\|^2}, \tag{4.23}$$

where $d(x_i, y_i, z_i), \vec{C}_{\text{targ}})$ denotes the euclidean distance from a vertex (x_i, y_i, z_i) of the approximated shape to the closest vertex of the target shape $\vec{C}_{\text{targ}}$, and v is the number of vertices of the approximated shape. Please note that the root-mean-square error from Eq. (4.23) contains a small bias that depends on the parameterization of the target shape, i.e. on the distance between neighboring vertices on the target shape. So, in order to reduce the bias in the root-mean-square error, we refine the mesh of the target shape prior to the evaluation of the error function by introducing new vertices in the middle of each triangle until the maximal edge length of each triangle is smaller than 0.5 mm. After this midpoint-refinement process each target shape $\vec{C}_{\text{targ}}$ consists of about 800,000 vertices and the expected mean bias in the root-mean-square error can be obtained via simulation to 0.1755 mm.[2] As mentioned above, for both implicit statistical shape models, the resulting shape approximations $\Phi_{\text{glob}}(\vec{w}_{\text{impl.}})$ and $\Phi_{\text{loc}}(\vec{\Psi}_{\text{loc.}})$ are converted to an explicit, mesh-based representation with the help of the *Marching Cubes* algorithm [81] in order to compute the error from Eq. (4.23).

The resulting approximation errors for the three investigated statistical shape models are depicted in Fig. 4.21. On the horizontal axis one can see the number of training shapes that have been used to train the statistical shape models, and on the vertical axis one can see the root-mean-square error between each model approximation and the target shape as defined in Eq. (4.23). The approximation errors for the ten target shapes are given as boxplots where the median error of all ten faces is depicted as a black dot. The bottom and top of the boxes denote the first and third quartiles of the errors for all ten faces, and the whiskers denote the minimal and the

[2] We sample uniformly distributed random points inside an equilateral triangle with an edge length of 0.5 mm and compute the root-mean-square error of all these points to the nearest vertex.

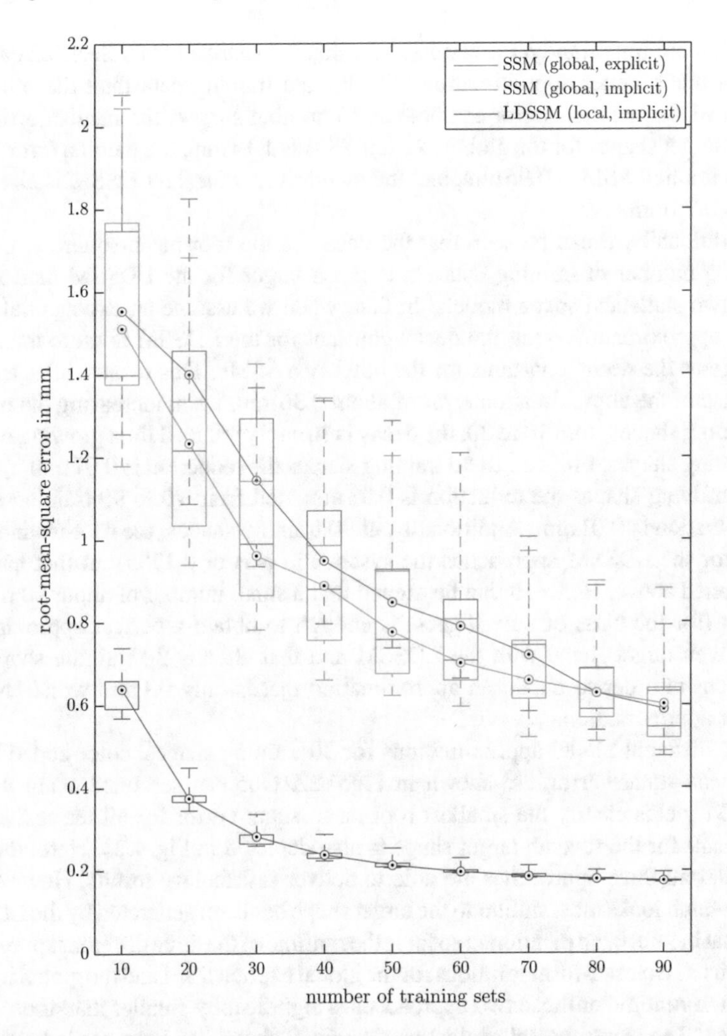

Fig. 4.21 Boxplots of the root-mean-square errors from shapes generated by various statistical shape models (SSMs) to ten given target shapes, plotted against the number of training shapes which have been used to train the SSMs. © Eurographics Association 2013. Reproduced from [6] by kind permission of the Eurographics Association

maximal errors along all ten faces, respectively. It is clearly visible how the root-mean-square error between the shape approximation and the target shape decreases with a growing number of training datasets for all three investigated statistical shape models. The results also show that the global implicit SSM from Sect. 2.3, which does not depend on any point-correspondences, can generate a shape approximation that is equal to (or even slightly better than) the solution that is generated by the global explicit SSM from Sect. 2.2, which depends on point correspondences.

But, more importantly, the results show that the LDSSM from Sect. 3.2 can generate a much better approximation with limited training data than the other two statistical shape models. For example at 30 training shapes, the median error over all ten target shapes for the global explicit SSM is 1.14 mm, the median error for the global implicit SSM is 0.96 mm, and the median error for the LDSSM is already as low as 0.28 mm.

Additionally, it can be seen that the decay of the root-mean-square error for a growing number of training datasets is much larger for the LDSSM than for the other two statistical shape models. In fact, when we assume an exponential decay of the approximation error, the decay constant for the LDSSM is more than twice as large as the decay constants for the other two SSMs. This means, after a strong decrease of the approximation error of about 0.36 mm, when increasing the number of training shapes from 10 to 30, the decay is strongly reduced for a growing number of training shapes. From 30 to 50 training shapes the reduction is 0.07 mm, from 50 to 70 training shapes the reduction is 0.03 mm, and from 70 to 90 training shapes the reduction is 0.01 mm. Additionally, at 90 training shapes, the root-mean-square error for the LDSSM approaches the systematic bias of 0.1755 mm that has been mentioned above. Hence, it can be argued that a small number of about 90 training shapes (for the class of face shapes) is enough to obtain a perfect approximation of a given target shape with the LDSSM and that already 30 training shapes are sufficient in order to obtain an approximation that is only 0.1 mm worse than the perfect approximation.

The different model approximations for 30 training shapes, color-coded by the root-mean-square error, are shown in Fig. 4.22. One can see once again that the LDSSM yields clearly the smallest root-mean-square error for all ten test shapes. The result for the seventh target shape is also depicted in Fig. 4.23. Here, it can be seen that all three approaches are able to deliver satisfactory results. However, the result which looks most similar to the target shape has been generated by the LDSSM.

What has not been mentioned so far is the runtime of the three different approaches. As their exist closed-form solutions for the global explicit SSM and the global implicit SSM, the runtime of those two approaches is significantly smaller than those of the LDSSM. The experimental evaluation described above has been carried out on an Intel Core i7 3770K quad core processor with 3.5 GHz clock speed. On this processor, the approximation of the target shape takes less than a minute for both global SSMs, and the runtime for the LDSSM can be seen in Fig. 4.24. The runtime for the LDSSM is significantly larger than for both global SSMs as the local approximation has to be obtained iteratively. For 10 training shapes, the runtime is approximately 5.2 min per target shape, and it increases up to 79 min for 90 training shapes. One can see that the runtime-complexity is quadratic $\mathcal{O}(cm^2)$ with regard to the number of SSM base vectors m, where $c \leq 1$ is the fraction of voxels that is considered in the weight update step in line 6 of Algorithm 3.1.

The quadratic time-complexity is due to the computation of the outer product in the Cauchy step size from Eq. (3.19). However, this computation is carried out

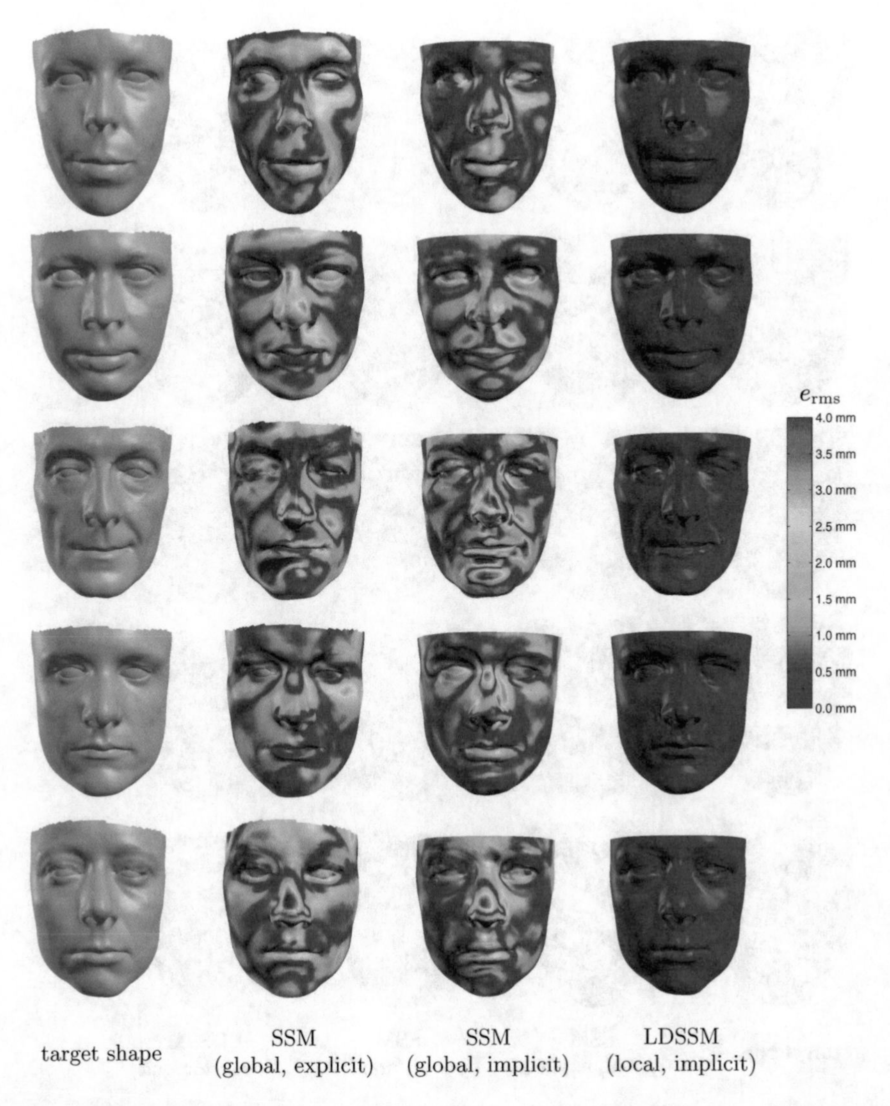

Fig. 4.22 Root-mean-square errors from shapes generated by various SSMs to ten given target shapes. 30 training shapes have been used to train the SSMs. The target shapes originate from the database that is described in [98]

only for a small fraction c of all voxels. Please note that the time complexity of the smoothing step in line 10 of Algorithm 3.1, which has to be carried out for all voxels, is linear with regard to the number of SSM base vectors m, as the smoothing is performed independently for each weight field $\Psi_i, i = 1 \ldots m$.

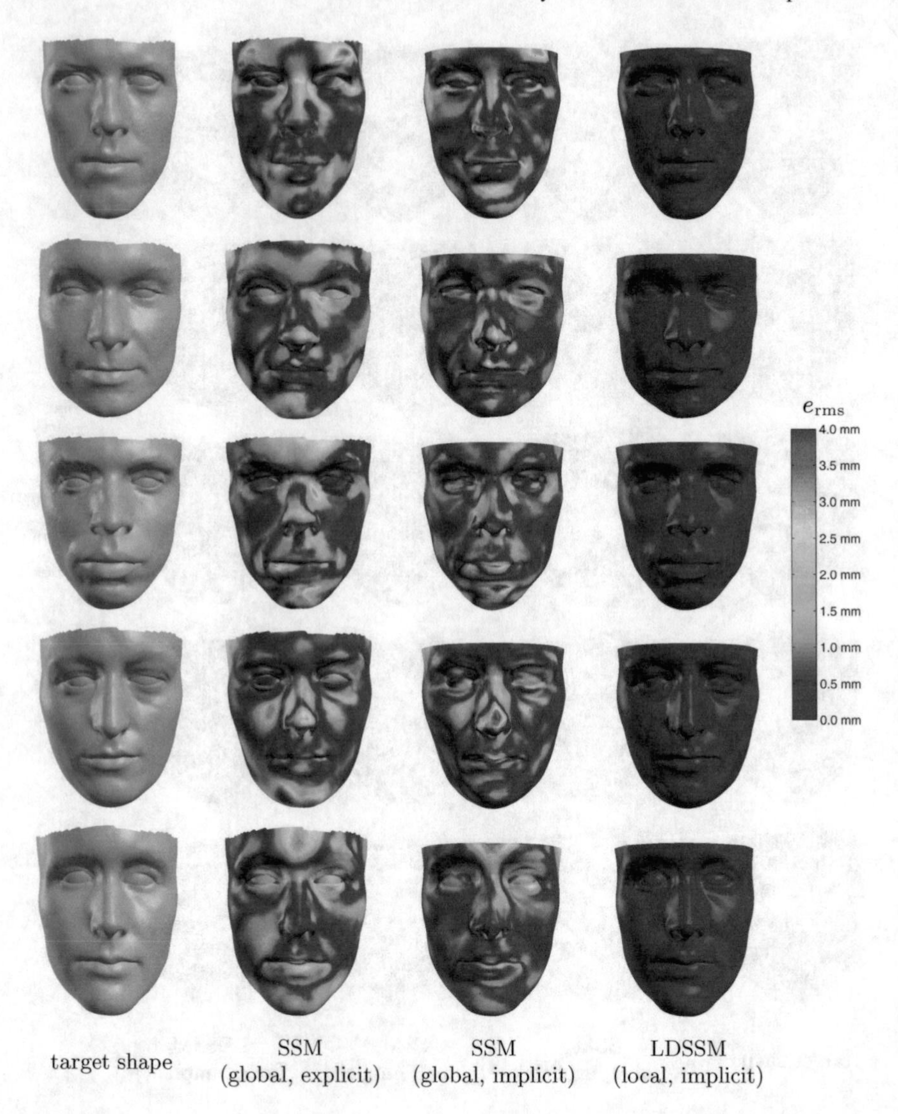

Fig. 4.22 (continued)

4.3.3 Application: Reconstructing Incomplete Face Scans

In the following, we would like to present a possible application for the iterative approximation of a partial target shape with the help of the LDSSM that has been presented in Sect. 3.3.3, namely the reconstruction of missing regions in 3D face scans. By 3D face scan we mean a 2-dimensional triangulated surface mesh in the 3-dimensional Euclidean space that represents the surface of a human face. It can

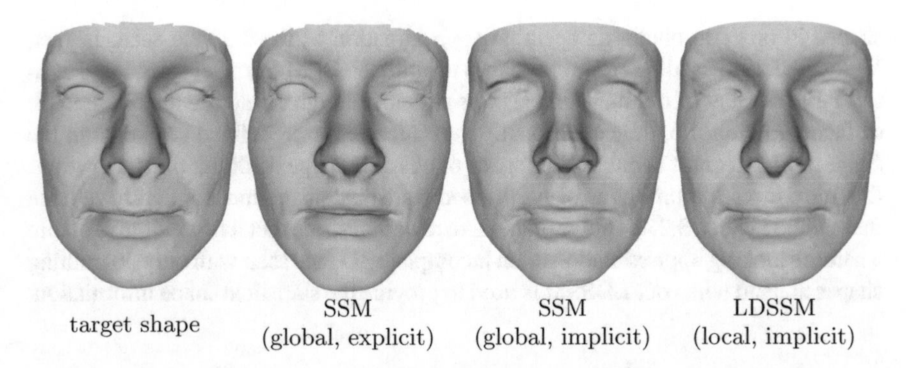

Fig. 4.23 Example for different approximations of a given target shape, generated by the global explicit SSM, the global implicit SSM, and the local implicit LDSSM. The SSMs have been trained using 30 data sets.

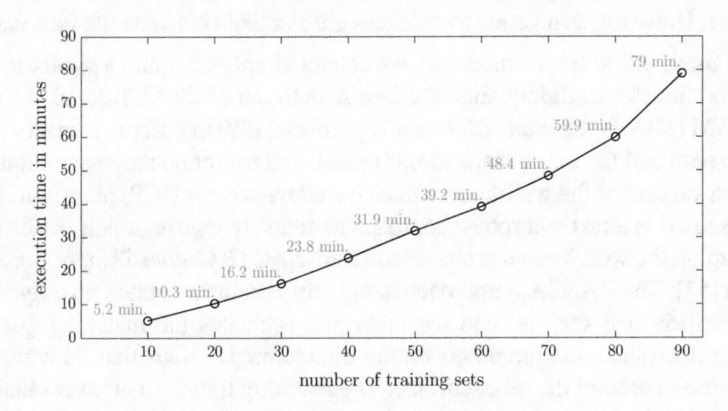

Fig. 4.24 Runtime for the LDSSM that is needed to approximate a given target shape, depending on the number of training shapes that are used to train the model

be acquired for example with the help of a laser range scanner [139] or a coded light approach [132] (see e.g. [17] for an overview). The problem is that an acquired 3D face scan almost always contains holes. Most of them are due to occlusions, but also other factors, like e.g. low reflectance or specular reflections, are plausible explanations [20]. Some examples can be seen in the first column of Fig. 4.27. Now, in order to use the face scan in an application, like e.g. facial animation [79], facial plastic surgery [62], or face recognition [20], usually a face scan without holes is required. So, one could try to close the holes using linear or polynomial interpolation techniques [89]. However, these simple approaches quickly come to their limits when interpolating the difficult structures of a human face.

Interpolating the missing regions with the help of statistical shape information has proven to produce much better results (see e.g. [98] or [79]). As shown in Sect. 4.3.2, a global SSM like the BFM from Sect. 4.3.1 requires a large amount of training shapes in order to obtain an accurate approximation of the 3D face scan. In fact, Vetter et al. [98] used 200 training shapes to build the BFM from Sect. 4.3.1 and even this large number of training samples was still not enough to capture the full amount of face variation. So, they additionally introduced four predefined segments in the shape model in order to enlarge the space of possible shape configurations. However, this introduces the problems of partitioned statistical shape models that have been mentioned in Sect. 3.2. Now, we would like to demonstrate that it is possible to obtain a natural-looking approximation of an incomplete 3D face scan with only 30 training shapes at hand when our LDSSM is used to provide the statistical shape information.

4.3.3.1 Rigid Alignment

As the LDSSM captures only nonrigid shape deformations, the mean shape of the LDSSM has to be rigidly aligned to the face scan with regard to rotation, translation, and scale. However, as no correspondences are available between the face scan $\vec{C}_{\text{targ}}$ and the mean shape of our model $\bar{\vec{C}}$, we cannot simply compute a similarity transformation in order to rigidly align the face scan to our LDSSM. Instead, we use the RANSAM (*RANdom SAmple Matching*) approach [139] in order to coarsely register the face scan and the mean shape of our model, and we refine the registration with a modified version of the well-known *Iterative Closest Point* (ICP) algorithm [15].

RANSAM is a fast and robust approach to coarsely register a pair of surfaces. It builds up on the well-known probabilistic RANSAC (*RANdom SAmple Consensus*) method [53]: The RANSAM approach iteratively generates random pose hypotheses that transform one surface onto the other and evaluates the matching quality by counting the inliers, i.e. the points on the transformed surface that lie within an ϵ-neighborhood around the other surface. The resulting transformation is obtained as the one with the highest matching quality. However, the RANSAM approach has only been designed to register two surfaces with regard to rotation and translation.

Now, in order to additionally obtain an estimate for the initial scale factor between the face scan $\vec{C}_{\text{targ}}$ and the mean shape of our model $\bar{\vec{C}}$, we use Algorithm 4.1: We iterate over a finite set of scale factors, scale the face scan by the given scale factor, and compute a RANSAM match for each scaled face scan. A depiction of this process can be seen in Fig. 4.25. The resulting initial scale factor is then simply the one that delivers the highest RANSAM matching quality and the resulting rotation and translation estimates are given by the RANSAM match at that scale.

This simple approach proved to perform very well in estimating the initial rotation, translation, and scale for the following fine registration step. In order to refine the initial rotation, translation, and scale estimates, we use a modified version of the well-known *Iterative Closest Point* (ICP) algorithm [15]. The ICP algorithm iteratively selects a random set of closest point pairs on two coarsely-registered surfaces and minimizes the mean squared distance between these two point sets with regard to rotation and translation. After a few iterations, this usually leads to a good registration of the two shapes.

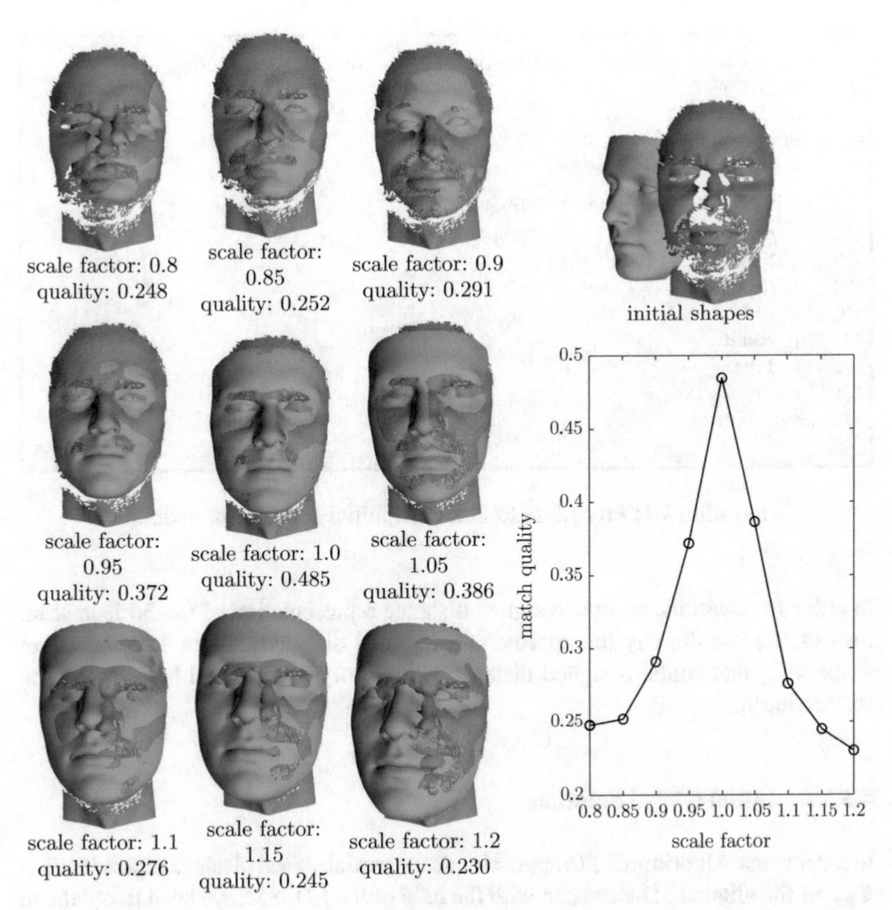

Fig. 4.25 Best RANSAM match and corresponding quality for various scale factors

Our modification now consists of including the scale factor in the minimization process. This is achieved by using the similarity transformation, introduced in Sect. 2.4.1, in order to minimize the mean squared distance between the two point sets in each ICP iteration. Again, this simple modification performed very well in registering the shapes with regard to rotation, translation, and scale.

After having rigidly aligned the 3D face scan $\vec{C}_{\text{targ}}$ to the mean shape of our model $\bar{\vec{C}}$, we need to convert it to an implicit signed distance representation Φ_{targ} in order to be able to close the holes with the help of the LDSSM. Like in Sect. 4.3.2, we use the approach from [42] for this purpose which approximates a signed distance representation in a small hull around the surface mesh. However, the difference to Sect. 4.3.2 is that the 3D face scan now contains holes.

Because of this, the interior and exterior of the 3D face scan is not clearly defined. So, we are not able to propagate the signed distance hull to the rest of the volume

```
1: procedure OBTAININITIALSCALEFACTOR(C̄, C_targ, minScale, maxScale)
2:     quality ← 0
3:     bestScale ← 0
4:     for f ← minScale : 0.05 : maxScale do
5:         C_scale ← SCALESHAPEBYFACTOR(C_targ, f)
6:         FINDOPTIMALRANSAMMATCH(C̄, C_scale)
7:         qTemp ← EVALUATEMATCHINGQUALITY(C̄, C_targ)
8:         if qTemp > quality then
9:             quality ← qTemp
10:            bestScale ← f
11:        end if
12:    end for
13:    return bestScale
14: end procedure
```

Algorithm 4.1: Procedure to obtain an initial scale factor estimate.

in order to obtain the complete signed distance representation of the 3d face scan. Instead, we use directly the approximated signed distance hull as a partial target shape Φ_{targ} that contains signed distance values only in the small hull around the surface mesh.

4.3.3.2 Initial Global Solution

In order to use Algorithm 3.1 to approximate the partial signed distance representation Φ_{targ} of the aligned 3D face scan with the help of the LDSSM, we need to obtain an initial solution $\vec{\Psi}_{\text{init}}$ for the field of weight vectors. In Sect. 4.3.2, we used the global closed-form solution $\vec{w}_{\text{impl.}}$ from Eq. (4.21) in order to initialize the field of weight vectors. The problem is that we do not have a complete signed distance representation of our acquired 3D face scan. So, we cannot calculate the closed form solution from Eq. (4.21). But, we can approximate the global closed-form solution by formulating it as an iterative fixed point problem.

Let Φ_{targ} be the signed distance representation of our 3D face scan $\vec{C}_{\text{targ}}$ that contains only distance values in a small hull around the scanned surface (further denoted by $\text{hull}(\vec{C})$). We start with initial weights $\vec{w}^0 = \vec{0}$ and calculate a first approximation of the target shape with the help of the global SSM from Eq. (2.34) as

$$\Phi_{\text{glob}}(\vec{w}^k) = \bar{\Phi} + \sum_{i=1}^{m} w_i^k \, \tilde{\Phi}_i \,, \tag{4.24}$$

with $k = 0$. Then, the signed distance values from the global model approximation are replaced by the known signed distance values in the hull around the scanned surface:

$$\Phi^{\star}_{\vec{w}^k}(\vec{x}) = \begin{cases} \Phi_{\text{targ}}(\vec{x}) & \text{, if } \vec{x} \in \text{hull}(\vec{C}) \\ \Phi_{\text{glob}}(\vec{w}^k)(\vec{x}) & \text{, else} \end{cases} . \tag{4.25}$$

Afterwards, a new weight vector $\vec{w}^{k+1}$ is obtained by inserting the vectorized level set $\vec{\Phi}^{\star}_{\vec{w}^k}(\vec{x})$ as the new target shape in Eq. (4.21):

$$\vec{w}^{k+1} = \left(\vec{\Phi}^{\star}_{\vec{w}^k} - \bar{\vec{\Phi}}\right) \tilde{\mathbf{\Phi}}^T . \tag{4.26}$$

By iterating Eqs. (4.24), (4.25), and (4.26) until convergence, one obtains an initial approximation of the face scan by the global SSM. Please note that it can be shown that this global model approximation approach is a specific instance of an approach which has been by presented by Tsai et al. in [130]. So, it has a sound mathematical foundation. The approach of Tsai et al. will be discussed in detail in Sect. 5.3. Additionally, in order to further refine the rigid transformation parameters, we perform one step of our modified ICP after each change in the global weight vector.

4.3.3.3 Experimental Setup and Results

In order to demonstrate the reconstruction of missing regions in 3D face scans with the help of the LDSSM, we train our LDSSM using 30 training faces as described in Sect. 4.3.2. This means, 30 explicit training shapes have been synthesized with the help of the BFM from Sect. 4.3.1 (see Fig. 4.18) and they have been converted to a signed distance representation in order to use them in the training process of the LDSSM. The target shapes have been acquired by ourselves with the help of the "DAVID-SLS-2 Structured Light 3D Scanner" [44]. We demonstrate our approach using 6 face scans of 3 different individuals. The scanned faces are depicted in the leftmost column of Fig. 4.27. The shapes in rows 3 and 5 have been acquired from the front and the shapes in rows 4 and 6 from the side. The shapes in rows 1 and 2 have been stitched together with the help of the RANSAM approach [139] using 3 overlapping scans. It can be seen that only the stitched scans contain almost no holes. All the other scans contain non-negligible holes and missing regions that have to be closed with the help of statistical shape information in order to reproduce a natural-looking face.

The reconstruction process is shown in Fig. 4.26. The scanned faces are rigidly aligned to the mean-shape $\bar{\vec{C}}$ of the LDSSM by the approach presented above. For Algorithm 4.1, we use a minimal scale factor of 0.8 and a maximal scale factor of 1.2, respectively, and we use 100 iterations of the ICP algorithm to refine the initial result.

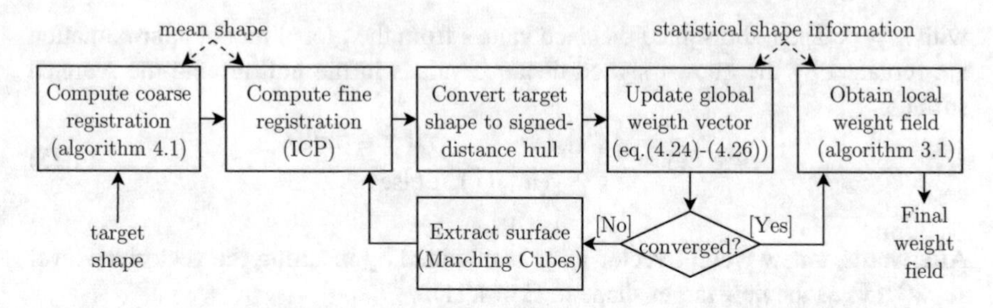

Fig. 4.26 Flowchart for the reconstruction of incomplete face scans with the LDSSM

Afterwards, an initial global shape approximation is obtained by the approach from the previous section. Thereby, the diameter of the signed distance hull is chosen to 10 mm, and Eqs. (4.24), (4.25), and (4.26) are iterated until the error between the weight vectors of two subsequent iterations, normalized with regard to the trained standard deviations, is less than 0.0012. With the thus obtained initial solution, we further approximate the 3D face scan with the help of the LDSSM by using Algorithm 3.1. The parameters of the algorithm are chosen exactly as in Sect. 4.3.2. This means, we choose an initial cubic averaging kernel with a size of $64 \times 64 \times 64$ mm and reduce the filter size every 10 iterations.

The only difference to Sect. 4.3.2 is that we use a minimal filter size of $16 \times 16 \times 16$ mm for the mostly incomplete face scans in lines 3–6 of Fig. 4.27. For the almost complete scans in lines 1 and 2 of Fig. 4.27, a minimal filter size of $2 \times 2 \times 2$ mm is used. So, for the face scans in rows 1 and 2 of Fig. 4.27, a total of 60 iterations are needed in order to obtain the local fitting result and for the shapes in rows 3–6, 30 iterations are needed.

The model approximations are depicted in the second column (initial global model approximation) and in the third column (local model approximation) of Fig. 4.27, respectively. The fourth and fifth columns show the modified 3D scans, where the holes have been filled using the local model approximations. In column four, the local model approximation is additionally highlighted in orange. It can be seen that the local model has been able to capture the almost complete scans very accurately. But, more importantly, also for the other shapes the local model approximation looks much more similar to the original scan than the global model approximation. The existing regions of the shapes have been precisely approximated and the missing regions have been nicely interpolated. However, in those cases where only one side of the face is present in the scan, the local model approximations do not look exactly symmetric. This is due to the fact that there exists no similarity criterion in the

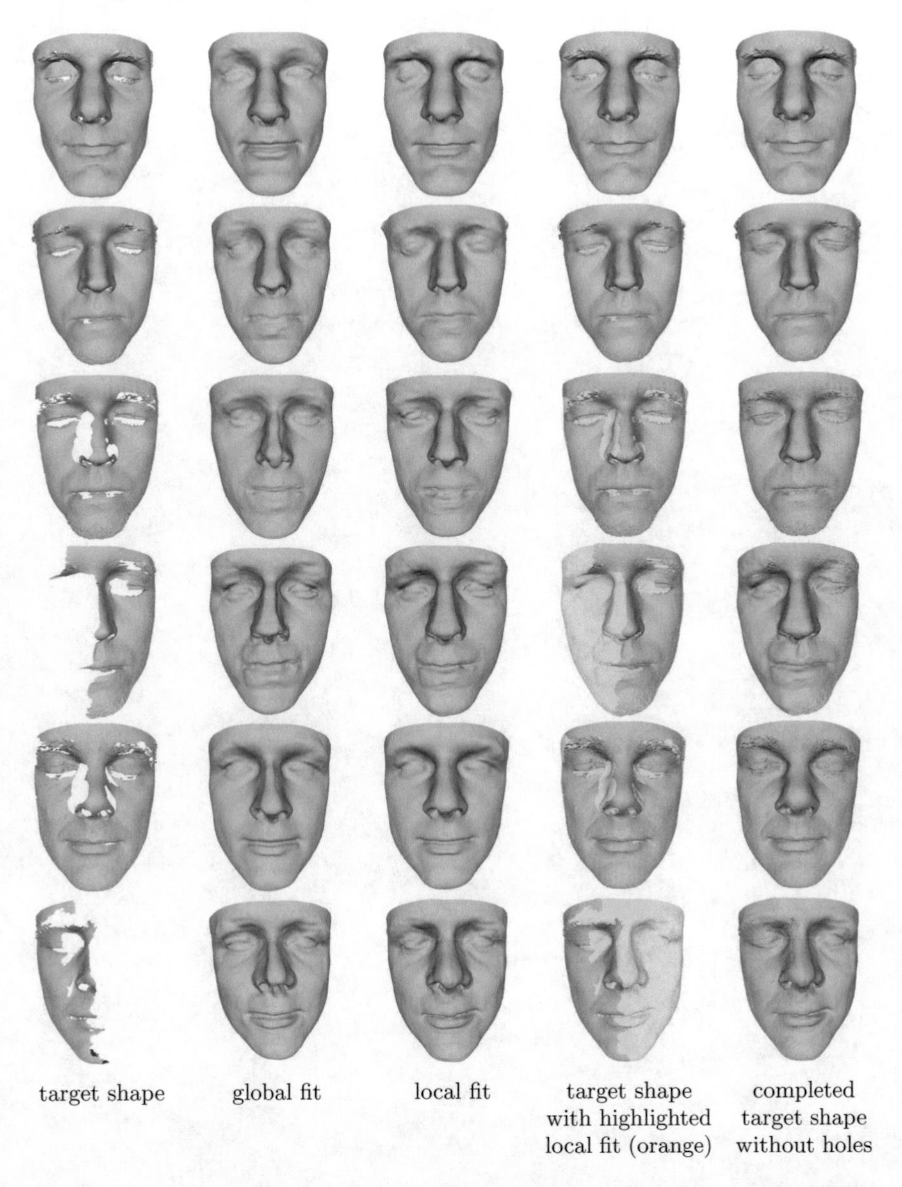

Fig. 4.27 Results of the 3D face scan completion experiment. ©Eurographics Association 2013. Reproduced from [6] by kind permission of the Eurographics Association

incorporated statistical shape model. So, the side with no data mostly resembles the global model approximation, whereas the other side is given as a precise fit of the local model. Nevertheless, also the reconstructed partial faces look all very natural.

Chapter 5
Global-To-Local Shape Priors for Variational Level Set Methods

In Chap. 3, we have presented our *Locally Deformable Statistical Shape Model* (LDSSM) together with an iterative approach that couples the shape information provided by our LDSSM with level set segmentation methods. The level set evolution is used to estimate a weight update target which then in turn is approximated by our LDSSM (c.f. Fig. 3.8). Through this model approximation, the evolved level set is restricted to the space of feasible shapes that is consistent with our LDSSM. The presented approach is straightforward to understand and can thus be easily used to extend classical level set segmentation methods with local statistical shape information.

However, one drawback of our iterative approach is that the statistical shape information is used only *after* the estimation of the weight update target. Therefore, the weight update target estimation step of our approach generally evolves the level set in a way that is not consistent with our LDSSM. As a consequence, the evolved level set has to be restricted back to the space of feasible shapes in the following model fitting step. To improve the segmentation process, it would be beneficial to incorporate the statistical shape information already in the weight update target estimation step. In order to achieve this, we need to provide our iterative approach from Chap. 3 with a sound mathematical foundation. For this purpose, we propose to embed our LDSSM into the variational image segmentation framework from Sect. 1.5. This will be the topic of the following sections.

The following sections are an extension of the work that we have published in [2]. In Sect. 5.1, we start by pointing out the limitations of our iterative segmentation framework from Sect. 3.3.4. Then, in Sect. 5.2, we briefly review existing approaches that try to integrate shape information into variational image segmentation approaches. Afterwards, in Sect. 5.3, we thoroughly review an approach that integrates the global *Statistical Shape Model* (SSM) from Sect. 2.3 into the variational image segmentation framework. Building on this, we finally present a new solution to integrate our LDSSM into the variational image segmentation framework in Sects. 5.4 and 5.5.

C. Last, *From Global to Local Statistical Shape Priors*, Studies in Systems, Decision and Control 98, DOI 10.1007/978-3-319-53508-1_5

5.1 Problems of the Iterative Segmentation Framework

We recall that the LDSSM has been used in Sect. 3.3.4 in an iterative framework for segmenting 2D images and 3D volume data. Our framework consists of four steps that have to be iterated in order to obtain the desired segmentation result. These four steps are shown in Fig. 3.8, and we will shortly recapitulate them here as they are needed for the upcoming illustrations.

In the first step, a weight update target is estimated based on image information around the current model contour. Please note that this weight update target is not required to lie in the space of feasible shapes which is consistent with our LDSSM as no model information is used in the estimation of the weight update target. So, the following three steps are responsible for restricting the estimated weight update target back to the space of feasible shapes: In step 2, updated weight fields are computed which reduce the element-wise squared distance between the weight update target and the LDSSM representation. To make sure that these updated weight fields are in accordance with the trained shape distribution in each element of the data domain, they are truncated to lie within three times the standard deviation of the trained Gaussian distribution in step 3. Finally, in step 4, the updated weight fields are convolved with a smoothing kernel in order to ensure that they satisfy the smoothness condition of our LDSSM.

A first connection between our iterative framework and level set methods has been made at the end of Sect. 3.3.4.3. We showed that our framework can be regarded as introducing an additional shape regularization component (steps 2–4 of our framework) in the level set evolution process (step 1 of our framework). More specifically, our iterative weight field update for a known target shape from Sect. 3.3.3 (steps 2–4 of our segmentation framework) minimizes the squared distance between the modeled shape and the target shape under the constraints that the resulting weight fields should be smooth and bounded by the trained shape distribution (c.f. Sect. 3.3.1).

Furthermore, we explained in Sect. 1.5 that in state of the art variational image segmentation methods the level set contour evolution (step 1 of our framework) is derived as a gradient flow which minimizes an appropriate image energy. So, our segmentation framework from Sect. 3.3.4, can be regarded as iteratively minimizing two coupled problems: an image energy and the constrained squared distance to an estimated weight update target.

This means, in the derivation of our iterative segmentation framework, we treated the shape approximation problem independently from the image segmentation problem, and we connected both problems by iteratively minimizing them until one obtains a stable solution. This formulation has the drawback that the statistical shape information is not considered in the energy minimization problem which is used to drive the level set contour evolution in step 1 of our framework. Hence, it leads to the above discussed problem that the evolved shape is not restricted to the space of feasible shapes. One might also say that the shape approximation step is extrinsic to the level set contour evolution as it is applied only after the level set has already evolved.

Now, as mentioned in the introduction to this chapter, we want to integrate the statistical shape information directly into the level set evolution process. This means, our goal is to obtain a mathematical formulation of the segmentation problem which integrates both problems, the image segmentation problem and the shape approximation problem, into one combined energy so that the shape regularization component becomes intrinsic to the level set evolution process and the level set is only allowed to evolve in a way which is consistent with our LDSSM.

5.2 Existing Variational Image Segmentation Approaches with Shape Priors

We are not the first to couple statistical shape models and variational image segmentation approaches. So, we will discuss the most prominent existing approaches in this section. For a recent review article, we additionally refer the reader to [40]. In addition to the nonrigid shape deformations, most of the following approaches also try to estimate the rigid pose parameters during the segmentation process. In order to achieve this, there exists two different ways: One can either estimate the rigid pose of the evolving level set with regard to the shape prior (e.g. [27, 77, 109]) or one can estimate the rigid pose of the evolving level set with regard to the image (e.g. [106, 108, 130]), respectively. Hence, in the second approach, one can always assume that the shape prior is rigidly aligned to the evolving level set. Because of this and since the rigid pose estimation is beyond the scope of this thesis, we will omit the rigid pose estimation in the description of the methods. However, one possible method to include the rigid pose estimation in the segmentation process will be presented later in Sect. 5.3.5.

For the following descriptions, we divide the approaches in two main categories: Approaches that allow only solutions which lie within the linear subspace of feasible shapes spanned by the eigenshapes of the global implicit linear parametric SSM from Sect. 2.3 and approaches with additional flexibility that allow also solutions which deviate from the subspace of feasible shapes. All discussed approaches are summarized in Table 5.1. Additionally, it is shown where the approach that we are about to introduce can be placed with regard to the existing approaches.

5.2.1 Linear Subspace-Constrained Approaches

In this section, we start by describing the linear subspace-constrained approaches. The approaches with additional flexibility will be described subsequently in Sect. 5.2.2, and we will discuss the drawbacks of the existing approaches in Sect. 5.2.3.

Table 5.1 Overview of existing approaches that have the objective to integrate statistical shape information into the variational image segmentation framework

Linear subspace-constrained approaches	Approaches with additional flexibility
Leventon et al. [77]	Chen et al. [27]
Tsai et al. [130]	Bresson et al. [22]
Rousson and Cremers [106]	Rousson et al. [109]
	Cremers et al. [39]
	Rousson and Paragios [108]
	Our approach [2]

5.2.1.1 Leventon et al. [77] (2000)

In 2000, Leventon et al. [77] proposed a first approach to incorporate statistical shape information in the evolution process of the *Geodesic Active Contours* (GAC) level set segmentation framework from Sect. 1.5.1. For this purpose, they used the global implicit linear parametric SSM from Sect. 2.3 in order to extend the evolution equation of the GAC approach (Eq. (1.56)) by an additional term that guides the level set contour evolution to *likely* shapes. In their approach, the likelihood is modeled by a *Maximum A Posteriori* (MAP) approach:

$$P(\vec{w}_{\text{MAP}}) = \arg\max_{\vec{w}} P(\vec{w}|\Phi, I) . \tag{5.1}$$

This means, one searches for the most likely weight vector $\vec{w}_{\text{MAP}}$ of the global implicit model given the current level set function Φ and the image data I. Using the laws of conditional probability and Bayes' theorem, the right hand side of Eq. (5.1) can be expanded to [77, Eq. (11)]

$$P(\vec{w}|\Phi, I) = \frac{P(\Phi|\vec{w})P(I|\vec{w}, \Phi)P(\vec{w})}{P(\Phi, I)} , \tag{5.2}$$

where the denominator may be dropped as it does not depend on $\vec{w}$. Leventon et al. used the term $P(\Phi|\vec{w})$ to favor evolving level set functions Φ with a zero level curve that lies completely inside the modeled shape $\Phi_{\text{glob}}(\vec{w})$ (they initialized the evolving contour inside the object of interest). The term $P(I|\vec{w}, \Phi)$ is used to model how good the modeled shape $\Phi_{\text{glob}}(\vec{w})$ approximates image edges and the term $P(\vec{w})$ models the likelihood of a shape according to Eq. (2.9).

Using the so obtained MAP shape estimate $\Phi_{\text{glob}}(\vec{w}_{\text{MAP}})$, Leventon et al. proposed to extend the evolution equation of the GAC approach (Eq. (1.56)) as

$$\Phi^{k+1} = \Phi^k + \lambda_1 \left(\delta_{\Phi^k} \left\langle \nabla g \frac{\nabla \Phi^k}{|\nabla \Phi^k|} \right\rangle + \delta_{\Phi^k} g \left[\operatorname{div}\left(\frac{\nabla \Phi^k}{|\nabla \Phi^k|} \right) + \nu \right] \right) + \lambda_2 \left(\Phi_{\text{glob}}(\vec{w}_{\text{MAP}}) - \Phi^k \right), \tag{5.3}$$

where the additional last term on the right hand side of Eq. (5.3) exerts a force on the evolving level set function in the direction of the MAP shape estimate. The weighting factors λ_1 and λ_2 are used to balance the influence of the shape model guided contour update and the original GAC evolution. The final segmentation result is then given as the maximum a posteriori shape model estimate $\Phi_{\text{glob}}(\vec{w}_{\text{MAP}})$ after the last iteration of Eq. (5.3).

When we have a closer look at Eq. (5.3), we can see that the second term on the right hand side (i.e. the GAC contour evolution) is derived without consideration of the additional shape constraint. So, like our iterative framework, the formulation of Leventon et al. has the drawback that the shape constraint is derived independently from the data-driven contour evolution.

5.2.1.2 Tsai et al. [130] (2003)

Motivated by the work of [77], Tsai et al. [130] proposed to directly optimize the parameter vector $\vec{w}$ of the global implicit SSM in the low-dimensional subspace of feasible shapes defined by the mean shape $\bar{\Phi}$ (Eq. (2.28)) and the m main modes of variation $\{\tilde{\Phi}_1, \ldots, \tilde{\Phi}_m\}$ (Eq. (2.32)). This leads to an energy $E(\vec{w})$ that depends only on the parameter vector $\vec{w}$ of the global implicit SSM from Eq. (2.34). So, the restriction of possible results to the subspace of feasible shapes is now intrinsic to the the energy $E(\vec{w})$ and the need to minimize two competing energies is completely removed. We will discuss the approach of Tsai et al. in more detail in Sect. 5.3.

5.2.1.3 Rousson and Cremers [106] (2005)

Both so far mentioned approaches assume that plausible shapes are distributed uniformly [130] or according to a linear multivariate Gaussian distribution [77] inside the low-dimensional subspace of feasible shapes. However, this assumption is not always suitable. Especially for modeling moving objects, e.g. the silhouette of a walking person [39] or the contours obtained from different views of the same three-dimensional object [38], it is more reasonable to describe the likelihood of plausible shapes by a nonlinear distribution. For more information about level set based tracking with shape priors we refer the reader to [35, 36, 112].

The need for a nonlinear distribution becomes obvious when we have a look at the projection of the motion capture data of a running human onto its two main modes of variation. The motion capture data is shown in Fig. 5.1a and the projection is shown in Fig. 5.1b, respectively. The motion capture data has been obtained from the database described in [90]. A density function that is appropriate to describe such

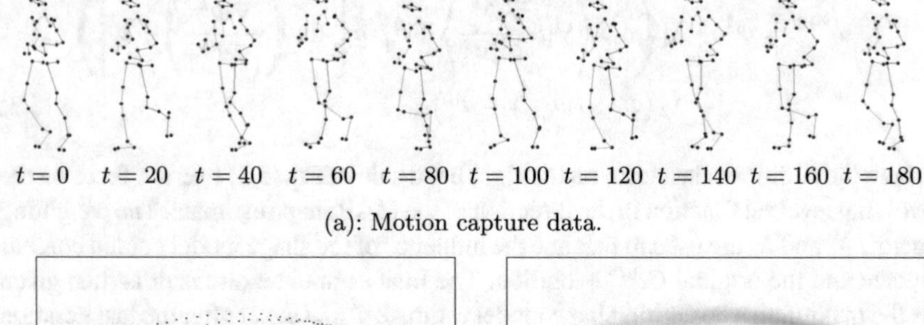

(a): Motion capture data.

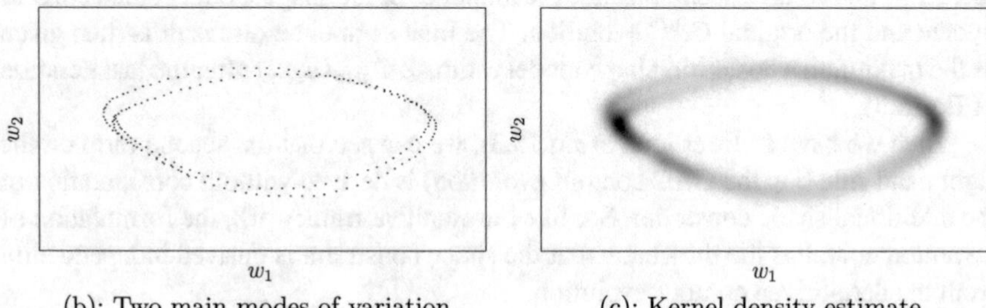

(b): Two main modes of variation. (c): Kernel density estimate.

Fig. 5.1 Motion capture data of a running human (taken from [90]) (**a**) and its projection onto the two main modes of variation after full generalized Procrustes analysis (**b**). The density function that describes likely shape weights (w_1, w_2) is shown in **c**. The darker the density, the more likely a pair of shape weights

a highly nonlinear shape distribution can be obtained by *Kernel Density Estimation* (KDE) methods [97, 105]. An example for an appropriate density function is shown in Fig. 5.1c.

In KDE methods, one tries to estimate an unknown density function $P(\vec{u})$ from a set of n samples $\{\vec{u}_1, \ldots, \vec{u}_n\}$ as

$$P(\vec{u}) = \frac{1}{n\sigma} \sum_{i=1}^{n} K\left(\frac{\vec{u} - \vec{u}_i}{\sigma}\right), \tag{5.4}$$

where K is the so-called kernel function and σ is a smoothing parameter called bandwidth. A common kernel function is e.g. the Gaussian kernel:

$$K(v) = \frac{1}{\sqrt{2\pi}} \mathrm{e}^{-\frac{v^2}{2}}. \tag{5.5}$$

In this case, the bandwidth parameter σ denotes the standard deviation of the (isotropic) Gaussian distribution and the density estimate is obtained as a mixture of k Gaussians with fixed standard deviations that are centered around the samples $\vec{u}_i$.

Rousson and Cremers [106] proposed to perform the KDE in the linear subspace of feasible shapes defined in Eq. (2.34). In this case, the data samples are given as the projections $\{\vec{w}_1, \ldots, \vec{w}_n\}$ of the implicit training shapes $\{\vec{\Phi}_1, \ldots, \vec{\Phi}_n\}$ into this subspace. They are obtained as (c.f. Eq. (2.8))

$$\vec{w}_i = \left(\vec{\Phi}_i - \bar{\vec{\Phi}} \right) \tilde{\mathbf{\Phi}}^T , \tag{5.6}$$

so that the density estimate becomes

$$P(\vec{w}) = \frac{1}{n\sigma} \sum_{i=1}^{n} K\left(\frac{\vec{w} - \vec{w}_i}{\sigma} \right) . \tag{5.7}$$

In Eq. (5.7), the variance σ^2 is given as the average squared distance from any training weight vector to its nearest neighbor:

$$\sigma^2 = \frac{1}{n} \sum_{i=1}^{n} \min_{i \neq j} \left|\left| \vec{w}_i - \vec{w}_j \right|\right|^2 . \tag{5.8}$$

This ensures that on average the next training shape is located within one standard deviation of each Gaussian distribution [39]. The so-obtained kernel density estimate of the running person can be seen in Fig. 5.1c.

Now, similar to the approach from [130], Rousson and Cremers define an energy that depends only on the weight vector $\vec{w}$. However, their energy consists of two distinct parts, a data term and a weighted shape term:

$$E(\vec{w}) = E_{\text{data}}(\vec{w}) + \lambda E_{\text{shape}}(\vec{w}) , \tag{5.9}$$

where the shape energy is given as the negative log-likelihood of the estimated shape density from Eq. (5.7):

$$E_{\text{shape}}(\vec{w}) = -\log P(\vec{w}) . \tag{5.10}$$

The gradient of the shape energy with regard to the parameter vector $\vec{w}$ is given as

$$\frac{\partial E_{\text{shape}}}{\partial \vec{w}} = \frac{1}{\sigma^2} \frac{\sum_{i=1}^{n} K\left(\frac{\vec{w} - \vec{w}_i}{\sigma} \right) (\vec{w} - \vec{w}_i)}{\sum_{i=1}^{n} K\left(\frac{\vec{w} - \vec{w}_i}{\sigma} \right)} . \tag{5.11}$$

Hence, the shape energy exerts a force on the evolving parameter vector $\vec{w}$ towards each training weight vector $\vec{w}_i$. The force in the direction of each training vector exponentially decays with the distance to the training weight vector. So, the evolving parameter vector is mostly drawn in the direction of the nearest training weight vector.

5.2.2 Approaches with Additional Flexibility

All approaches discussed so far have in common that the final segmentation result has to lie in the low-dimensional subspace spanned by the training shapes. Other results are *not* possible. As a result, the approaches of [77, 106, 130] work pretty well when the number of training shapes is large enough to capture the full intra-class variation of a particular class of training shapes. However, as already mentioned several times, in many real-world applications the number of available training shapes does not suffice in order to reproduce the full subspace of feasible shapes. In these cases, the approaches based on the assumption that the final segmentation can fully be described by the implicit statistical shape model from Eq. (2.34) will fail. To deal with this problem, other variational approaches have been proposed where the shape prior is included as a weighted regularization term in the energy functional that has to be minimized, thus allowing solutions that deviate from the trained shape distribution.

5.2.2.1 Chen et al. [27] (2002)

For example, Chen et al. [27] proposed to include a shape term in the Geodesic Active Contours energy from Eq. (1.53) that penalizes the squared distance to the zero level curve $\bar{C}$ of the mean shape from Eq. (2.28)[1]:

$$E(\Phi) = \int_{\Omega} \delta_{\Phi(\vec{x})} \left[g(I(\vec{x})) + \frac{\lambda}{2} d^2(\Phi(\vec{x}), \bar{C}) \right\} |\nabla\Phi(\vec{x})| \, \mathrm{d}\vec{x} \,, \tag{5.12}$$

where $d(\Phi(\vec{x}), \bar{C})$ denotes the closest distance from a point $\vec{x}$ on the zero level curve of Φ to any point on the zero level curve $\bar{C}$ of the mean shape. λ is a weighting factor that balances the influence of the shape term. The corresponding level set evolution equation can be obtained as

$$\frac{\mathrm{d}}{\mathrm{d}t}\Phi = \delta_\Phi \left[\left\langle \nabla \left(g + \frac{\lambda}{2} d^2 \right) \frac{\nabla\Phi}{|\nabla\Phi|} \right\rangle + \left(g + \frac{\lambda}{2} d^2 \right) \operatorname{div}\left(\frac{\nabla\Phi}{|\nabla\Phi|} \right) \right] . \tag{5.13}$$

Clearly, the approach by Chen et al. is also not well suited for shape classes with high intra-class shape variability where the shapes may differ significantly from the average zero level curve. In this case, one has to choose the weighting factor λ fairly small so that the segmentation result is mostly determined by the original GAC terms and not by the new shape terms. Chen et al. somehow addressed this issue by clustering the training shapes. This leads to an artificial reduction of the intra-class variance at the price of additional shape classes. However, they did not mention how the correct shape class is selected prior to the segmentation process.

---

[1] In fact they used a slightly different approach to obtain an estimate of the mean shape. For details we refer the reader to [27].

5.2.2.2 Bresson et al. [22] (2003)

The approach from [27] has later been extended by Bresson et al. [22]. They replaced the squared distance to the zero level curve of the mean shape by the squared level set values of the global implicit SSM from Eq. (2.34):

$$E(\Phi, \vec{w}) = \int_{\Omega} \delta_{\Phi(\vec{x})} \left[g(I(\vec{x})) + \frac{\lambda}{2} \Phi_{\text{glob}}(\vec{w})(\vec{x}) \right] |\nabla \Phi(\vec{x})| \, \mathrm{d}\vec{x} \,, \tag{5.14}$$

which leads to the following evolution equations that have to be iterated in turn in order to obtain the segmentation result:

$$\frac{\mathrm{d}}{\mathrm{d}t} \Phi = \delta_{\Phi} \left[\left\langle \nabla \left(g + \frac{\lambda}{2} \Phi^2_{\text{glob}}(\vec{w}) \right) \frac{\nabla \Phi}{|\nabla \Phi|} \right\rangle + \left(g + \frac{\lambda}{2} \Phi^2_{\text{glob}}(\vec{w}) \right) \operatorname{div} \left(\frac{\nabla \Phi}{|\nabla \Phi|} \right) \right], \tag{5.15}$$

$$\frac{\mathrm{d}}{\mathrm{d}t} \vec{w} = -\lambda \int_{\Omega} \delta_{\Phi(\vec{x})} \Phi_{\text{glob}}(\vec{w})(\vec{x}) \tilde{\Phi}_i(\vec{x}) \, |\nabla \Phi(\vec{x})| \, \mathrm{d}\vec{x} \,. \tag{5.16}$$

Integrating over the squared level set values of the global implicit SSM along the zero level curve of Φ is supposed to favor those segmentation results that have a similar shape as a certain modeled shape $\Phi_{\text{glob}}(\vec{w})$. However, this approach also has two drawbacks:

1. The space of signed distance functions is not a linear space [40], i.e. linear combinations of two signed distance functions yield in general no signed distance function. This means that the shape term introduced by Bresson et al. does not exactly penalize the squared distance to the zero level curve of a certain modeled shape.
2. Integrating over the squared level set values of the global implicit SSM along the zero level curve of Φ favors short zero level curves which has no theoretical justification. In fact, the shape term should favor those curves with a length comparable to the zero level curve of the modeled shape.

5.2.2.3 Rousson et al. [109] (2004)

An approach that addresses the two above-mentioned drawbacks has been proposed by Rousson et al. [109]. Like in the approaches from [22, 27], their energy functional consists of two parts, a data term and a shape term:

$$E(\Phi, \vec{w}) = \lambda E_{\text{data}}(\Phi) + (1 - \lambda) E_{\text{shape}}(\Phi, \vec{w}) \,, \tag{5.17}$$

where λ is again a weighting factor that balances the influence of the shape term and $\vec{w}$ is the parameter vector of the global implicit SSM from Eq. (2.34). The data term can for example be the GAC energy from Eq. (1.53) but also any other energy is possible like for example the *Active Contours Without Edges* approach from Eq. (1.58) or the *Geodesic Active Regions* model from [94].

The crucial part of the work by Rousson et al. consists of the shape energy. Instead of penalizing the squared distance from the zero level curve of Φ to the zero level curve $\bar{C}$ of the mean shape (as in [27]) or penalizing the squared level set values on the zero level curve of the global implicit SSM (as in [22]), they penalize the squared distance from the zero level curve of Φ to the modeled shape $\Phi_{\text{glob}}(\vec{w})$:

$$E_{\text{shape}}(\Phi, \vec{w}) = \int_{\Omega} \delta_{\Phi(\vec{x})} \left[\Phi(\vec{x}) - \Phi_{\text{glob}}(\vec{w})(\vec{x})\right]^2 \, \mathrm{d}\vec{x} \,. \tag{5.18}$$

This energy neither favors those level sets Φ with a short zero level curve nor does it assume implicitly that the modeled shape is a signed distance function. The functional derivative of this shape energy with regard to the level set function Φ is given as

$$\frac{\partial E(\Phi)}{\partial \Phi(\vec{x})} = 2\delta_{\Phi(\vec{x})} \left[\Phi(\vec{x}) - \Phi_{\text{glob}}(\vec{w})(\vec{x})\right] . \tag{5.19}$$

Now, in order to minimize Eq. (5.18), one needs to determine an appropriate weight vector of the global model approximation $\vec{w}$ in addition to the level set function Φ. This can be achieved by solving the linear system $\vec{w}\,\mathbf{U} = \vec{b}$ in each iteration of the level set evolution, where the entries of the $m \times m$ matrix $\mathbf{U}$ and the m-dimensional vector $\vec{b}$ are defined as [109]

$$U(i, j) = \int_{\Omega} \delta_{\Phi(\vec{x})} \tilde{\Phi}_i(\vec{x}) \tilde{\Phi}_j(\vec{x}) \, \mathrm{d}\vec{x} \,, \tag{5.20}$$

$$b(i) = \int_{\Omega} \delta_{\Phi(\vec{x})} \left(\Phi(\vec{x}) - \bar{\Phi}(\vec{x})\right) \tilde{\Phi}_i(\vec{x}) \, \mathrm{d}\vec{x} \,. \tag{5.21}$$

The $\tilde{\Phi}_i$ are the main modes of variation (Eq. (2.32)) and $\bar{\Phi}$ is the mean shape (Eq. (2.28)) of the global implicit model, respectively.

The level set functions Φ which minimize the shape energy from Eq. (5.18) are located within the subspace of feasible shapes spanned by the global implicit SSM from Eq. (2.34) as for those functions the shape energy yields zero. So, for λ close to 1 the solution space of the approach by Rousson et al. becomes equal to the subspace of feasible shapes like for the approaches from Sect. 5.2.1. Yet, when $\lambda < 1$ the shape energy by Rousson et al. provides more flexibility as the solutions are allowed to lie outside the subspace of feasible shapes.

5.2.2.4 Cremers et al. [39] (2006)

The approaches mentioned so far in this subsection have in common that the penalization terms depend on the distance of the segmentation result to linear combinations of the training shapes. However, as already mentioned in the description of the approach from Rousson and Cremers, this is not always a good choice. So, Cremers et al. proposed an approach that is closely related to the approach from [106]. However, instead of performing the KDE in the low-dimensional subspace of feasible shapes, they proposed to perform the KDE directly in the infinite-dimensional space that contains all level set functions. For this purpose, Cremers et al. defined the squared distance between two level set functions Φ_1 and Φ_2 as the squared area of their set symmetric difference, i.e. the area where the interiors of the represented shapes do not overlap:

$$d^2(\Phi_1, \Phi_2) = \int_\Omega \left[H(-\Phi_1(\vec{x}) - H(-\Phi_2(\vec{x})\right]^2 \, \mathrm{d}\vec{x} \,. \tag{5.22}$$

With the help of this distance, they obtain the kernel density estimate of a density function that describes plausible shapes as

$$P(\Phi) = \frac{1}{n\sigma} \sum_{i=1}^{n} K\left(\frac{d(\Phi, \Phi_i)}{\sigma}\right), \tag{5.23}$$

where the Φ_i denote the training shapes.

Similar to the approach by Rousson and Cremers, the variance σ^2 is given as the average squared distance from any training shape to its nearest neighbor:

$$\sigma^2 = \frac{1}{n} \sum_{i=1}^{n} \min_{i \neq j} d^2(\Phi_i, \Phi_j) \,. \tag{5.24}$$

Now, one can again define an energy that consists of two distinct parts, a data term and a weighted shape term:

$$E(\Phi) = E_{\text{data}}(\Phi) + \lambda E_{\text{shape}}(\Phi) \,. \tag{5.25}$$

However, this time the energy does not depend on any parameter vector but only on the evolving level set function Φ itself. Consequently, one is not bound to the linear subspace of feasible shapes. As before, the shape energy is given as the negative log-likelihood of the estimated density function:

$$E_{\text{shape}}(\Phi) = -\log P(\vec{\Phi}) \,, \tag{5.26}$$

and its functional derivative can be obtained as

$$\frac{\partial E_{\text{shape}}(\Phi)}{\partial \Phi(\vec{x})} = \frac{1}{\sigma^2} \frac{\sum_{i=1}^{n} K\left(\frac{d(\Phi,\Phi_i)}{\sigma}\right) \delta_\Phi(\vec{x}) \left[H(-\Phi(\vec{x}) - H(-\Phi_i(\vec{x})\right]}{\sum_{i=1}^{n} K\left(\frac{d(\Phi,\Phi_i)}{\sigma}\right)} \,. \tag{5.27}$$

Now, similar to the previous approach, it can be seen that the shape energy exerts a force on the evolving level set Φ towards each training shape Φ_i that exponentially decays with regard to the distance to each training shape. So, the approach by Cremers et al. also favors those shapes that are similar to any training shape. However, it is more flexible than the approach presented in [106] since one is no more tied to the linear subspace of feasible shapes.

5.2.2.5 Rousson and Paragios [108] (2008)

All previously mentioned approaches have the drawback that their penalization terms have a global influence on the shape evolution. Hence, if the variations of a specific shape class in a local region of the shape are not accurately represented through the model, the whole segmentation result will strongly deviate from the subspace of feasible shapes. So, the segmentation result will mostly be determined through image features and will no longer be guided by the trained shape distribution.

In order to address this issue, Rousson and Paragios proposed another approach that explicitly models the *local confidence* in the training shapes [107, 108]. This is achieved by keeping the mean shape $\bar{\Phi}$ as a global shape descriptor but instead of extracting the global linear main modes of variation from the training shapes, they combine it with spatially varying shape variances $\bar{\sigma}^2(\vec{x})$. A large variance in an element $\vec{x}$ corresponds to large variations in the training data in this point. An additional smoothness criterion for the spatially varying shape variances ensures that the shape variability fluctuates only slightly in local regions of the image domain Ω. Based on these two parameters (mean shape and local variances), one can define an element-wise shape probability according to

$$\bar{p}(\Phi(\vec{x})) = \frac{1}{\sqrt{2\pi}\bar{\sigma}(\vec{x})} \exp\left(-\frac{(\Phi(\vec{x}) - \bar{\Phi}(\vec{x}))^2}{2\bar{\sigma}^2(\vec{x})}\right). \tag{5.28}$$

The corresponding shape energy is given as the average negative log-likelihood of Eq. (5.28) on the zero level curve of the evolving level set function:

$$\begin{aligned} E_{\text{shape}}(\Phi) &= \int_\Omega \delta_{\Phi(\vec{x})} \left(-\log \bar{p}(\Phi(\vec{x}))\right) \mathrm{d}\vec{x} \\ &\sim \int_\Omega \delta_{\Phi(\vec{x})} \left(\log \bar{\sigma}(\vec{x}) + \frac{(\Phi(\vec{x}) - \bar{\Phi}(\vec{x}))^2}{2\bar{\sigma}^2(\vec{x})}\right) \mathrm{d}\vec{x}\,. \end{aligned} \tag{5.29}$$

So, the shape energy from Eq. (5.29) basically enhances the shape energy from [27] by spatially varying shape variances. This means, like in [27], the shape energy from Eq. (5.29) becomes minimal when the segmentation result equals the mean shape. However, this time the segmentation result is being less penalized when it deviates from the mean shape in image regions with large variations in the training shapes than in those regions with small variations in the training data. Hence, the shape prior is assumed to be more significant in those regions with small variations in the training shapes, whereas in image regions with large shape variations the image data is considered to be more reliable than the shape information. The functional derivative of Eq. (5.29) can be obtained as

$$\frac{\partial E(\Phi)}{\partial \Phi(\vec{x})} = 2\delta_{\Phi(\vec{x})} \frac{\left(\Phi(\vec{x}) - \bar{\Phi}(\vec{x})\right)}{\bar{\sigma}^2(\vec{x})} . \tag{5.30}$$

It can be seen one more time that the evolving level set is forced stronger towards the mean shape in regions with small shape variances. Rousson and Paragios denoted this term the *shape consistency force*.

The spatially varying local shape variances have the advantage that even with a small training set, one can construct a strong shape prior in local regions with small shape variations inside the considered shape class. However, in those regions where the shapes differ significantly from the mean shape, the segmentation result is still determined mostly by the image features.

5.2.3 Drawbacks of Existing Approaches

All discussed approaches have in common that their penalization terms have a global influence on the shape evolution. This means that their individual penalization terms depend either on the distance of the segmentation result to the training shapes or on the distance of the segmentation result to linear combinations of those training shapes. None of the proposed penalization terms considers the fact that the desired segmentation result might consist of local, i.e. nonlinear, combinations of the training shapes (c.f. Sect. 3.1). Such a case would be penalized by all presented penalization terms with a large value.

The approach by Rousson and Paragios [108] thereby represents a single exception. However, their approach bases on the assumption that the segmentation result resembles the mean shape in large parts. So, when the intra-class variance of the training shapes is high so that many shapes greatly differ from the mean shape, the shape prior by Rousson and Paragios has no statistical significance and the segmentation result will almost completely be determined by the image features.

We clarify these explanations with the help of the hand-example from Fig. 3.1. It can be seen that the desired segmentation result differs from the mean shape, which in this case is a hand with medium-length fingers. So, the penalization term from Eq. (5.12) will result in a large value and the shape prior from Eq. (5.29) has no statistical significance. Furthermore, the penalization terms from Eqs. (5.14) and (5.18), which consider the distance from the desired segmentation result to a linear combination of the training shapes, will also result in large values as the desired segmentation result differs from all linear combinations of the training shapes (c.f. the best possible linear approximation in Fig. 3.1c). This is because the desired result is a local, i.e. nonlinear, combination of the training shapes. Additionally, it can be seen

Table 5.2 Properties of existing variational image segmentation approaches that include shape priors. The assumed shape prior is indicated in columns 2–5 and columns 6–8 show the additional flexibility with regard to the shape prior (subspace constrained means no extra flexibility)

	Shape prior				Extra flexibility		
	Uniform	Gaussian	Nonlinear	Fixed	Subspace constrained	Global	Local
Leventon et al. [77]		x			x		
Tsai et al. [130]	x				x		
Rousson and Cremers [106]			x		x		
Bresson et al. [22]	x					x	
Rousson et al. [109]	x					x	
Cremers et al. [39]			x			x	
Chen et al. [27]				x		x	
Rousson and Paragios [108]				x			x
Our approach [2]	x	(x)					x

that the desired segmentation result also differs from all the training shapes. Thus, the penalization term from Eq. (5.26), which includes the exponentially weighted mean distance to all training shapes, would result in a large value for the desired result too.

This example clearly shows that a segmentation result which consists of local combinations of training shapes is not intended in the above mentioned approaches (an overview can also be seen in Table 5.2). Consequently, the extension of the variational image segmentation framework by our LDSSM from Chap. 3, which explicitly models these local combinations, will be the topic of the following sections.

5.3 Variational Formulation for a Global Shape Prior

Our approach to integrate our LDSSM into the level set evolution process is mostly inspired by the work which has been presented by Tsai et al. in [130]. The goal of their approach is to integrate the global implicit *Statistical Shape Model* (SSM) from Eq. (2.34) into the level set segmentation framework. We will show later that our approach can be regarded as an extension of their method by replacing the SSM through our more general LDSSM. As a consequence, we first need to review their approach before we can continue with the description of our method.

As detailed in Sect. 1.5, the objective in state of the art variational level set methods is to minimize an energy functional $E(\Phi)$ which exactly describes the segmentation problem. The minimization of such an energy functional is obtained by taking its functional derivative with regard to the level set function Φ and evolving the level set according to the thus obtained gradient descent equation until one reaches a stable solution.

Now, in order to integrate the shape information provided by the SSM from Sect. 2.3, one can replace the level set function Φ in the formulation of the energy functional by the global model $\Phi_{\text{glob}}(\vec{w})$:

$$E(\Phi) \quad \Longrightarrow \quad E(\Phi_{\text{glob}}(\vec{w})) \,. \tag{5.31}$$

By doing this, the energy $E(\Phi_{\text{glob}}(\vec{w}))$ now depends on the modeled shape $\Phi_{\text{glob}}(\vec{w})$ instead of an arbitrary level set function Φ. This brings two main advantages:

1. The segmentation result is guaranteed to reside in the space of feasible shapes spanned by the eigenshapes of the SSM.
2. The segmentation result can be obtained by minimizing the energy directly with regard to the weight vector $\vec{w}$ of the SSM.

In the following section, we will review the just mentioned approach by Tsai et al. in detail. Moreover, we will make their idea more clear by discussing the solution for an exemplary segmentation problem (Sect. 5.3.2). Afterwards, we extend the work of Tsai et al. by providing a general solution for an arbitrary segmentation problem (Sect. 5.3.3).

We further present an extension of the approach by Tsai et al. that incorporates the trained shape distribution in the formulation of the energy functional (Sect. 5.3.4). Finally, in Sect. 5.3.5, we briefly discuss how the rigid transformation from the SSM to the data under consideration can be considered in the segmentation process.

5.3.1 Problem Formulation

As pointed out in Sect. 2.3, each shape which can be generated by the SSM from Eq. (2.34) is completely described by a vector $\vec{w} \in \mathbb{R}^m$. The modeled shape $\Phi_{\text{glob}}(\vec{w})$ is given as the mean shape $\bar{\Phi}$ plus a weighted sum of the m most important eigenshapes $\tilde{\Phi}_i$ of a set of training shapes:

$$\Phi_{\text{glob}}(\vec{w}) = \bar{\Phi} + \sum_{i=1}^{m} w_i \, \tilde{\Phi}_i \,.$$

A main property of this shape representation is that the eigenshapes $\tilde{\Phi}_i$ span a low-dimensional linear subspace where all feasible shapes reside in. The vector $\vec{w}$ is the representation of each modeled shape in this subspace.

Tsai et al. proposed in [130] that one can formulate a segmentation problem, which makes use of the statistical shape information provided by the SSM, by defining an appropriate energy $E(\Phi_{\text{glob}}(\vec{w}))$ that depends on the modeled shape $\Phi_{\text{glob}}(\vec{w})$ and which should be minimized in order to obtain the segmentation result. They presented the solution for three exemplary energies in their work. We will recapitulate one of these energies in Sect. 5.3.2 and the other two segmentation problems can be found in Appendix E. However, we will extend their work by providing a general solution in Sect. 5.3.3 which enables us to use a large number of energy functionals $E(\Phi)$ that have been proposed in the level set literature for level set segmentation approaches without shape information. The only modification which has to be made is that the level set function Φ has to be replaced by the modeled shape $\Phi_{\text{glob}}(\vec{w})$.

By replacing the level set function Φ through the modeled shape $\Phi_{\text{glob}}(\vec{w})$, the segmentation problem remains the same but the solution space is restricted to the subspace of feasible shapes spanned by the SSM. Mathematically speaking, one reduces an infinite dimensional optimization problem, where one searches for a function $\Phi : \Omega \to \mathbb{R}$ which minimizes an energy functional $E : (\Omega \to \mathbb{R}) \to \mathbb{R}$, to a low-dimensional optimization problem, where one searches for a vector $\vec{w} \in \mathbb{R}^m$ which minimizes an energy function $E_{\text{glob}} : \mathbb{R}^m \to \mathbb{R}$. The energy function E_{glob} is thereby given as the composition of the original energy functional E and the statistical shape model $\Phi_{\text{glob}} : \mathbb{R}^m \to (\Omega \to \mathbb{R})$ so that

$$E_{\text{glob}}(\vec{w}) = E(\Phi_{\text{glob}}(\vec{w})). \tag{5.32}$$

In order to obtain the desired segmentation result, Tsai et al. proposed to derive a level set evolution equation for the energy function $E(\Phi_{\text{glob}}(\vec{w}))$ by performing gradient descent with regard to the parameter vector $\vec{w}$ of the statistical shape model:

$$\vec{w}^{k+1} = \vec{w}^k - \gamma^k \, \nabla E(\Phi_{\text{glob}}(\vec{w}^k)) \,, \tag{5.33}$$

where γ^k denotes the step size in each iteration k and the gradient of the energy function is given as

$$\nabla E(\Phi_{\text{glob}}(\vec{w})) = \left(\frac{\partial E(\Phi_{\text{glob}}(\vec{w}))}{\partial w_1}, \ldots, \frac{\partial E(\Phi_{\text{glob}}(\vec{w}))}{\partial w_m} \right). \tag{5.34}$$

This means, instead of varying the level set function Φ explicitly, the modeled shape $\Phi_{\text{glob}}(\vec{w})$ now evolves implicitly by varying the parameter vector $\vec{w}$ of the SSM. Note that, in contrast to the standard narrow-band level set evolution, no re-initialization of the modeled shape $\Phi_{\text{glob}}(\vec{w})$ to a signed distance function after each few iterations is necessary. This is due to the fact that $\Phi_{\text{glob}}(\vec{w})$ is evolved on the whole data domain Ω.

5.3.2 An Exemplary Segmentation Problem

In order to illustrate the approach of Tsai et al., we now consider a specific segmentation problem. For this purpose, we use the region-based energy functional by Chan and Vese [26] which has been defined in Eq. (1.58) as

$$E(\Phi) = \lambda_1 \int_\Omega (I - c_1)^2 H(-\Phi)\, d\vec{x} + \lambda_2 \int_\Omega (I - c_2)^2 H(\Phi)\, d\vec{x} \\ +\mu \int_\Omega \delta_\Phi ||\nabla \Phi||\, d\vec{x} + \nu \int_\Omega H(-\Phi)\, d\vec{x}\,.$$

Now, as stated in Sect. 5.3.1, in order to introduce statistical shape information into the problem, the level set function Φ is replaced by the modeled shape $\Phi_{\text{glob}}(\vec{w})$ so that Eq. (1.58) modifies to

$$\begin{aligned} E(\Phi_{\text{glob}}(\vec{w})) = \lambda_1 &\int_\Omega (I - c_1)^2 H(-\Phi_{\text{glob}}(\vec{w}))\, \mathrm{d}\vec{x} \\ +\lambda_2 &\int_\Omega (I - c_2)^2 H(\Phi_{\text{glob}}(\vec{w}))\, \mathrm{d}\vec{x} \\ +\mu &\int_\Omega \delta_{\Phi_{\text{glob}}(\vec{w})} |\nabla \Phi_{\text{glob}}(\vec{w})|\, \mathrm{d}\vec{x} \\ +\nu &\int_\Omega H(-\Phi_{\text{glob}}(\vec{w}))\, \mathrm{d}\vec{x}\,. \end{aligned} \tag{5.35}$$

Having a closer look at Eq. (5.35), one can see that the first two terms on the right hand side are the so-called *fitting* terms which connect the model $\Phi_{\text{glob}}(\vec{w})$ and the image I. The third and the fourth term are regularizing terms which prefer modeled shapes having a short contour or a small region inside the contour, respectively. These terms are necessary in order to obtain satisfactory results for the original energy functional (1.58) but they can be dropped in the modified energy (5.35).

This is because the set of possible functions which may represent a minimum of the modified energy (5.35) is already severely restricted by the SSM so that an additional regularization is not required. Furthermore, in the approach of Tsai et al. it is not intended to neither prefer modeled shapes having a short contour nor a small region inside the contour. All modeled shapes are considered equally likely instead. So, the simplified energy with no additional regularization terms is given as

$$\begin{aligned} E(\Phi_{\text{glob}}(\vec{w})) =\lambda_1 &\int_\Omega (I - c_1)^2 H(-\Phi_{\text{glob}}(\vec{w}))\, \mathrm{d}\vec{x} \\ +\lambda_2 &\int_\Omega (I - c_2)^2 H(\Phi_{\text{glob}}(\vec{w}))\, \mathrm{d}\vec{x}\,. \end{aligned} \tag{5.36}$$

Please note that it sometimes may still be beneficial to prefer some modeled shapes over the others. How this can be achieved will be discussed in Sect. 5.3.4.

As already mentioned above, in order to minimize the energy function in Eq. (5.36), Tsai et al. proposed to perform gradient descent with regard to the parameter vector $\vec{w}$ of the SSM. Thus, one needs to calculate the gradient of the energy function. Here, the i-th component of the gradient vector is given as

$$\begin{aligned} \frac{\partial E(\Phi_{\text{glob}}(\vec{w}))}{\partial w_i} = \; & \lambda_1 \frac{\partial}{\partial w_i} \left[\int_\Omega (I - c_1)^2 H(-\Phi_{\text{glob}}(\vec{w})) \, \mathrm{d}\vec{x} \right] \\ & + \lambda_2 \frac{\partial}{\partial w_i} \left[\int_\Omega (I - c_2)^2 H(\;\; \Phi_{\text{glob}}(\vec{w})) \, \mathrm{d}\vec{x} \right] . \end{aligned} \tag{5.37}$$

As the integration domain does not depend on w_i, the Leibniz integral rule allows us to interchange the order of integration and differentiation:

$$\begin{aligned} \frac{\partial E(\Phi_{\text{glob}}(\vec{w}))}{\partial w_i} = \; & \lambda_1 \int_\Omega \frac{\partial}{\partial w_i} \left[(I - c_1)^2 H(-\Phi_{\text{glob}}(\vec{w})) \right] \mathrm{d}\vec{x} \\ & + \lambda_2 \int_\Omega \frac{\partial}{\partial w_i} \left[(I - c_2)^2 H(\;\; \Phi_{\text{glob}}(\vec{w})) \right] \mathrm{d}\vec{x} . \end{aligned} \tag{5.38}$$

Now, we can apply the chain rule in order to obtain the derivative of the Heaviside step function:

$$\begin{aligned} \frac{\partial E(\Phi_{\text{glob}}(\vec{w}))}{\partial w_i} = & - \lambda_1 \int_\Omega \delta_{\Phi_{\text{glob}}(\vec{w})} (I - c_1)^2 \frac{\partial \Phi_{\text{glob}}(\vec{w})}{\partial w_i} \, \mathrm{d}\vec{x} \\ & + \lambda_2 \int_\Omega \delta_{\Phi_{\text{glob}}(\vec{w})} (I - c_2)^2 \frac{\partial \Phi_{\text{glob}}(\vec{w})}{\partial w_i} \, \mathrm{d}\vec{x} , \end{aligned} \tag{5.39}$$

where $\delta_{\Phi_{\text{glob}}(\vec{w})}$ is the Dirac delta function that is nonzero only at the zero-crossings of $\Phi_{\text{glob}}(\vec{w})$. With the help of Eq. (2.34), the derivative of the shape model $\Phi_{\text{glob}}(\vec{w})$ with regard to the weight vector w_i is given as

$$\frac{\partial \Phi_{\text{glob}}(\vec{w})}{\partial w_i} = \frac{\partial}{\partial w_i} \left[\bar{\Phi} + \sum_{u=1}^{m} w_u \tilde{\Phi}_u \right] = \sum_{u=1}^{m} \frac{\partial w_u}{\partial w_i} \tilde{\Phi}_u , \tag{5.40}$$

where

$$\frac{\partial w_u}{\partial w_i} = \begin{cases} 1, \text{ if } i = u \\ 0, \text{ else} \end{cases} . \tag{5.41}$$

So, Eq. (5.40) simplifies to

$$\frac{\partial \Phi_{\text{glob}}(\vec{w})}{\partial w_i} = \tilde{\Phi}_i . \tag{5.42}$$

By inserting Eq. (5.42) into Eq. (5.39), the i-th component of the gradient vector finally reads as

$$\begin{aligned}\frac{\partial E(\Phi_{\text{glob}}(\vec{w}))}{\partial w_i} = &-\lambda_1 \int_\Omega \delta_{\Phi_{\text{glob}}(\vec{w})}(I - c_1)^2 \tilde{\Phi}_i \, \mathrm{d}\vec{x} \\ &+ \lambda_2 \int_\Omega \delta_{\Phi_{\text{glob}}(\vec{w})}(I - c_2)^2 \tilde{\Phi}_i \, \mathrm{d}\vec{x} \,. \end{aligned} \tag{5.43}$$

This result (up to a term that does not depend on the evolving shape) can also be found in [130]. However, Tsai et al. did not mention that Eq. (5.43) is the rate of change of the fitting terms of the original energy functional (c.f. Eq. (1.59) when the regularization terms are neglected) along the i-th eigenshape $\tilde{\Phi}_i$ of the SSM:

$$\frac{\partial E(\Phi_{\text{glob}}(\vec{w}))}{\partial w_i} = dE(\Phi, \tilde{\Phi}_i) \,. \tag{5.44}$$

This can be seen when we compare the right hand side of Eq. (5.43) with the functional derivative of the fitting terms from the original energy functional:

$$\frac{\partial E(\Phi)}{\partial \Phi(\vec{x})} = -\lambda_1 \delta_\Phi (I - c_1)^2 + \lambda_2 \delta_\Phi (I - c_2)^2 \,, \tag{5.45}$$

and when we have a look at the definition of the functional differential in Eq. (1.31).

As the eigenshapes $\tilde{\Phi}_i$ define an orthonormal basis of the SSM subspace, the just derived gradient from Eq. (5.43) can be interpreted as the orthogonal projection of the functional derivative from Eq. (5.45) into the low-dimensional SSM subspace.

5.3.3 General Solution for Arbitrary Segmentation Problems

For three popular segmentation problems, the derivation of the gradient with regard to the weight vector $\vec{w}$ of the global model $\Phi_{\text{glob}}(\vec{w})$ is shown in Sect. 5.3.2 and in Appendix E. The results presented therein can also be found in [130]. However, the question arises whether there exists a general solution so that the gradient from Eq. (5.34) can easily be derived for an arbitrary segmentation problem from the level set literature, e.g. the *Geodesic Active Contours* approach from Sect. 1.5.1. The answer to this question cannot be found in [130].

So, we derive such a general solution ourselves. For this purpose, we remind us that the general idea of the approach by Tsai et al. is to narrow down the domain of a given energy functional $E(\Phi)$ to the subspace of feasible shapes of the SSM by replacing the level set function Φ through the modeled shape $\Phi_{\text{glob}}(\vec{w})$. So, the modified energy $E(\Phi_{\text{glob}}(\vec{w}))$ is a composition of the original energy functional E and the global model Φ_{glob}. Additionally, we use the fact that the weight vector $\vec{w} = (w_1, \ldots, w_m)$ of the SSM can be interpreted as a function which is defined on the discrete domain $\{1, \ldots, m\}$ via

$$\vec{w} : \{1, \ldots, m\} \rightarrow \mathbb{R}$$
$$\vec{w}(i) \mapsto w_i \,. \tag{5.46}$$

Thus, in each element $\vec{x}$ of the data domain Ω, the SSM can be interpreted as a functional that maps the (discrete) weight function $\vec{w}$ to a real value $\Phi_{\text{glob}}(\vec{w})(\vec{x})$.

The composed energy $E(\Phi_{\text{glob}}(\vec{w}))$ is then a functional of a functional that depends on the discrete weight function $\vec{w}$. In order to differentiate this energy E with regard to the weight w_i, we can apply the chain rule for the functional derivative from Eq. (1.24) (by making the following substitutions: $F \equiv E$, $G \equiv \Phi$, and $k \equiv \vec{w}$ so that $k(y) \equiv \vec{w}(i) = w_i$):

$$\frac{\partial E(\Phi_{\text{glob}}(\vec{w}))}{\partial w_i} = \int_\Omega \frac{\partial E(\Phi_{\text{glob}}(\vec{w}))}{\partial \Phi_{\text{glob}}(\vec{w})(\vec{x})} \frac{\partial \Phi_{\text{glob}}(\vec{w})(\vec{x})}{\partial w_i} \, \mathrm{d}\vec{x} \,. \tag{5.47}$$

With the help of Eq. (5.42), this simplifies to

$$\frac{\partial E(\Phi_{\text{glob}}(\vec{w}))}{\partial w_i} = \int_\Omega \frac{\partial E(\Phi_{\text{glob}}(\vec{w}))}{\partial \Phi_{\text{glob}}(\vec{w})(\vec{x})} \tilde{\Phi}_i(\vec{x}) \, d\vec{x} \,. \tag{5.48}$$

So, the observation that the gradient with regard to the weight vector $\vec{w}$ can be interpreted as the orthogonal projection of the functional derivative into the SSM subspace is not only valid for the exemplary energies from Sect. 5.3.2 and Appendix E. It also holds for all energies that are of the form $E(\Phi_{\text{glob}}(\vec{w}))$, i.e. where the level set function Φ has been replaced by the global model $\Phi_{\text{glob}}(\vec{w})$. With the help of Eq. (5.48), the gradient from Eq. (5.34) can thus be written as

$$\nabla E(\Phi_{\text{glob}}(\vec{w})) = \Bigg(\int_\Omega \frac{\partial E(\Phi_{\text{glob}}(\vec{w}))}{\partial \Phi_{\text{glob}}(\vec{w})(\vec{x})} \tilde{\Phi}_1(\vec{x}) \, \mathrm{d}\vec{x}, \ldots$$
$$\ldots, \int_\Omega \frac{\partial E(\Phi_{\text{glob}}(\vec{w}))}{\partial \Phi_{\text{glob}}(\vec{w})(\vec{x})} \tilde{\Phi}_m(\vec{x}) \, \mathrm{d}\vec{x} \Bigg) \,. \tag{5.49}$$

Consequently, one can easily derive the gradient for an existing energy functional by taking the published functional derivative from the level set literature and inserting it into Eq. (5.49).

5.3.4 Extension of the Approach by the Trained Weight Distribution

The approach by Tsai et al., to incorporate statistical shape information in the level set evolution process, relies on the assumption that all acceptable solutions lie in the low-dimensional linear subspace of feasible shapes spanned by the eigenshapes of the SSM. Searching for a solution in this subspace greatly reduces the complexity of a given segmentation problem as only an unknown m-dimensional weight vector $\vec{w}$ has to be obtained instead of an unknown level set function Φ.

However, what has been neglected in the approach of Tsai et al. is that not all weights in the subspace spanned by the eigenshapes of the SSM are equally likely. Instead, it follows from the construction of the subspace that the weights are distributed according to a Gaussian distribution (c.f. Eq. (2.9)):

$$P(\vec{w}) = \frac{1}{\sqrt{(2\pi)^m |\tilde{\mathbf{\Sigma}}|}} \exp\left(-\frac{1}{2}\vec{w}\tilde{\mathbf{\Sigma}}^{-1}\vec{w}^T\right).$$

In order to incorporate this additional information in the solution process of a given segmentation problem, we propose to add a regularization term $E_{\text{shape}}(\vec{w})$ to the energy function $E(\Phi_{\text{glob}}(\vec{w}))$ so that the modified segmentation problem reads as

$$E_{\text{glob}}(\vec{w}) = E(\Phi_{\text{glob}}(\vec{w})) + \alpha\, E_{\text{shape}}(\vec{w})\,, \tag{5.50}$$

and the modified gradient descent iteration is given as

$$\vec{w}^{k+1} = \vec{w}^k - \gamma^k\, \nabla E_{\text{glob}}(\vec{w}^k)\,. \tag{5.51}$$

In Eq. (5.50), α is a scalar weighting factor that determines the influence of the regularization term on the segmentation result and $E_{\text{shape}}(\vec{w})$ is chosen to be the negative logarithm of the Gaussian weight distribution:

$$E_{\text{shape}}(\vec{w}) = -\ln P(\vec{w}) \tag{5.52}$$

$$= \ln\left((2\pi)^{\frac{m}{2}}|\tilde{\mathbf{\Sigma}}|^{\frac{1}{2}}\right) + \frac{1}{2}\vec{w}\tilde{\mathbf{\Sigma}}^{-1}\vec{w}^T\,. \tag{5.53}$$

We call the regularization term *shape energy* because the weight vector $\vec{w}$ corresponds to the low-dimensional representation of the modeled shape $\Phi_{\text{glob}}(\vec{w})$ in the SSM subspace. Thus, imposing constraints on the weights $\vec{w}$ directly corresponds to imposing constraints on the shapes which can be generated by the SSM.

The first term on the right hand side of Eq. (5.53) is constant with regard to the weight vector $\vec{w}$. Hence, it has no influence on the segmentation result. With the help of $\Sigma = \text{diag}(\sigma_1^2, \ldots, \sigma_m^2)$, where σ_i^2 is the variance of the Gaussian distribution along the i-th axis of the SSM subspace, Eq. (5.53) can thus be written as

$$E_{\text{shape}}(\vec{w}) = \frac{1}{2}\sum_{i=1}^{m}\frac{w_i^2}{\sigma_i^2} + \text{const.} \tag{5.54}$$

So, the shape energy is given by the squared Mahalanobis distance of the weights w_i from the origin of the SSM subspace, which means that solutions that are close to the mean shape are preferred over those that are far away from the mean shape. This has been our intention by introducing the shape energy, as the mean shape is the most likely shape in our Gaussian framework from Eq. (2.9). Computing the gradient of the shape energy yields

$$\frac{\partial E_{\text{shape}}(\vec{w})}{\partial w_i} = \frac{w_i}{\sigma_i^2} \tag{5.55}$$

for the i-th component, and so the gradient of the modified segmentation problem from Eq. (5.50) reads as

$$\nabla E_{\text{glob}}(\vec{w}) = \nabla E(\Phi_{\text{glob}}(\vec{w})) + \alpha \left(\frac{w_1}{\sigma_1^2}, \ldots, \frac{w_m}{\sigma_m^2} \right), \tag{5.56}$$

where $\nabla E(\Phi_{\text{glob}}(\vec{w}))$ is given in Eq. (5.49).

Now, when we perform gradient descent in order to find a minimum of the modified energy function from Eq. (5.50), the weights w_i are additionally forced towards zero by the term $-\frac{w_i}{\sigma_i^2}$. The additional force on the individual weights w_i thereby depends on the trained shape variance σ_i^2 along the i-th axis of the SSM subspace. So, those weights w_i with small variances σ_i^2 are forced stronger towards the mean shape than weights with large variances.

This is comparable to the third step in our iterative approach from Sect. 3.3.4, where we truncated the weights to three standard deviations of the trained Gaussian distribution in order to stick to likely shapes. However, by doing so, we implicitly assumed that the weights are equally distributed within three standard deviations of the Gaussian distribution, whereas we have now considered directly the Gaussian distribution.

5.3.5 Consideration of Rigid Transformations

In the previous sections, we have shown how to find a minimum of a given energy function by performing gradient descent with regard to the weight vector $\vec{w}$ of the SSM. However, as the SSM captures only the nonrigid variations in the training shapes (c.f. Chap. 2), the SSM has to be rigidly aligned to the data under consideration with regard to pose and scale.

The goal of this thesis is to improve the nonrigid adaptability of a given SSM by replacing the weight vector $\vec{w}$ through a field of weight vectors $\vec{\Psi}$ in order to allow local adaptations of the model to the data under consideration. Therefore, we treat the determination of the rigid transformation from the shape model to the data as a preprocessing step so that it does not influence the shape fitting process. This means, throughout this thesis we always assume that the shape model and the data under consideration are rigidly aligned at the beginning of the shape fitting process. Some further remarks on the rigid-alignment step can be found in Appendix F.

5.4 Variational Formulation for a Local Shape Prior

Now that we have recapitulated the approach of Tsai et al. and extended it to include the trained weight distribution, we continue to describe our approach of integrating our LDSSM into the level set evolution process. The following sections are organized as follows: In Sect. 5.4.1, we will give the motivation for our approach and in Sect. 5.4.2 we will formulate an energy functional that integrates the LDSSM into the level set evolution process. Afterwards, in Sect. 5.4.3, we will extend our local energy by the trained weight distribution similar to the extension of the global energy (c.f. Sect. 5.3.4). Having formulated the energy for our local approach, we will continue by discussing how to obtain a minimum of this energy in Sect. 5.4.4. Finally, we will give the functional gradient that is needed for the solution in Sect. 5.4.5.

5.4.1 Motivation

The approach of Tsai et al. provides an elegant method to obtain segmentation results that reside in the low-dimensional subspace spanned by the eigenshapes of the global SSM. However, as a consequence, the SSM is also the main bottleneck of their approach when the dimension of the shape space is not large enough to cover the whole amount of variation inside a specific shape class. This is especially the case when only a small number of training shapes is available, as the dimension of the shape space cannot exceed the number of training shapes minus one (c.f. Sect. 2.2). So, the approach of Tsai et al. might not be able to provide a satisfying segmentation result with only a limited amount of training shapes at hand.

To circumvent this problem, one could use more training shapes in order to increase the dimension of the SSM subspace. However, the attainment of a sufficient amount of training shapes is one of the main problems in statistical shape models (see Sect. 2.5). Therefore, the utilization of more training shapes in order to increase the dimension of the SSM subspace is not really an option. In fact, how to deal with a limited amount of training shapes is the main point which is addressed in this thesis. So, one has to find another way of how to obtain better segmentation results with the available training data.

Some approaches that try to address this problem have already been discussed in Sect. 5.2. For example, in the approach of Rousson et al. [109] the solution space is extended by demanding that the segmentation result has to be close to the SSM subspace instead of demanding that is has to be inside the SSM subspace. The closeness of the segmentation result to the SSM subspace is thereby ensured by penalizing the mean square error between the segmentation result and its projection into the SSM subspace (c.f. Eq. (5.18)). Similar penalization terms that have also been discussed in Sect. 5.2 are the mean square error to the mean shape (c.f. Eq. (5.12)) or the exponentially weighted mean distance to all training shapes (c.f. Eq. (5.23)). However, as already mentioned, none of the existing approaches takes explicitly into account that the segmentation result may consist of a nonlinear combination of training shapes. This is where our LDSSM comes into play.

As introduced in Chap. 3, the LDSSM extends the subspace of feasible shapes from the SSM by considering also nonlinear combinations of the main modes of variation. This is achieved by replacing the scalar shape weights $\vec{w}$, which describe a linear combination of the main modes of variation, through a vector of smooth weight functions $\vec{\Psi}$ so that the modeled shape is now described by a vector of local weights $\vec{\Psi}(\vec{x})$ in each element $\vec{x}$ of the data domain Ω. So, with the help of the LDSSM, we can extend the solution space by taking explicitly those segmentation results into account which consist of a local, i.e. nonlinear, combination of training shapes.

In the approaches presented in Sect. 5.2, the shape information has been added as an additional regularization term to an energy functional that depends on the level set function Φ. In contrast, we follow the idea of Tsai et al. and replace the level set function Φ through our local model $\Phi_{\text{loc}}(\vec{\Psi})$ in the formulation of the energy functional. By doing this, we obtain the desired segmentation result directly by minimizing the modified energy functional with regard to the vector of weight fields $\vec{\Psi}$. This means, in our approach only those shapes are allowed as a segmentation result which can be generated by our LDSSM. So, what we propose in the following sections is an extension of the approach by Tsai et al. where our LDSSM is used instead of the global SSM to define the solution space.

As mentioned above, the subspace of the LDSSM is much larger than the subspace of the SSM because also local combinations of the training shapes are allowed. How much the LDSSM subspace differs from the SSM subspace thereby depends on the smoothness of the weight functions. Anyhow, we consider the LDSSM subspace as general enough to contain all the desired segmentation results and we do not think that it is necessary to allow results which do not lie inside this LDSSM subspace. This is why we will formulate our segmentation problem in the style of Tsai et al. and not in the style of the approaches from Sect. 5.2.

5.4.2 *Problem Formulation*

As discussed in the previous section, our approach to integrate our LDSSM into the level set evolution process builds up on the work of Tsai et al. They proposed to search for the solution of a given segmentation problem directly in the SSM subspace. We extend this idea by replacing the SSM through our more general LDSSM:

$$\Phi_{\text{glob}}(\vec{w}) = \bar{\Phi} + \sum_{i=1}^{m} w_i \, \tilde{\Phi}_i \quad \Longrightarrow \quad \Phi_{\text{loc}}(\vec{\Psi}) = \bar{\Phi} + \sum_{i=1}^{m} \Psi_i \times \tilde{\Phi}_i \,, \tag{5.57}$$

and we search for the solution of a given segmentation problem in the LDSSM subspace instead of the SSM subspace. We recall that the major difference between our LDSSM and the SSM is that each shape in our LDSSM is completely defined by a vector of smooth weight functions $\vec{\Psi}$ instead of a vector of scalar weights $\vec{w}$. Thus, by replacing the SSM through our more general LDSSM, we now have to search for a vector of smooth weight functions $\vec{\Psi} = (\Psi_1, \ldots, \Psi_m)$, $\Psi_i : \Omega \to \mathbb{R}$, in order to find the solution to a given segmentation problem instead of searching for a vector of scalar weights $\vec{w} = (w_1, \ldots, w_m)$, $w_i \in \mathbb{R}$.

More specifically, instead of describing a segmentation problem by a global energy function $E_{\text{glob}} : \mathbb{R}^m \to \mathbb{R}$ which depends on the vector of scalar weights $\vec{w}$ (c.f. Sect. 5.3.1), we describe it by a local energy functional $E_{\text{loc}} : (\Omega \to \mathbb{R})^m \to \mathbb{R}$ that depends on the vector of smooth weight fields:

$$E_{\text{glob}}(\vec{w}) \quad \Longrightarrow \quad E_{\text{loc}}(\vec{\Psi})\,. \tag{5.58}$$

As for the global energy function $E_{\text{glob}}(\vec{w})$ from Eq. (5.32), the energy functional $E_{\text{loc}}(\vec{\Psi})$ can be obtained by taking any energy functional $E(\Phi)$ from the level set literature and replacing the unconstrained level set function Φ by those shapes $\Phi_{\text{loc}}(\vec{\Psi})$ which can be generated by our LDSSM. However, such an energy functional of the form

$$E_{\text{loc}}(\vec{\Psi}) = E(\Phi_{\text{loc}}(\vec{\Psi})) \tag{5.59}$$

is alone not totally sufficient to constrain the segmentation results to the LDSSM subspace.

As we have already said in Sect. 3.2, the weight functions of our LDSSM have to be smooth. This smoothness is not ensured in Eq. (5.59). So, arbitrary steep weight functions are allowed as a solution of the segmentation problem which is described by the energy functional $E(\Phi_{\text{loc}}(\vec{\Psi}))$. In order to integrate the demanded smoothness constraint on the weight functions Ψ_i in the definition of our segmentation problem, we formulate the constraint as an additional regularization term $E^{\text{loc}}_{\text{smooth}}(\vec{\Psi})$ which is added to the energy functional $E(\Phi_{\text{loc}}(\vec{\Psi}))$. So, we finally obtain our formulation of an energy functional which constrains the segmentation results to the LDSSM subspace as

$$E_{\text{loc}}(\vec{\Psi}) = E(\Phi_{\text{loc}}(\vec{\Psi})) + \beta E^{\text{loc}}_{\text{smooth}}(\vec{\Psi})\,, \tag{5.60}$$

where $\beta \in \mathbb{R}$ is a weighting factor that determines the influence of the smoothing term on the segmentation result. The larger the weighting factor β, the smoother the weight functions Ψ_i which are allowed as a solution to the segmentation problem. For $\beta = 0$, no additional constraints are added to the segmentation problem. However, as we are searching only for smooth functions as a solution to the segmentation problem, β always has to be greater than zero.

We formulate the smoothness constraint in our approach as the integral over the squared gradient magnitudes of the weight functions:

$$E^{\text{loc}}_{\text{smooth}}(\vec{\Psi}) = \frac{1}{2} \int_{\Omega} ||\nabla \Psi_1(\vec{x})||^2 + \cdots + ||\nabla \Psi_m(\vec{x})||^2 \, d\vec{x}\,. \tag{5.61}$$

By doing this, discontinuities in the weight fields Ψ_i are penalized. This is a frequently used regularization method in order to obtain smooth results to a given problem. It has been first used in the context of image processing in the famous paper of Horn and Schunck about determining smooth optical flow fields [63].

5.4.3 Extension of Our Approach by the Trained Weight Distribution

In Sect. 5.3.4, we have shown that it is possible to extend the approach of Tsai et al. by adding an additional regularization term $E_{\text{shape}}(\vec{w})$ to the energy function $E(\Phi_{\text{glob}}(\vec{w}))$ (Eq. (5.50)):

$$E_{\text{glob}}(\vec{w}) = E(\Phi_{\text{glob}}(\vec{w})) + \alpha\, E_{\text{shape}}(\vec{w}) .$$

This regularization term takes into account the Gaussian distribution of plausible weights $P(\vec{w})$ from Eq. (2.9). It has the effect that shapes with likely model weights $\vec{w}$, according to the trained weight distribution $P(\vec{w})$, are preferred in the solution of the segmentation problem over those shapes with unlikely model weights. The distribution of plausible weights $P(\vec{w})$ is thereby estimated from the set of training shapes $\{\Phi_1, \ldots, \Phi_n\}$ during the construction phase of the SSM (c.f. Chap. 2).

Now, we extend our local problem formulation from Eq. (5.60) by a similar regularization term $E_{\text{shape}}^{\text{loc}}(\vec{\Psi})$ which favors shapes that have been generated by likely fields of weight vectors $\vec{\Psi}$ over those shapes which have been generated by unlikely fields of weight vectors. In order to do this, we first have to define how *likely* a field of weight vectors $\vec{\Psi}$ is. This means, we need to determine a probability distribution $P_{\text{loc}}(\vec{\Psi})$ that assigns a probability to each field of weight vectors $\vec{\Psi}$.

When we take a look back at the introduction of the LDSSM in Sect. 3.2, we remember that it has been developed on the basis of the SSM by introducing an individual weight vector $\vec{\Psi}(\vec{x})$ in each element $\vec{x}$ of the data domain Ω. This allows local combinations of the main modes of variation. So, as the LDSSM is an extension of the SSM, it is a reasonable assumption that each of the local weight vectors $\vec{\Psi}(\vec{x})$ is distributed according to the same trained weight distribution as the global weight vector $\vec{w}$. This means, the weight probability distribution $P(\vec{w})$ from Eq. (2.9) can be used in order to obtain an element-wise weight probability distribution $P(\vec{\Psi}(\vec{x}))$ for each weight vector $\vec{\Psi}(\vec{x})$ of our weight field. For this purpose, we take the weight probability $P(\vec{w})$ from Eq. (2.9) and replace the global weight vector $\vec{w}$ by a local weight vector $\vec{\Psi}(\vec{x})$. Thus, we obtain

$$P(\vec{\Psi}(\vec{x})) = \frac{1}{(2\pi)^{\frac{m}{2}} |\Sigma|^{\frac{1}{2}}} \mathrm{e}^{-\frac{1}{2}\vec{\Psi}(\vec{x})\Sigma^{-1}\vec{\Psi}(\vec{x})^T} , \tag{5.62}$$

with $\Sigma = \mathrm{diag}(\sigma_1^2, \ldots, \sigma_m^2)$, as our element-wise weight probability distribution. Since the distribution in Eq. (5.62) is the same as in Eq. (2.9), the standard deviations σ_1 to σ_m along the individual dimensions of the distribution are also identical. They are obtained from the PCA of the training shapes as explained in Sect. 2.2.

Next, the element-wise weight probabilities from Eq. (5.62) have to be combined into one total probability $P_{\mathrm{loc}}(\vec{\Psi})$ for the field of weight vectors $\vec{\Psi}$. In order to achieve this, we assume that the element-wise probabilities $P(\vec{\Psi}(\vec{x}))$ are statistically independent. This assumption is not totally correct as the field of weight vectors should be smooth (c.f. Sect. 3.2) which introduces a statistical dependency between adjacent elements $\vec{x}$. However, as this smoothness condition has already been explicitly addressed in the smoothness energy from Eq. (5.61), we neglect it in the derivation of our shape energy. Thus, we can compute the probability distribution $P_{\mathrm{loc}}(\vec{\Psi})$ for the field of weight vectors $\vec{\Psi}$ as the product of all element-wise probabilities $P(\vec{\Psi}(\vec{x}))$ according to the multiplication rule for independent events [23, p. 810, Eq. (16.41b)]:

$$P_{\mathrm{loc}}(\vec{\Psi}) = \prod_{\vec{x} \in \Omega} P(\vec{\Psi}(\vec{x})) . \tag{5.63}$$

Now, having determined the probability $P_{\mathrm{loc}}(\vec{\Psi})$ of a field of weight vectors $\vec{\Psi}$, we define the local shape energy $E_{\mathrm{shape}}^{\mathrm{loc}}(\vec{\Psi})$ as the negative log-likelihood of the weight field probability:

$$E_{\mathrm{shape}}^{\mathrm{loc}}(\vec{\Psi}) = -\ln P_{\mathrm{loc}}(\vec{\Psi}) . \tag{5.64}$$

This is identical to the definition of the shape energy for the global approach in Eq. (5.52). With the help of Eq. (5.63), Eq. (5.64) can be written as the negative logarithm of the product of the element-wise shape probabilities:

$$E_{\mathrm{shape}}^{\mathrm{loc}}(\vec{\Psi}) = -\ln \prod_{\vec{x} \in \Omega} P(\vec{\Psi}(\vec{x})) . \tag{5.65}$$

Considering the logarithmic identities on the continuous domain Ω, we can further rewrite our shape energy as

$$E_{\mathrm{shape}}^{\mathrm{loc}}(\vec{\Psi}) = \int_{\Omega} -\ln P(\vec{\Psi}(\vec{x}))\, d\vec{x} . \tag{5.66}$$

The integrand of Eq. (5.66) can be considered as an element-wise shape energy. In fact, when we compare it to Eq. (5.52), we can see that the integrand is identical to the global shape energy $E_{\mathrm{shape}}(\vec{w})$ where the global weight vector $\vec{w}$ has been replaced by the local weight vector $\vec{\Psi}(\vec{x})$:

$$E_{\mathrm{shape}}(\vec{\Psi}(\vec{x})) = -\ln P(\vec{\Psi}(\vec{x})) . \tag{5.67}$$

So, the local shape energy $E^{\mathrm{loc}}_{\mathrm{shape}}(\vec{\Psi})$ can be interpreted as evaluating the global shape energy independently for each local weight vector $\vec{\Psi}(\vec{x})$ and accumulating the obtained results:

$$E^{\mathrm{loc}}_{\mathrm{shape}}(\vec{\Psi}) = \int_{\Omega} E_{\mathrm{shape}}(\vec{\Psi}(\vec{x}))\,\mathrm{d}\vec{x}\,. \tag{5.68}$$

This is not surprising as we have assumed statistical independence between the individual elements $\vec{x}$.

In summary, we have defined a local shape energy $E^{\mathrm{loc}}_{\mathrm{shape}}(\vec{\Psi}(\vec{x}))$ which is based on the trained shape distribution from Eq. (2.9): The element-wise weight probabilities are combined into one shape energy by integrating over their negative log-likelihoods. This local shape energy can now be integrated into our energy functional from Eq. (5.60) in order to prefer shapes which have been generated by *likely* weight fields:

$$E_{\mathrm{loc}}(\vec{\Psi}) = E(\Phi_{\mathrm{loc}}(\vec{\Psi})) + \alpha E^{\mathrm{loc}}_{\mathrm{shape}}(\vec{\Psi}) + \beta E^{\mathrm{loc}}_{\mathrm{smooth}}(\vec{\Psi})\,, \tag{5.69}$$

where α is another weighting factor which controls the influence of the shape energy on the total energy.

5.4.4 Finding a Solution to the Problem by Functional Gradient Descent

Now that we have defined an energy functional which allows only those segmentation results that are constraint to the LDSSM subspace (Eq. (5.69)), the goal is to find a field of weight vectors $\vec{\Psi} : \Omega \to \mathbb{R}^m$ that minimizes the given energy functional. As explained in Sect. 1.5, a solution can be obtained via functional gradient descent. Please note that the main difference between our approach and other state of the art level set segmentation approaches is that the energy functional in Eq. (5.69) depends on the field of weight vectors $\vec{\Psi}$ instead of the level set function Φ (c.f. Sect. 5.4.1). This means, we have to perform functional gradient descent with regard to the field of weight vectors $\vec{\Psi}$ and not with regard to the level set function Φ as in the approaches from Sect. 1.5.

So, in order to minimize the energy functional from Eq. (5.69), we start with an initial guess $\vec{\Psi}^0$ for the field of weight vectors and take repeated steps in the direction of the negative functional gradient:

$$\vec{\Psi}^{k+1}(\vec{x}) = \vec{\Psi}^{k}(\vec{x}) - \gamma^k \frac{\partial E_{\mathrm{loc}}(\vec{\Psi}^k)}{\partial \vec{\Psi}^k(\vec{x})}\,, \tag{5.70}$$

where $k \in \mathbb{N}$ denotes the k-th iteration and $\gamma^k \in \mathbb{R}$ is the step size in each iteration. Equation (5.70) is iterated until a stationary point of the given energy functional is reached. This point most likely indicates a local minimum as we were moving in the direction of steepest descent of the energy functional (c.f. Sect. 1.2.1.5). Now, in order to perform the gradient descent iteration which is defined in Eq. (5.70), we need to calculate the functional derivative of the energy functional $E_{\text{loc}}(\vec{\Psi})$ with regard to the field of smooth weight vectors $\vec{\Psi}$. For this purpose, we recall from Sect. 3.2 that the field of weight vectors $\vec{\Psi} : \Omega \rightarrow \mathbb{R}^m$ can be written as

$$\vec{\Psi} = (\Psi_1, \ldots, \Psi_m) , \tag{5.71}$$

where each of the m components Ψ_i is itself a scalar weight field $\Psi_i : \Omega \rightarrow \mathbb{R}$. Moreover, we make use of Eq. (1.29) which says that the functional derivative of a functional that depends on multiple functions can be obtained as a vector of all partial functional derivatives $\frac{\partial E(\vec{\Psi})}{\partial \Psi_i(\vec{x})}$:

$$\frac{\partial E_{\text{loc}}(\vec{\Psi})}{\partial \vec{\Psi}(\vec{x})} = \left(\frac{\partial E_{\text{loc}}(\vec{\Psi})}{\partial \Psi_1(\vec{x})}, \ldots, \frac{\partial E_{\text{loc}}(\vec{\Psi})}{\partial \Psi_m(\vec{x})} \right) . \tag{5.72}$$

Now, with the help of Eqs. (5.71) and (5.72), the gradient descent iteration from Eq. (5.70) can be rewritten as

$$\begin{aligned} \left(\Psi_1^{k+1}(\vec{x}), \ldots, \Psi_m^{k+1}(\vec{x})\right) = & \left(\Psi_1^k(\vec{x}), \ldots, \Psi_m^k(\vec{x})\right) \\ & - \gamma^k \left(\frac{\partial E_{\text{loc}}(\vec{\Psi}^k)}{\partial \Psi_1^k(\vec{x})}, \ldots, \frac{\partial E_{\text{loc}}(\vec{\Psi}^k)}{\partial \Psi_m^k(\vec{x})} \right) . \end{aligned} \tag{5.73}$$

This is basically a system of m coupled gradient descent equations, one for each weight field Ψ_i.

Then, as stated in Eq. (1.30), a necessary condition for a minimum of Eq. (5.69) is a system of m Euler–Lagrange equations, one for each function Ψ_i:

$$\begin{aligned} \frac{\partial E_{\text{loc}}(\Psi_1, \ldots, \Psi_m)}{\partial \Psi_1(\vec{x})} &= 0 \\ &\vdots \\ \frac{\partial E_{\text{loc}}(\Psi_1, \ldots, \Psi_m)}{\partial \Psi_m(\vec{x})} &= 0 . \end{aligned} \tag{5.74}$$

In practical implementations, Eq. (5.74) will never be exactly zero due to noise in the data or numerical rounding errors. So, it is a common practice to either iterate Eq. (5.70) for a fixed number of iterations or to check whether two consecutive solutions $\vec{\Psi}^k$ and $\vec{\Psi}^{k+1}$ lie within a predefined convergence tolerance ϵ, i.e. stop iterating Eq. (5.70) in the case that the following termination condition is fulfilled:

$$||\vec{\Psi}^{k+1} - \vec{\Psi}^k||^2 = \sum_{i=i}^{m} \int_{\Omega} \left(\Psi_i^{k+1}(\vec{x}) - \Psi_i^k(\vec{x})\right)^2 \, \mathrm{d}\vec{x} < \epsilon . \tag{5.75}$$

However, this means that in order to evaluate the termination condition one has to store m weight fields from the previous iteration. In order to reduce the memory requirements and to speed up the evaluation of the termination condition, we evaluate a termination condition of the form

$$||\Phi_{\text{loc}}(\vec{\Psi}^{k+1}) - \Phi_{\text{loc}}(\vec{\Psi}^{k})||^2 = \int_{\Omega} \left(\Phi_{\text{loc}}(\vec{\Psi}^{k+1})(\vec{x}) - \Phi_{\text{loc}}(\vec{\Psi}^{k})(\vec{x})\right)^2 \, \mathrm{d}\vec{x} < \epsilon \, . \tag{5.76}$$

This means, instead of directly evaluating the squared error between two consecutive fields of weight vectors, we evaluate the squared error between the corresponding level set representations that can be obtained with the help of Eq. (3.3).

5.4.5 Obtaining the Functional Gradient

In the previous section, we have discussed how the solution to the segmentation problem defined by the energy functional from Eq. (5.69) can be obtained via functional gradient descent. However, so far we have not mentioned how to obtain the partial functional derivatives $\frac{\partial E_{\text{loc}}(\vec{\Psi})}{\partial \Psi_i(\vec{x})}$ of the energy functional with regard to the i-th weight field that are needed for this approach.

In order to obtain those partial functional derivatives, we make use of the fact that the energy functional $E_{\text{loc}}(\vec{\Psi})$ from Eq. (5.69) consists of three distinct terms which are linearly combined: a fitting energy $E(\Phi_{\text{loc}}(\vec{\Psi}))$ that connects the LDSSM to the data under consideration, a shape energy $E^{\text{loc}}_{\text{shape}}(\vec{\Psi})$ that favors likely weight fields, and a smoothing energy $E^{\text{loc}}_{\text{smooth}}(\vec{\Psi})$ which ensures that the resulting weight fields are smooth. Now, it follows from Eq. (1.22) that like in ordinary vector calculus the derivative of the functional $E_{\text{loc}}(\vec{\Psi})$ with regard to the i-th weight field can be obtained as the linear combination of the functional derivatives of its individual terms:

$$\frac{\partial E_{\text{loc}}(\vec{\Psi})}{\partial \Psi_i(\vec{x})} = \frac{\partial E(\Phi_{\text{loc}}(\vec{\Psi}))}{\partial \Psi_i(\vec{x})} + \alpha \frac{\partial E^{\text{loc}}_{\text{shape}}(\vec{\Psi})}{\partial \Psi_i(\vec{x})} + \beta \frac{\partial E^{\text{loc}}_{\text{smooth}}(\vec{\Psi})}{\partial \Psi_i(\vec{x})} \, . \tag{5.77}$$

In the following subsections, we will show how to obtain those functional derivatives.

5.4.5.1 Smoothing Energy Gradient

We start with the computation of the functional derivative of the smoothing energy functional $E^{\text{loc}}_{\text{smooth}}$. For this purpose, we recall that it has been defined in Eq. (5.61) as

$$E^{\text{loc}}_{\text{smooth}}(\vec{\Psi}) = \frac{1}{2} \int_{\Omega} ||\nabla \Psi_1(\vec{x})||^2 + \cdots + ||\nabla \Psi_m(\vec{x})||^2 \, d\vec{x} \, .$$

When we denote the integrand of the smoothing energy functional as $F^{\text{loc}}_{\text{smooth}}$, Eq. (5.61) can be written as

$$E^{\text{loc}}_{\text{smooth}}(\vec{\Psi}) = \int_{\Omega} F^{\text{loc}}_{\text{smooth}}\left(\frac{\partial \Psi_1}{\partial x_1}, \ldots, \frac{\partial \Psi_1}{\partial x_n}, \ldots\ldots, \frac{\partial \Psi_m}{\partial x_1}, \ldots, \frac{\partial \Psi_m}{\partial x_n}\right) d\vec{x}\,, \quad (5.78)$$

where the integrand $F^{\text{loc}}_{\text{smooth}}$ is given as

$$F^{\text{loc}}_{\text{smooth}} = \frac{1}{2}\Bigg[\left(\frac{\partial \Psi_1}{\partial x_1}\right)^2 + \cdots + \left(\frac{\partial \Psi_1}{\partial x_n}\right)^2 + \cdots \\ \cdots + \left(\frac{\partial \Psi_m}{\partial x_1}\right)^2 + \cdots + \left(\frac{\partial \Psi_m}{\partial x_n}\right)^2\Bigg]. \quad (5.79)$$

Now, in order to compute the functional derivative of the smoothing energy $E^{\text{loc}}_{\text{smooth}}$ with regard to the weight field Ψ_i, we can apply Eq. (1.26) for the functional derivative in the case of functions with multiple arguments. By doing this, we obtain

$$\frac{\partial E^{\text{loc}}_{\text{smooth}}(\vec{\Psi})}{\partial \Psi_i(\vec{x})} = -\sum_{u=1}^{n} \frac{\partial}{\partial x_u} \frac{\partial F^{\text{loc}}_{\text{smooth}}}{\partial(\frac{\partial \Psi_i}{\partial x_u})}\,. \quad (5.80)$$

The first term on the right hand side of Eq. (1.26) vanishes because the integrand $F^{\text{loc}}_{\text{smooth}}$ depends only on the partial derivatives of the functions Ψ_i but not on the functions Ψ_i itself.

We can continue by computing the inner derivative of Eq. (5.80). It results with the help of Eq. (5.79) to

$$\frac{\partial F^{\text{loc}}_{\text{smooth}}}{\partial(\frac{\partial \Psi_i}{\partial x_u})} = \frac{1}{2}\,2\left(\frac{\partial \Psi_i}{\partial x_u}\right) = \frac{\partial \Psi_i}{\partial x_u}\,. \quad (5.81)$$

This can then be used in order to compute the outer derivative of Eq. (5.80) as

$$\frac{\partial}{\partial x_u} \frac{\partial F^{\text{loc}}_{\text{smooth}}}{\partial(\frac{\partial \Psi_i}{\partial x_u})} = \frac{\partial}{\partial x_u} \frac{\partial \Psi_i}{\partial x_u} = \frac{\partial^2 \Psi_i}{\partial x_u^2}\,. \quad (5.82)$$

When we reinsert this result into Eq. (5.80), the functional derivative of the smoothing energy with regard to the weight field Ψ_i is finally obtained as the negative sum of the second order partial derivatives of the weight field Ψ_i. This can be written as

$$\frac{\partial E^{\text{loc}}_{\text{smooth}}(\vec{\Psi})}{\partial \Psi_i(\vec{x})} = -\Delta \Psi_i(\vec{x})\,, \quad (5.83)$$

where Δ denotes the Laplace operator:

$$\Delta\Psi_i = \frac{\partial^2\Psi_i}{\partial x_1^2} + \cdots + \frac{\partial^2\Psi_i}{\partial x_n^2} . \tag{5.84}$$

That the smoothing energy $E^{\mathrm{loc}}_{\mathrm{smooth}}$ has indeed the desired smoothing effect on the weight fields Ψ_i can be seen when we insert its functional derivative from Eq. (5.83) into our gradient descent Eq. (5.73). By doing this, we obtain

$$\Psi_i^{k+1} = \Psi_i^k + \gamma^k \, \Delta\Psi_i^k . \tag{5.85}$$

This is the Euler forward discretization of the heat diffusion equation, which is well-known for its smoothing properties [63].

5.4.5.2 Shape Energy Gradient

We continue by describing the functional derivative of the shape energy $E^{\mathrm{loc}}_{\mathrm{shape}}$. Equation (5.66) defines $E^{\mathrm{loc}}_{\mathrm{shape}}$ as the integral over the negative logarithms of the element-wise weight probabilities:

$$E^{\mathrm{loc}}_{\mathrm{shape}}(\vec{\Psi}) = \int_{\Omega} -\ln P(\vec{\Psi}(\vec{x})) \, d\vec{x} \, ,$$

where the probability of each weight vector $\vec{\Psi}(\vec{x})$ is computed as (see Eq. (5.62))

$$P(\vec{\Psi}(\vec{x})) = \frac{1}{(2\pi)^{\frac{m}{2}} |\Sigma|^{\frac{1}{2}}} \mathrm{e}^{-\frac{1}{2}\vec{\Psi}(\vec{x})\Sigma^{-1}\vec{\Psi}(\vec{x})^T} .$$

When we insert Eq. (5.62) into Eq. (5.66), we obtain

$$E^{\mathrm{loc}}_{\mathrm{shape}}(\vec{\Psi}) = \int_{\Omega} \left[\frac{1}{2}\vec{\Psi}(\vec{x})\Sigma^{-1}\vec{\Psi}(\vec{x})^T + \ln\left((2\pi)^{\frac{m}{2}} |\Sigma|^{\frac{1}{2}} \right) \right] \mathrm{d}\vec{x} \, . \tag{5.86}$$

In Eq. (5.86), the integrand consists of two terms. The second term does neither depend on the field of weight vectors $\vec{\Psi}$ nor on the variable of integration $\vec{x}$. So, Eq. (5.86) can be rewritten as

$$E^{\mathrm{loc}}_{\mathrm{shape}}(\vec{\Psi}) = \int_{\Omega} \frac{1}{2}\vec{\Psi}(\vec{x})\Sigma^{-1}\vec{\Psi}(\vec{x})^T \, \mathrm{d}\vec{x} + \underbrace{\ln\left((2\pi)^{\frac{m}{2}} |\Sigma|^{\frac{1}{2}} \right) |\Omega|}_{\text{const.}} , \tag{5.87}$$

where $|\Omega|$ denotes the cardinality of the set Ω. It can be seen that the second term is constant with regard to the field of weight vectors $\vec{\Psi}$ so that it vanishes when we take the functional derivative. Now, when we additionally denote the remaining term under the integral as $F^{\mathrm{loc}}_{\mathrm{shape}}$, Eq. (5.87) further simplifies to

$$E_{\text{shape}}^{\text{loc}}(\vec{\Psi}) = \int_{\Omega} F_{\text{shape}}^{\text{loc}}(\Psi_1(\vec{x}), \ldots, \Psi_m(\vec{x}))\, d\vec{x} + \text{const.}, \tag{5.88}$$

where $F_{\text{shape}}^{\text{loc}}$ is given as

$$F_{\text{shape}}^{\text{loc}} = \frac{1}{2}\vec{\Psi}(\vec{x})\Sigma^{-1}\vec{\Psi}(\vec{x})^T = \frac{1}{2}\left(\frac{\Psi_1(\vec{x})^2}{\sigma_1^2} + \cdots + \frac{\Psi_m(\vec{x})^2}{\sigma_m^2}\right). \tag{5.89}$$

Now, we can apply Eq. (1.26) and obtain

$$\frac{\partial E_{\text{shape}}^{\text{loc}}(\vec{\Psi})}{\partial \Psi_i(\vec{x})} = \frac{\partial F_{\text{shape}}^{\text{loc}}(\Psi_1(\vec{x}), \ldots, \Psi_m(\vec{x}))}{\partial \Psi_i(\vec{x})} \tag{5.90}$$

for the functional derivative of the shape energy $E_{\text{shape}}^{\text{loc}}$. In this case, all terms but the first on the right hand side of Eq. (1.26) vanish because the integrand $F_{\text{shape}}^{\text{loc}}$ depends only on the functions Ψ_i itself but not on their partial derivatives.

The partial derivative of $F_{\text{shape}}^{\text{loc}}$ with regard to Ψ_i on the right hand side of Eq. (5.90) can be calculated with the help of Eq. (5.89). So, we finally obtain the functional derivative of the shape energy $E_{\text{shape}}^{\text{loc}}$ with regard to the function Ψ_i as

$$\frac{\partial E_{\text{shape}}^{\text{loc}}(\vec{\Psi})}{\partial \Psi_i(\vec{x})} = \frac{\Psi_i(\vec{x})}{\sigma_i^2}. \tag{5.91}$$

When we insert this result into our gradient descent Eq. (5.73), we get

$$\begin{aligned} \Psi_i^{k+1} &= \Psi_i^k - \gamma^k \frac{\Psi_i^k}{\sigma_i^2} \\ &= \Psi_i^k \left(1 - \frac{\gamma^k}{\sigma_i^2}\right). \end{aligned} \tag{5.92}$$

Now, we choose $\gamma^k < \sigma_i^2$ such that this gradient descent equation forces the elements of the weight fields towards zero, which corresponds to forcing the modeled shape towards the mean shape. This makes sense, as the mean shape is the most likely one in our Gaussian framework from Eq. (5.63). How strong the elements of each weight field Ψ_i are forced towards zero depends on the trained shape variance σ_i^2 along the corresponding axis of the SSM subspace. The elements of those weight fields Ψ_i with small variances σ_i^2 are forced stronger towards the mean shape than the elements of those weight fields with large variances. This also makes sense, as larger deviations shall be allowed for the elements of those weight fields having large variances in the training data than for the elements of those weight fields with small variances.

5.4.5.3 Exemplary Fitting Energy Gradient

Now, what remains to derive is the functional gradient of the fitting energy. However, as the fitting energy is problem dependent, there exists not one special fitting energy but a variety of different fitting energies. For a better understanding, we will derive the functional gradient of an exemplary fitting energy in the remainder of this subsection. Then, we will give a general solution for arbitrary fitting energies in the next subsection.

We discuss the same example that has already been discussed for the global approach of Tsai et al. in Sect. 5.3.2, namely the region-based segmentation problem by Chan and Vese [26]. However, this time we replace the level set function Φ by our local model $\Phi_{\text{loc}}(\vec{\Psi})$ instead of the global model $\Phi_{\text{glob}}(\vec{w})$ so that the energy from Eq. (5.36) changes to

$$E(\Phi_{\text{loc}}(\vec{\Psi})) = \lambda_1 \int_{\Omega} (I - c_1)^2 H(-\Phi_{\text{loc}}(\vec{\Psi}))\, \mathrm{d}\vec{x} + \lambda_2 \int_{\Omega} (I - c_2)^2 H(\Phi_{\text{loc}}(\vec{\Psi}))\, \mathrm{d}\vec{x}. \tag{5.93}$$

From Eq. (1.26) follows that the functional derivative of Eq. (5.93) with regard to the weight field Ψ_i can be obtained as

$$\begin{aligned} \frac{\partial E(\Phi_{\text{loc}}(\vec{\Psi}))}{\partial \Psi_i(\vec{x})} &= \lambda_1 (I(\vec{x}) - c_1)^2 \frac{\partial H(-\Phi_{\text{loc}}(\vec{\Psi}(\vec{x})))}{\partial \Psi_i(\vec{x})} \\ &\quad + \lambda_2 (I(\vec{x}) - c_2)^2 \frac{\partial H(\Phi_{\text{loc}}(\vec{\Psi}(\vec{x})))}{\partial \Psi_i(\vec{x})}. \end{aligned} \tag{5.94}$$

Again, all terms but the first on the right-hand side of Eq. (1.26) vanish as Eq. (5.93) depends only on the weight field Ψ_i but not on its derivatives. Like in Sect. 5.3.2, the derivative of the Heaviside step function can be obtained as the Dirac delta function that is nonzero only at the zero-crossings of $\Phi_{\text{loc}}(\vec{\Psi})$. So, Eq. (5.94) simplifies to

$$\begin{aligned} \frac{\partial E(\Phi_{\text{loc}}(\vec{\Psi}))}{\partial \Psi_i(\vec{x})} &= \left[-\delta_{\Phi_{\text{loc}}(\vec{\Psi}(\vec{x}))} \lambda_1 (I(\vec{x}) - c_1)^2 + \delta_{\Phi_{\text{loc}}(\vec{\Psi}(\vec{x}))} \lambda_2 (I(\vec{x}) - c_2)^2 \right] \\ &\quad \times \frac{\partial \Phi_{\text{loc}}(\vec{\Psi}(\vec{x}))}{\partial \Psi_i(\vec{x})}. \end{aligned} \tag{5.95}$$

Now, in order to further simplify Eq. (5.95) we determine the derivative of the LDSSM. With the help of Eq. (3.3) it reads as

$$\begin{aligned}\frac{\partial \Phi_{\text{loc}}(\vec{\Psi}(\vec{x}))}{\partial \Psi_i(\vec{x})} &= \frac{\partial}{\partial \Psi_i(\vec{x})}\left[\bar{\Phi}(\vec{x}) + \sum_{u=1}^{m} \Psi_u(\vec{x})\tilde{\Phi}_u(\vec{x})\right] \\ &= \sum_{u=1}^{m} \frac{\partial \Psi_u(\vec{x})}{\partial \Psi_i(\vec{x})}\tilde{\Phi}_u(\vec{x}) \\ &= \tilde{\Phi}_i(\vec{x}) \,.\end{aligned} \tag{5.96}$$

So, we finally obtain the functional derivative of the exemplary fitting energy as

$$\begin{aligned}\frac{\partial E(\Phi_{\text{loc}}(\vec{\Psi}))}{\partial \Psi_i(\vec{x})} = \Big[&- \delta_{\Phi_{\text{loc}}(\vec{\Psi}(\vec{x}))}\lambda_1 (I(\vec{x}) - c_1)^2 \\ &+ \delta_{\Phi_{\text{loc}}(\vec{\Psi}(\vec{x}))}\lambda_2 (I(\vec{x}) - c_2)^2\Big]\tilde{\Phi}_i(\vec{x}) \,.\end{aligned} \tag{5.97}$$

This result is identical to the functional derivative of the original energy functional by Chan and Vese (c.f. Eq. (5.45)) up to the additional factor $\tilde{\Phi}_i(\vec{x})$. This additional factor results from the application of the chain rule as $\tilde{\Phi}_i(\vec{x})$ is the derivative of the LDSSM with regard to the local weight vector $\Psi_i(\vec{x})$. We will show in the following section that this additional factor not only appears in this specific energy functional but in all energy functionals that are of the form $E(\Phi_{\text{loc}}(\vec{\Psi}))$.

The difference between the derivative of the global problem $E(\Phi_{\text{glob}}(\vec{w}))$ with regard to the global weight w_i in Eq. (5.43) and the functional derivative of the local problem $E(\Phi_{\text{loc}}(\vec{\Psi}))$ with regard to the weight field Ψ_i in Eq. (5.97) is the missing integral over the whole data domain Ω. Thus, the global weight update in Eq. (5.43) can be interpreted as averaging the local weight updates in all elements $\vec{x}$ of the data domain Ω.

5.4.5.4 General Fitting Energy Gradient

Having derived the functional gradient of an exemplary fitting energy in the previous section, we will continue by providing a general solution for arbitrary fitting energies. For this purpose, we make for the moment use of the fact that many fitting energies are of the form

$$\begin{aligned}E(\Phi_{\text{loc}}(\vec{\Psi})) = \int_{\Omega} F\Big(&\Phi_{\text{loc}}(\vec{\Psi}(\vec{x})), \frac{\partial \Phi_{\text{loc}}(\vec{\Psi}(\vec{x}))}{\partial x_1}, \ldots \\ &\ldots, \frac{\partial \Phi_{\text{loc}}(\vec{\Psi}(\vec{x}))}{\partial x_n}, x_1, \ldots, x_n\Big)\, \mathrm{d}\vec{x} \,.\end{aligned} \tag{5.98}$$

For example the region-based segmentation problem by Chan and Vese or the *Geodesic Active Contours* approach, which have both been defined in Sect. 1.5, are of this form when we replace the level set function Φ by our LDSSM $\Phi_{\text{loc}}(\vec{\Psi})$.

The solution for the functional derivative of Eq. (5.98) with regard to the i-th weight field Ψ_i has been defined in Eq. (1.26) as

$$\frac{\partial E(\Phi_{\text{loc}}(\vec{\Psi}))}{\partial \Psi_i(\vec{x})} = \frac{\partial F}{\partial \Psi_i(\vec{x})} - \sum_{u=1}^{n} \frac{\partial}{\partial x_u} \frac{\partial F}{\partial \left(\frac{\partial \Psi_i(\vec{x})}{\partial x_u}\right)} . \tag{5.99}$$

However, as F does not depend directly on $\Psi_i(\vec{x})$, we need to use the chain rule in order to extend Eq. (5.99) to

$$\begin{aligned}\frac{\partial E(\Phi_{\text{loc}}(\vec{\Psi}))}{\partial \Psi_i(\vec{x})} =& \frac{\partial F}{\partial \Phi_{\text{loc}}(\vec{\Psi}(\vec{x}))} \frac{\partial \Phi_{\text{loc}}(\vec{\Psi}(\vec{x}))}{\partial \Psi_i(\vec{x})} \\ &+ \sum_{k=1}^{n} \frac{\partial F}{\partial \left(\frac{\partial \Phi_{\text{loc}}(\vec{\Psi}(\vec{x}))}{\partial x_k}\right)} \frac{\partial \left(\frac{\partial \Phi_{\text{loc}}(\vec{\Psi}(\vec{x}))}{\partial x_k}\right)}{\partial \Psi_i(\vec{x})} \\ &- \sum_{u=1}^{n} \frac{\partial}{\partial x_u} \left[\frac{\partial F}{\partial \Phi_{\text{loc}}(\vec{\Psi}(\vec{x}))} \frac{\partial \Phi_{\text{loc}}(\vec{\Psi}(\vec{x}))}{\partial \left(\frac{\partial \Psi_i(\vec{x})}{\partial x_u}\right)} \right. \\ &\left. + \sum_{k=1}^{n} \frac{\partial F}{\partial \left(\frac{\partial \Phi_{\text{loc}}(\vec{\Psi}(\vec{x}))}{\partial x_k}\right)} \frac{\partial \left(\frac{\partial \Phi_{\text{loc}}(\vec{\Psi}(\vec{x}))}{\partial x_k}\right)}{\partial \left(\frac{\partial \Psi_i(\vec{x})}{\partial x_u}\right)} \right] . \end{aligned} \tag{5.100}$$

The additional sums in Eq. (5.100) occur as not only $\Phi_{\text{loc}}(\vec{\Psi}(\vec{x}))$ but also each term $\frac{\partial \Phi_{\text{loc}}(\vec{\Psi}(\vec{x}))}{\partial x_u}$ may depend directly on the weight field $\Psi_i(\vec{x})$ or its partial derivatives. As $\Phi_{\text{loc}}(\vec{\Psi}(\vec{x}))$ does not depend directly on $\frac{\partial \Psi_i(\vec{x})}{\partial x_u}$, the first term in the square brackets on the right-hand side of Eq. (5.100) vanishes. Additionally, when we use Eq. (3.3) in order to obtain

$$\begin{aligned}\frac{\partial \Phi_{\text{loc}}(\vec{\Psi}(\vec{x}))}{\partial x_k} &= \frac{\partial}{\partial x_k} \left[\bar{\Phi}(\vec{x}) + \sum_{i=1}^{m} \Psi_i(\vec{x}) \tilde{\Phi}_i(\vec{x}) \right] \\ &= \frac{\partial \bar{\Phi}(\vec{x})}{\partial x_k} + \sum_{i=1}^{m} \left[\frac{\partial \Psi_i(\vec{x})}{\partial x_k} \tilde{\Phi}_i(\vec{x}) + \Psi_i(\vec{x}) \frac{\partial \tilde{\Phi}_i(\vec{x})}{\partial x_k} \right] , \end{aligned} \tag{5.101}$$

Eq. (5.100) simplifies to

$$\begin{aligned}\frac{\partial E(\Phi_{\text{loc}}(\vec{\Psi}))}{\partial \Psi_i(\vec{x})} =& \frac{\partial F}{\partial \Phi_{\text{loc}}(\vec{\Psi}(\vec{x}))} \frac{\partial \Phi_{\text{loc}}(\vec{\Psi}(\vec{x}))}{\partial \Psi_i(\vec{x})} + \sum_{k=1}^{n} \frac{\partial F}{\partial \left(\frac{\partial \Phi_{\text{loc}}(\vec{\Psi}(\vec{x}))}{\partial x_k}\right)} \frac{\partial \tilde{\Phi}_i(\vec{x})}{\partial x_k} \\ &- \sum_{u=1}^{n} \frac{\partial}{\partial x_u} \left[\frac{\partial F}{\partial \left(\frac{\partial \Phi_{\text{loc}}(\vec{\Psi}(\vec{x}))}{\partial x_u}\right)} \tilde{\Phi}_i(\vec{x}) \right] . \end{aligned} \tag{5.102}$$

This is because

$$\frac{\partial\left(\frac{\partial \Phi_{\text{loc}}(\vec{\Psi}(\vec{x}))}{\partial x_k}\right)}{\partial \Psi_i(\vec{x})} = \frac{\partial \tilde{\Phi}_i(\vec{x})}{\partial x_k} \quad \text{and} \quad \frac{\partial\left(\frac{\partial \Phi_{\text{loc}}(\vec{\Psi}(\vec{x}))}{\partial x_k}\right)}{\partial\left(\frac{\partial \Psi_i(\vec{x})}{\partial x_u}\right)} = \begin{cases} \tilde{\Phi}_i(\vec{x}) & \text{, if } k = u \\ 0 & \text{, otherwise} \end{cases} . \tag{5.103}$$

Equation (5.102) can be further expanded with the help of the product rule to

$$\begin{aligned} \frac{\partial E(\Phi_{\text{loc}}(\vec{\Psi}))}{\partial \Psi_i(\vec{x})} &= \frac{\partial F}{\partial \Phi_{\text{loc}}(\vec{\Psi}(\vec{x}))} \frac{\partial \Phi_{\text{loc}}(\vec{\Psi}(\vec{x}))}{\partial \Psi_i(\vec{x})} + \sum_{k=1}^{n} \frac{\partial F}{\partial\left(\frac{\partial \Phi_{\text{loc}}(\vec{\Psi}(\vec{x}))}{\partial x_k}\right)} \frac{\partial \tilde{\Phi}_i(\vec{x})}{\partial x_k} \\ &- \sum_{u=1}^{n} \frac{\partial}{\partial x_u} \frac{\partial F}{\partial\left(\frac{\partial \Phi_{\text{loc}}(\vec{\Psi}(\vec{x}))}{\partial x_u}\right)} \tilde{\Phi}_i(\vec{x}) - \sum_{u=1}^{n} \frac{\partial F}{\partial\left(\frac{\partial \Phi_{\text{loc}}(\vec{\Psi}(\vec{x}))}{\partial x_u}\right)} \frac{\partial \tilde{\Phi}_i(\vec{x})}{\partial x_u} . \end{aligned} \tag{5.104}$$

The second and the last term on the right-hand side of Eq. (5.104) cancel each other out. So, we finally obtain the functional derivative of the energy functional from Eq. (5.98), with regard to the i-th weight field Ψ_i, as

$$\begin{aligned} \frac{\partial E(\Phi_{\text{loc}}(\vec{\Psi}))}{\partial \Psi_i(\vec{x})} &= \frac{\partial F}{\partial \Phi_{\text{loc}}(\vec{\Psi}(\vec{x}))} \tilde{\Phi}_i(\vec{x}) - \sum_{u=1}^{n} \frac{\partial}{\partial x_u} \frac{\partial F}{\partial\left(\frac{\partial \Phi_{\text{loc}}(\vec{\Psi}(\vec{x}))}{\partial x_u}\right)} \tilde{\Phi}_i(\vec{x}) \\ &= \left[\frac{\partial F}{\partial \Phi_{\text{loc}}(\vec{\Psi}(\vec{x}))} - \sum_{u=1}^{n} \frac{\partial}{\partial x_u} \frac{\partial F}{\partial\left(\frac{\partial \Phi_{\text{loc}}(\vec{\Psi}(\vec{x}))}{\partial x_u}\right)} \right] \tilde{\Phi}_i(\vec{x}) . \end{aligned} \tag{5.105}$$

When we compare the term in square brackets on the right-hand side of Eq. (5.105) with the right hand side of Eq. (1.26) (functional derivative in the case of functions with multiple arguments), we can see that it is the functional derivative of the energy $E(\Phi_{\text{loc}}(\vec{\Psi}))$ with regard to the level set function Φ_{loc}. The second term on the right-hand side of Eq. (5.105) is the derivative of the LDSSM $\Phi_{\text{loc}}(\vec{\Psi})$ with regard to the i-th weight field Ψ_i. This result is similar to the results that we expect from applying the chain rule in order to obtain a derivative of composed functions.

Indeed, when we have a closer look at our energy functional $E(\Phi_{\text{loc}}(\vec{\Psi}))$, we can see that it is a functional of a function of a function. So, we can apply the chain rule for functional derivatives from Eq. (1.25) in order to obtain the same result as in Eq. (5.105):

$$\begin{aligned} \frac{\partial E(\Phi_{\text{loc}}(\vec{\Psi}))}{\partial \Psi_i(\vec{x})} &= \frac{\partial E(\Phi_{\text{loc}}(\vec{\Psi}))}{\partial \Phi_{\text{loc}}(\vec{\Psi}(\vec{x}))} \frac{\partial \Phi_{\text{loc}}(\vec{\Psi}(\vec{x}))}{\partial \Psi_i(\vec{x})} \\ &= \frac{\partial E(\Phi_{\text{loc}}(\vec{\Psi}))}{\partial \Phi_{\text{loc}}(\vec{\Psi}(\vec{x}))} \tilde{\Phi}_i(\vec{x}) . \end{aligned} \tag{5.106}$$

Furthermore, the formulation from Eq. (5.106) is more general than Eq. (5.105). So, we can drop the requirement that the fitting energy has to be of the form as in Eq. (5.98) and consider even more complex energy functionals that may contain for example higher order partial derivatives of the LDSSM.

The benefit of the formulation in Eq. (5.106) is that for any given energy functional $E(\Phi)$ from the level set literature, we can look up the functional derivative $\frac{\partial E(\Phi)}{\partial \Phi(\vec{x}))}$ from the literature. Then, we simply append the base function $\tilde{\Phi}_i(\vec{x})$ as another factor in order to obtain the functional derivative of the modified energy functional $\frac{\partial E(\Phi_{\mathrm{loc}}(\vec{\Psi})}{\partial \Psi_i(\vec{x})}$, where the level set function Φ has been replaced by our local model $\Phi_{\mathrm{loc}}(\vec{\Psi})$.

5.5 Global-to-Local Variational Formulation

The approach presented in Sect. 5.4 enables us to obtain those results to a given segmentation problem that are guaranteed to lie within the generalized subspace of feasible shapes which is described by our LDSSM. This means, all possible segmentation results are given as local combinations of some given training shapes. Since the LDSSM subspace allows much more possible shape configurations than the SSM subspace, it is possible to obtain much more accurate segmentation results with our local approach from Sect. 5.4 than with the global approach by Tsai et al. that has been presented in Sect. 5.3. This is due to the fact that, in our local approach, the shape constraint is only enforced locally in a neighborhood around each element $\vec{x}$ of the data domain Ω instead of enforcing a common global shape constraint for all elements of the data domain.

However, as in many engineering problems, the increase in segmentation accuracy comes at the price of reduced robustness to noise. This means, our local gradient descent method (Eq. (5.73)) is more likely to get trapped inside local minima due to missing or wrong image information than the global descent method (Eq. (5.33)). In order to avoid those local minima, we need to strengthen the influence of the shape model in the segmentation process. This means that the resulting weight fields Ψ_i have to be as smooth as possible. As explained in Sect. 3.2, smooth weight fields correspond to a strong shape prior which allows us to guide the modeled shape also in those regions with missing or wrong image information.

Solutions with smooth weight fields Ψ_i can be obtained by raising the influence of the smoothing energy $E^{\mathrm{loc}}_{\mathrm{smooth}}$ in our variational formulation from Eq. (5.69). This can be done by increasing the value of the weighting factor β. Raising the influence of the smoothing energy $E^{\mathrm{loc}}_{\mathrm{smooth}}$ has the effect that the fitting energy $E(\Phi_{\mathrm{loc}}(\vec{\Psi}))$ becomes less important in the determination of the segmentation result. So, the weight fields Ψ_i will be influenced by the fitting energy $E(\Phi_{\mathrm{loc}}(\vec{\Psi}))$ mostly in those image regions where strong image information is present. In the other regions, where the image information is more ambiguous, the resulting weight fields Ψ_i will mostly be determined by the smoothing energy $E^{\mathrm{loc}}_{\mathrm{smooth}}$.

However, as nice as this sounds in theory, there exist some practical limitations in our gradient descent formulation from Eq. (5.73) that prevent us from obtaining an arbitrary large influence of our LDSSM on the segmentation results. We will address these limitations in Sect. 5.5.1. Afterwards, in Sect. 5.5.2, we will build a connection between our local approach from Sect. 5.4 and the global approach from Sect. 5.3. This connection will then be utilized in Sect. 5.5.3 in order to derive a new global-to-local approach for minimizing the energy functional from Eq. (5.69). This new approach removes the limitations of the gradient descent formulation from Eq. (5.73) and makes our local segmentation approach robust against local minima caused by noise and missing data.

5.5.1 Problems of the Variational Gradient Descent Approach

A property of common fitting energies is that their functional gradients differ from zero only on the zero level contour of the evolving level set (see e.g. Eq. (5.97)). This means, in each iteration of the gradient descent approach from Eq. (5.73), the gradient of the fitting energy $E(\Phi_{\text{loc}}(\vec{\Psi}))$ affects only those elements $\Psi_i(\vec{x})$ of the weight fields where the zero level contour is currently located. Consequently, the weight field elements in the largest area of the data domain Ω will only by altered by the gradients of the shape energy $E^{\text{loc}}_{\text{shape}}$ and the smoothness energy $E^{\text{loc}}_{\text{smooth}}$, respectively.

As the gradient of the smoothness energy is given by the heat equation (c.f. Eq. (5.83)), it has the effect that local changes in the weight fields Ψ_i are propagated to neighboring elements of the data domain. When we iterate our gradient descent approach from Eq. (5.73) until a stationary point of the energy functional from Eq. (5.69) is reached, this propagation of local weight changes has the following effects:

1. Wrong weight field updates, due to noise in the data that leads to local errors in the estimated gradient of the fitting energy $E(\Phi_{\text{loc}}(\vec{\Psi}))$, will be canceled out by neighboring weight field updates.
2. Missing weight field updates, due to missing or ambiguous image information, will be interpolated from neighboring weight field updates.

The size of the neighborhood in which weight field updates have an influence on each other thereby depends on the influence of the smoothness term in the energy functional from Eq. (5.69). So, if the data is highly corrupted or large areas of the data are missing, it is a good choice to use a large weight β for the smoothness energy term $E^{\text{loc}}_{\text{smooth}}$.

However, if the continuous Laplace operator from the smoothness energy gradient (c.f. Eq. (5.83)) is implemented via a central differencing scheme (c.f. Chap. 6), the value of β in Eq. (5.69) is limited by

$$\beta \leq \frac{1}{2^n} \frac{(\Delta x)^2}{\gamma^k}, \tag{5.107}$$

where Δx is the spatial resolution (same resolution across all dimensions) and n is the number of dimensions (i.e $n = 2$ or $n = 3$). Equation (5.107) has to be satisfied, otherwise the gradient descent evolution is not guaranteed to be stable [101]. So, as the spatial resolution Δx is fixed, we have to decrease our step size γ_k in order to obtain an arbitrarily large influence of the smoothing term $E^{\text{loc}}_{\text{smooth}}$ in our combined energy functional from Eq. (5.69). This leads to an impractically slow convergence rate of the gradient descent approach from Eq. (5.73) if local weight field updates are supposed to be propagated over large areas of the data domain Ω. As a result, the influence range of the fitting energy $E(\Phi_{\text{loc}}(\vec{\Psi}))$ on the weight fields is effectively limited to neighborhoods of a few elements in diameter around each element $\vec{x}$ of the data domain Ω.

This limitation on the value of β is not a problem if a good initial solution exists that should only be locally refined. However, it becomes a problem when the initial solution is far from the desired result so that large parts of the evolving contour may be located in regions where image information is missing, ambiguous, or highly corrupted by noise. As discussed above, in this case, instead of a local shape prior, a more global shape prior would be required that guides the evolving contour through those difficult regions. For the LDSSM, this means that the weight fields should be as smooth as possible.

Now, as it is not possible to enforce smooth weight fields over large regions due to the limitation on the value of β from Eq. (5.107), those parts of the contour that are located in difficult image regions are likely to get stuck in local minima during the gradient descent iteration from Eq. (5.73). In order to counteract this, i.e. to provide guidance to the segmentation contour also in those image regions with missing, ambiguous, or highly corrupted data, we need to find another way of how to propagate reliable local image information to larger parts of the data domain Ω.

5.5.2 *Connection Between the Global Approach and Our Local Approach*

An efficient way of propagating local shape changes to the rest of the data domain in order to obtain globally consistent weights can be observed in the global approach from Sect. 5.3. We recall that the partial derivative of the fitting energy $E(\Phi_{\text{glob}}(\vec{w}))$ with regard to the global weight w_i has been given as (Eq. (5.48))

$$\frac{\partial E(\Phi_{\text{glob}}(\vec{w}))}{\partial w_i} = \int_{\Omega} \frac{\partial E(\Phi_{\text{glob}}(\vec{w}))}{\partial \Phi_{\text{glob}}(\vec{w})(\vec{x})} \tilde{\Phi}_i(\vec{x})\, d\vec{x}\,.$$

In Eq. (5.48), a local change in the level set function Φ_{glob} is projected on the base vector $\tilde{\Phi}_i$ of the SSM in order to obtain the change in the weight w_i. The partial derivative of the fitting energy $E(\Phi_{\text{glob}}(\vec{w}))$ with regard to the level set function Φ_{glob} is different from zero only directly on the zero level contour of Φ_{glob} (c.f. Sect. 5.5.1). So, it represents a local change in the level set function Φ_{glob}. The weight w_i, however, has a global influence on the modeled shape through Eq. (2.34).

Now, we compare this result with our local approach from Sect. 5.4. The derivative of the fitting energy $E(\Phi_{\text{loc}}(\vec{\Psi}))$ with regard to the weight field Ψ_i at point $\vec{x}$ has been defined in Eq. (5.106) as

$$\frac{\partial E(\Phi_{\text{loc}}(\vec{\Psi}))}{\partial \Psi_i(\vec{x})} = \frac{\partial E(\Phi_{\text{loc}}(\vec{\Psi}))}{\partial \Phi_{\text{loc}}(\vec{\Psi}(\vec{x}))} \tilde{\Phi}_i(\vec{x}) .$$

It can be seen that the difference between the global and the local weight update is the missing integral over the whole data domain Ω for our local approach. This means, performing gradient descent for the global approach using Eq. (5.33) can be interpreted as averaging all functional derivatives over the whole data domain Ω in order to obtain the global weight update. This is what makes the global approach robust against noise and missing image information.

In order to transfer this robustness to our local approach, we can define a descent method so that it exactly behaves like the global approach. For this purpose, we change the descent direction for our local fitting energy $E(\Phi_{\text{loc}}(\vec{\Psi}))$. So far, it is given by its negative functional gradient (c.f. Eq. (5.106)):

$$\Psi_i^{k+1} = \Psi_i^k - \frac{\partial E(\Phi_{\text{loc}}(\vec{\Psi}))}{\partial \Phi_{\text{loc}}(\vec{\Psi}(\vec{x}))} \tilde{\Phi}_i(\vec{x}) , \tag{5.108}$$

which means that we descend in the direction of steepest descent of the fitting energy. Now, we can replace this descent direction by projecting each component of the negative functional gradient on a function $h : \Omega \rightarrow \mathbb{R}$, with $||h|| = 1$. The projection of the i-th component of the negative functional gradient onto such a function h is identical to the directional derivative along the function h [124]. So, it can be expressed via the functional differential from Eq. (1.31) as

$$- dE(\Phi_{\text{loc}}(\Psi_i), h)\, h , \tag{5.109}$$

where the functional differential is given as

$$\begin{aligned} dE(\Phi_{\text{loc}}(\Psi_i), h) &= \int_\Omega \frac{\partial E(\Phi_{\text{loc}}(\vec{\Psi}))}{\partial \Psi_i(\vec{x})} h(\vec{x})\, \mathrm{d}\vec{x} \\ &= \int_\Omega \frac{\partial E(\Phi_{\text{loc}}(\vec{\Psi}))}{\partial \Phi_{\text{loc}}(\vec{\Psi}(\vec{x}))} \tilde{\Phi}_i(\vec{x})\, h(\vec{x})\, \mathrm{d}\vec{x} . \end{aligned} \tag{5.110}$$

That $-dE(\Phi_{\text{loc}}(\Psi_i), h)\, h$ is indeed a descent direction for the i-th component of our fitting energy requires that the directional derivative along $-dE(\Phi_{\text{loc}}(\Psi_i), h)\, h$ has to be negative (c.f. Eq. (1.9)). In our case, this means that

$$dE\big(\Phi_{\text{loc}}(\Psi_i), -dE(\Phi_{\text{loc}}(\Psi_i), h)\, h\big) < 0 \tag{5.111}$$

has to be fulfilled, which can also be written as

$$-\int_\Omega \frac{\partial E(\Phi_{\text{loc}}(\vec{\Psi}))}{\partial \Psi_i(\vec{x})}\, dE(\Phi_{\text{loc}}(\Psi_i), h)\, h(\vec{x})\, \mathrm{d}\vec{x} < 0\,. \tag{5.112}$$

Since $dE(\Phi_{\text{loc}}(\Psi_i), h)$ does not depend on $\vec{x}$, it can be pulled out from the integral, and we obtain

$$\begin{aligned} &- dE(\Phi_{\text{loc}}(\Psi_i), h) \int_\Omega \frac{\partial E(\Phi_{\text{loc}}(\vec{\Psi}))}{\partial \Psi_i(\vec{x})}\, h(\vec{x})\, \mathrm{d}\vec{x} \\ &= -\big(dE(\Phi_{\text{loc}}(\Psi_i), h)\big)^2 \le 0\,. \end{aligned} \tag{5.113}$$

So, $-dE(\Phi_{\text{loc}}(\Psi_i), h)\, h$ is a descent direction if the functional differential along h differs from zero (which is true if we have not found a local minimum) and h is chosen in such a way that it is not zero everywhere on the data domain Ω.

Now, we choose the function h to be the normalized characteristic function $\frac{\mathbf{1}_\Omega}{||\mathbf{1}_\Omega||}$ of our data domain Ω, where $\mathbf{1}_\Omega$ is defined as

$$\mathbf{1}_\Omega(\vec{x}) = \begin{cases} 1 & \text{, if } \vec{x} \in \Omega \\ 0 & \text{, else} \end{cases}\,. \tag{5.114}$$

In this case, Eq. (5.108) modifies to

$$\Psi_i^{k+1}(\vec{y}) = \Psi_i^k(\vec{y}) - \gamma^k\, dE\left(\Phi_{\text{loc}}(\Psi_i^k), \frac{\mathbf{1}_\Omega}{||\mathbf{1}_\Omega||}\right) \frac{\mathbf{1}_\Omega(\vec{y})}{||\mathbf{1}_\Omega||} \tag{5.115}$$

for each element $\vec{y} \in \Omega$. In Eq. (5.115), the functional differential along the normalized characteristic function $\frac{\mathbf{1}_\Omega}{||\mathbf{1}_\Omega||}$ reads as

$$\begin{aligned} dE\left(\Phi_{\text{loc}}(\Psi_i), \frac{\mathbf{1}_\Omega}{||\mathbf{1}_\Omega||}\right) &= \int_\Omega \frac{\partial E(\Phi_{\text{loc}}(\vec{\Psi}))}{\partial \Phi_{\text{loc}}(\vec{\Psi}(\vec{x}))} \tilde{\Phi}_i(\vec{x})\, \frac{\mathbf{1}_\Omega(\vec{x})}{||\mathbf{1}_\Omega||}\, \mathrm{d}\vec{x} \\ &= \frac{1}{||\mathbf{1}_\Omega||} \int_\Omega \frac{\partial E(\Phi_{\text{loc}}(\vec{\Psi}))}{\partial \Phi_{\text{loc}}(\vec{\Psi}(\vec{x}))} \tilde{\Phi}_i(\vec{x})\, \mathrm{d}\vec{x}\,. \end{aligned} \tag{5.116}$$

By inserting Eq. (5.116) into Eq. (5.115), the weight field update for each element $\vec{y} \in \Omega$ can thus be written as

$$\Psi_i^{k+1}(\vec{y}) = \Psi_i^k(\vec{y}) - \gamma^k \frac{1}{||\mathbf{1}_\Omega||^2} \int_\Omega \frac{\partial E(\Phi_{\text{loc}}(\vec{\Psi}^k))}{\partial \Phi_{\text{loc}}(\vec{\Psi}^k(\vec{x}))} \tilde{\Phi}_i(\vec{x}) \, \mathrm{d}\vec{x} \, \mathbf{1}_\Omega(\vec{y}) \,. \quad (5.117)$$

Because $\mathbf{1}_\Omega(\vec{y}) = 1$ for all $y \in \Omega$ (c.f. Eq. (5.114)), this update rule simplifies to

$$\Psi_i^{k+1}(\vec{y}) = \Psi_i^k(\vec{y}) - \gamma^k \frac{1}{||\mathbf{1}_\Omega||^2} \int_\Omega \frac{\partial E(\Phi_{\text{loc}}(\vec{\Psi}^k))}{\partial \Phi_{\text{loc}}(\vec{\Psi}^k(\vec{x}))} \tilde{\Phi}_i(\vec{x}) \, \mathrm{d}\vec{x} \,, \ \forall \vec{y} \in \Omega \,. \quad (5.118)$$

It can be seen in Eq. (5.118) that by projecting the i-th component of the negative fitting energy gradient onto the characteristic function of our domain Ω, the update for each local weight $\Psi_i(\vec{y})$ is given by averaging the functional derivative along all elements $\vec{x}$ of the domain Ω. This is exactly what we wanted to achieve.

Furthermore, by setting $\Psi_i^0(\vec{x}) = w_i^0$ for all $\vec{x} \in \Omega$ so that $\Phi_{\text{loc}}(\vec{\Psi}^0(\vec{x})) = \Phi_{\text{glob}}(\vec{w}^0)(\vec{x})$, Eq. (5.118) becomes identical to Eq. (5.33) (where the gradient is given in Eq. (5.48)) up to the constant factor $||\mathbf{1}_\Omega||^{-2}$. This factor can be eliminated by choosing an appropriate step size $\tilde{\gamma}^k = \gamma^k \xi$, with $\xi = ||\mathbf{1}_\Omega||^2$:

$$\begin{aligned} \Psi_i^{k+1}(\vec{y}) &= \Psi_i^k(\vec{y}) - \tilde{\gamma}^k \frac{\xi}{||\mathbf{1}_\Omega||^2} \int_\Omega \frac{\partial E(\Phi_{\text{loc}}(\vec{\Psi}^k))}{\partial \Phi_{\text{loc}}(\vec{\Psi}^k(\vec{x}))} \tilde{\Phi}_i(\vec{x}) \, \mathrm{d}\vec{x} \\ &= \Psi_i^k(\vec{y}) - \tilde{\gamma}^k \int_\Omega \frac{\partial E(\Phi_{\text{loc}}(\vec{\Psi}^k))}{\partial \Phi_{\text{loc}}(\vec{\Psi}^k(\vec{x}))} \tilde{\Phi}_i(\vec{x}) \, \mathrm{d}\vec{x} \,. \end{aligned} \quad (5.119)$$

So, by iterating Eq. (5.119) until convergence, we obtain results with our local model that are identical to the global approach from Eq. (5.33). This is due to the fact that the weight fields Ψ_i are initialized to constant functions and each local weight is updated by the same expression (last term of Eq. (5.119)) so that the weight fields remain constant throughout all iterations.

Now, we also consider the gradients of the shape and the smoothing energy, given in Eqs. (5.91) and (5.83), respectively, which we have left out so far. By renaming $\tilde{\gamma}^k$ to γ^k, the weight update becomes (c.f. Eqs. (5.70) and (5.77))

$$\begin{aligned} \Psi_i^{k+1}(\vec{y}) = \Psi_i^k(\vec{y}) - \gamma^k \Bigg[& \frac{\xi}{||\mathbf{1}_\Omega||^2} \int_\Omega \frac{\partial E(\Phi_{\text{loc}}(\vec{\Psi}^k))}{\partial \Phi_{\text{loc}}(\vec{\Psi}^k(\vec{x}))} \tilde{\Phi}_i(\vec{x}) \, \mathrm{d}\vec{x} \\ & + \alpha \frac{\Psi_i^k(\vec{y})}{\sigma_i^2} - \beta \Delta \Psi_i^k(\vec{y}) \Bigg] \,. \end{aligned} \quad (5.120)$$

However, when we consider only constant weight fields Ψ_i^k, as explained above, the gradient of the smoothing term vanishes ($\Delta \Psi_i^k(\vec{y}) = 0$), and Eq. (5.120) simplifies to

$$\Psi_i^{k+1}(\vec{y}) = \Psi_i^k(\vec{y}) - \gamma^k \left[\frac{\xi}{||\mathbf{1}_\Omega||^2} \int_\Omega \frac{\partial E(\Phi_{\text{loc}}(\vec{\Psi}^k))}{\partial \Phi_{\text{loc}}(\vec{\Psi}^k(\vec{x}))} \tilde{\Phi}_i(\vec{x}) \, \mathrm{d}\vec{x} + \alpha \frac{\Psi_i^k(\vec{y})}{\sigma_i^2} \right] . \quad (5.121)$$

By iterating this equation until convergence, we obtain results with our local model that are identical to the global approach which has been extended by the trained weight distribution (c.f. Eq. (5.51) with gradients given in Eqs. (5.56) and (5.49)). In particular, it can be seen that the gradient of the local shape energy affects each local weight $\Psi_i^k(\vec{y})$ in the same way as the gradient of the global shape energy affects the global weight w_i in Eq. (5.56).

In summary, the global approach from Sect. 5.3 can be regarded as a special case of the local approach from Sect. 5.4 where all weight fields Ψ_i are initialized to a constant value w_i^0 and the descent direction of the fitting energy is chosen to $-dE(\Phi_{\text{loc}}(\Psi_i), \frac{\mathbf{1}_\Omega}{||\mathbf{1}_\Omega||})\frac{\mathbf{1}_\Omega}{||\mathbf{1}_\Omega||}$. This gives rise to the approach shown in Algorithm 5.1 in order to make the solution to our image segmentation problem from Eq. (5.69) more robust against noise and missing data.

1. Initialize all weight fields to zero and find a good initial solution to the problem by using the global weight field update from Eq. (5.121) (or equivalently Eq. (5.120), as $\Delta\Psi_i^k(\vec{y}) = 0$).
2. Refine the result using the local weight field update given in Eq. (5.73).

Algorithm 5.1: Necessary steps for the local solution with global initialization.

5.5.3 *Combining the Global Approach and Our Local Approach*

In the previous section, we have shown that the global approach from Sect. 5.3 is identical to our local approach from Sect. 5.4 when we choose the descent direction of the fitting energy for each weight field Ψ_i to $-dE(\Phi_{\text{loc}}(\Psi_i), \frac{\mathbf{1}_\Omega}{||\mathbf{1}_\Omega||})\frac{\mathbf{1}_\Omega}{||\mathbf{1}_\Omega||}$ (the projection of the i-th component of the fitting energy gradient onto the characteristic function $\mathbf{1}_\Omega$ of the data domain Ω). This descent direction has the property that the gradient of the fitting energy is averaged over the whole data domain Ω so that all elements $\Psi_i(\vec{y})$ of the weight field Ψ_i are updated to the same extent while the fitting energy gradient differs from zero only in a few elements $\vec{y}$ of the data domain Ω.

So, choosing $-dE(\Phi_{\text{loc}}(\Psi_i), \frac{\mathbf{1}_\Omega}{||\mathbf{1}_\Omega||})\frac{\mathbf{1}_\Omega}{||\mathbf{1}_\Omega||}$ as the descent direction of the fitting energy for each weight field Ψ_i makes the local approach from Sect. 5.4 robust against noise and missing data while completely loosing its key ability to locally adapt to the image structures. On the other hand, the descent direction from Eq. (5.108) is completely local because it is given for each element $\Psi_i(\vec{y})$ of the weight field Ψ_i as the i-th component of the negative functional gradient in the actually considered point $\vec{y}$:

$$-\frac{\partial E(\Phi_{\text{loc}}(\vec{\Psi}))}{\partial \Phi_{\text{loc}}(\vec{\Psi}(\vec{y}))}\tilde{\Phi}_i(\vec{y})\,. \tag{5.122}$$

This direction enables the LDSSM to locally adapt to the image structures but makes the local approach from Sect. 5.4 also more error-prone to noise (for which reason we need the smoothness energy in Eq. (5.69)). The locality in the descent direction becomes even more clear when we rewrite Eq. (5.122) in terms of the functional differential. In this case, it is given as

$$-\,d\,E_{\mathrm{img}}\left(\Phi_{\mathrm{loc}}(\Psi_i), \mathbf{1}_{\vec{y}}\right)\,\mathbf{1}_{\vec{y}}(\vec{y})\,, \tag{5.123}$$

where $\mathbf{1}_{\vec{y}}$ is the characteristic function of $\vec{y}$, i.e. it is nonzero only in the point $\vec{y}$:

$$\mathbf{1}_{\vec{y}}(\vec{x}) = \begin{cases} 1 & , \text{if } \vec{x} = \vec{y} \\ 0 & , \text{else} \end{cases}\,. \tag{5.124}$$

That Eqs. (5.122) and (5.123) are identical can easily be shown: The functional differential along $\mathbf{1}_{\vec{y}}$ can be written as (c.f. Eq. (5.110))

$$\begin{aligned} d\,E_{\mathrm{img}}\left(\Phi_{\mathrm{loc}}(\Psi_i), \mathbf{1}_{\vec{y}}\right) &= \int_{\Omega} \frac{\partial E(\Phi_{\mathrm{loc}}(\vec{\Psi}))}{\partial \Phi_{\mathrm{loc}}(\vec{\Psi}(\vec{x}))} \tilde{\Phi}_i(\vec{x})\,\mathbf{1}_{\vec{y}}(\vec{x})\,\mathrm{d}\vec{x} \\ &= \frac{\partial E(\Phi_{\mathrm{loc}}(\vec{\Psi}))}{\partial \Phi_{\mathrm{loc}}(\vec{\Psi}(\vec{y}))} \tilde{\Phi}_i(\vec{y})\,, \end{aligned} \tag{5.125}$$

and when we insert Eq. (5.125) into Eq. (5.123), we exactly obtain the descent direction from Eq. (5.122) as $\mathbf{1}_{\vec{y}}(\vec{y}) = 1$.

So, as a consequence, we can say that we can achieve a more global (and thus more robust) weight field update by widening the support of the function h on which we project the i-th component of our fitting energy gradient (from the actually considered point $\vec{y}$ in the case of $h = \mathbf{1}_{\vec{y}}$ to the whole data domain Ω in the case of $h = \frac{\mathbf{1}_{\Omega}}{||\mathbf{1}_{\Omega}||}$). Accordingly, we can obtain a more local (but possibly more error-prone) weight field update by narrowing down the support of the function h.

This finding gives rise to the following global-to-local weight field update approach: Start with a function h that has a large support and then subsequently limit the support to a smaller and smaller area so that the weight field update becomes more and more local. Please note that this is comparable to hierarchical image registration approaches which are well-known to be less prone to get stuck in local minima than other non-hierarchical approaches (see e.g. [76] or [142] for an overview).

To make it more specific, we define the set $\Theta_d(\vec{y})$ which contains all elements $\vec{x}$ of the data domain Ω that are located within a predefined radius d around the actually considered point $\vec{y}$:

$$\Theta_d(\vec{y}) = \left\{\vec{x} \in \Omega \mid ||\vec{y} - \vec{x}|| \le d\right\}, \tag{5.126}$$

and we choose the descent direction of the fitting energy for each weight field Ψ_i to

$$- dE\left(\Phi_{\text{loc}}(\Psi_i), \frac{\mathbf{1}_{\Theta_d(\vec{y})}}{||\mathbf{1}_{\Theta_d(\vec{y})}||}\right) \frac{\mathbf{1}_{\Theta_d(\vec{y})}}{||\mathbf{1}_{\Theta_d(\vec{y})}||} . \tag{5.127}$$

This defines the projection of the i-th component of the fitting energy gradient onto the characteristic function $\mathbf{1}_{\Theta_d(\vec{y})}$ of the just defined set $\Theta_d(\vec{y})$:

$$\mathbf{1}_{\Theta_d(\vec{y})}(\vec{x}) = \begin{cases} 1 & \text{, if } \vec{x} \in \Theta_d(\vec{y}) \\ 0 & \text{, else} \end{cases} . \tag{5.128}$$

Choosing the descent direction given by Eq. (5.127) has the effect that the weight update in each location $\vec{y}$ of the data domain Ω is obtained by averaging all local fitting energy gradients in a region with radius d around the currently considered point $\vec{y}$. This can be seen by rewriting the functional differential as

$$\begin{aligned} dE\left(\Phi_{\text{loc}}(\Psi_i), \frac{\mathbf{1}_{\Theta_d(\vec{y})}}{||\mathbf{1}_{\Theta_d(\vec{y})}||}\right) &= \int_\Omega \frac{\partial E(\Phi_{\text{loc}}(\vec{\Psi}))}{\partial \Phi_{\text{loc}}(\vec{\Psi}(\vec{x}))} \tilde{\Phi}_i(\vec{x}) \frac{\mathbf{1}_{\Theta_d(\vec{y})}(\vec{x})}{||\mathbf{1}_{\Theta_d(\vec{y})}||} \,\mathrm{d}\vec{x} \\ &= \frac{1}{||\mathbf{1}_{\Theta_d(\vec{y})}||} \int_{\Theta_d(\vec{y})} \frac{\partial E(\Phi_{\text{loc}}(\vec{\Psi}))}{\partial \Phi_{\text{loc}}(\vec{\Psi}(\vec{x}))} \tilde{\Phi}_i(\vec{x}) \,\mathrm{d}\vec{x} . \end{aligned} \tag{5.129}$$

If we choose d large enough so that $\Omega \subseteq \Theta_d(\vec{y})$, we obtain the global weight update from Eq. (5.115). If we choose $d = 0$, the characteristic function of the set $\Theta_d(\vec{y})$ becomes $\mathbf{1}_{\vec{y}}$ so that we obtain the local weight update from Eq. (5.108).

Now, when we again consider the gradients of the shape and the smoothness energy, which have been left out so far, our global-to-local descent iteration for the energy from Eq. (5.69) is given as (c.f. Eq. (5.120))

$$\begin{aligned} \Psi_i^{k+1}(\vec{y}) = \Psi_i^k(\vec{y}) - \gamma^k \Bigg[&\frac{\xi}{||\mathbf{1}_{\Theta_d(\vec{y})}||^2} \int_{\Theta_d(\vec{y})} \frac{\partial E(\Phi_{\text{loc}}(\vec{\Psi}^k))}{\partial \Phi_{\text{loc}}(\vec{\Psi}^k(\vec{x}))} \tilde{\Phi}_i(\vec{x}) \,\mathrm{d}\vec{x} \\ &+ \alpha \frac{\Psi_i^k(\vec{y})}{\sigma_i^2} - \beta \Delta \Psi_i^k(\vec{y}) \Bigg] . \end{aligned} \tag{5.130}$$

We empirically determined that it is a good approach to choose the weighting factor ξ (c.f. Eq. (5.119)) to $\xi = ||\mathbf{1}_\Omega|| \, ||\mathbf{1}_{\Theta_d(\vec{y})}||$ which is consistent with the explanations in Sect. 5.5.2. Then, our proposed global-to-local segmentation approach reads as shown in Algorithm 5.2.

As mentioned above, this approach is robust against local minima in the energy from Eq. (5.69) as the first term of the weight field update from Eq. (5.130) (i.e. the part that is based on the fitting energy) is now intrinsically smooth for large values of d in contrast to the weight field update from Eq. (5.73). We have already discussed this in detail for the case that $\Omega \subseteq \Theta_d(\vec{y})$. Now, when we halve the value of d, the weight update based on the fitting energy remains intrinsically smooth as the regions $\Theta_d(\vec{y})$ still highly overlap for neighboring points $\vec{y}$.

1. Choose an appropriate initial global weight vector $\vec{w}^{\text{init}} \in \mathbb{R}^m$ (e.g. $\vec{w}^{\text{init}} = \vec{0}$).
2. Initialize all weight fields Ψ_i, $i = 1 \ldots m$, to a constant value w_i^{init} so that

$$\Psi_i(\vec{y}) = w_i^{\text{init}} \,, \forall \vec{y} \in \Omega \,.$$

3. Choose a sufficiently large value for the radius d such that $\Omega \subseteq \Theta_d(\vec{y})$ for all $\vec{y} \in \Omega$, where d is further required to be a power of 2 (i.e. $d = 2^p$, $p \in \mathbb{N}$).

do

4. Iterate Eq. (5.130) until convergence for all weight fields Ψ_i, $i = 1 \ldots m$.
5. Reduce the value of d, where

$$d \leftarrow \begin{cases} \frac{d}{2} & \text{, if } d > 1 \\ 0 & \text{, if } d = 1 \\ -1 & \text{, else} \end{cases} \,.$$

while $d \geq 0$

Algorithm 5.2: Procedure of our global-to-local segmentation approach.

However, when we decrease the radius d of the regions $\Theta_d(\vec{y})$ during the fitting process, the individual weight field updates in each element $\vec{y}$ of the data domain Ω become more and more independent, allowing the model to adapt to more and more local image structures. In the final iteration, for $d = 0$, the smoothness of the weight fields is solely determined by the gradient of smoothness energy, as the weight

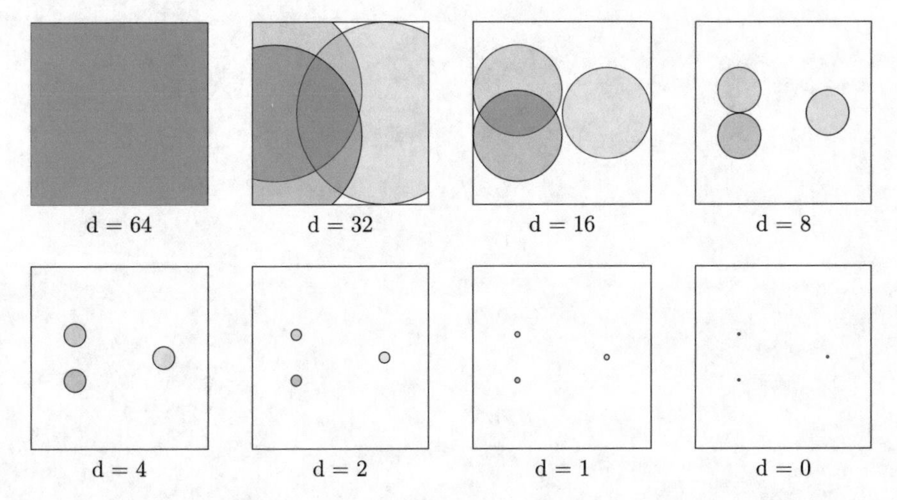

Fig. 5.2 Illustration of the sets $\Theta_d(16, 24)$ (*blue circle*), $\Theta_d(16, 40)$ (*red circle*), and $\Theta_d(48, 32)$ (*green circle*) for different values of d. The size of the data domain Ω is 64×64 pixels. It can be seen that the regions highly overlap for large d

updates due to the fitting energy are now completely independent. This is identical to our local approach from Eq. (5.73). A depiction that illustrates the subsequent size reduction of the regions $\Theta_d(\vec{y})$ can be seen in Fig. 5.2.

Chapter 6
Evaluation of the Global-To-Local Variational Formulation

In Chap. 5, we have presented an elegant formulation in order to integrate our proposed *Locally Deformable Statistical Shape Model* (LDSSM) from Chap. 3 into a variational level set segmentation framework. In the following, we will show that this new formulation improves the segmentation results obtained with our former heuristic approach (Sect. 3.3) by incorporating the statistical shape information already in the weight update target estimation step of our iterative processing chain (Fig. 3.8). For this purpose, we will consider one more time the extraction of the combined outer bony boundary of the nasal cavity and the paranasal sinuses from *Computed Tomography* (CT) data which has been already addressed with our former heuristic local approach in Sect. 4.1.

In Sect. 6.1, we will show the potential of our new global-to-local variational formulation by extracting the paranasal sinuses from two-dimensional CT slices, and in Sect. 6.2 we will go one step further by segmenting the paranasal sinuses directly from three-dimensional CT data. The results of the two-dimensional paranasal sinuses segmentation are an extended version of the results that we have already published in [2]. The results of the three-dimensional paranasal sinuses segmentation have not been published before.

6.1 Extracting the Nasal Cavity and the Paranasal Sinuses from Two-Dimensional CT Slices

We start our evaluation by comparing the results of our new global-to-local variational formulation (Sect. 5.5) against the results of the global variational formulation by Tsai et al. (Sect. 5.3). This will be done in Sect. 6.1.1. For comparison, we will also provide the results of our former heuristic local approach (Sect. 3.3.4). Afterwards, in Sect. 6.1.2, we will additionally evaluate the influence of our initial global solution

C. Last, *From Global to Local Statistical Shape Priors*, Studies in Systems, Decision and Control 98, DOI 10.1007/978-3-319-53508-1_6

(Sect. 4.1.2) on the results that we obtain with our new global-to-local approach as well as with our former heuristic local approach.

6.1.1 New Global-To-Local Approach Versus Global Approach by Tsai et al.

In the following, we compare the global approach by Tsai et al. against our new global-to-local approach and our former local heuristic approach. For this purpose, we first describe the experimental setup and subsequently we will discuss the results.

6.1.1.1 Experimental Setup

As explained in Sect. 4.1.2.1, we are interested in extracting the combined outer bony boundary of the nasal cavity and the paranasal sinuses. The available training data consists of 49 hand-segmented two-dimensional CT slices of the paranasal sinuses. For the evaluation of our former heuristic local approach (Sect. 4.1.3.2), it has been described how these slices have been extracted from the database that has been introduced in Sect. 4.1.1. The procedure of the following evaluation is identical to Sect. 4.1.3.2: In a leave-one-out approach, we always use the information from $n = 48$ slices to train the shape models that are used in the segmentation of the remaining slice. We utilize the $m = 10$ most important eigenshapes in order to define the SSM subspace. The same eigenshapes are also used by our LDSSM (Chap. 3).

In order to apply our new global-to-local approach, we need to find a proper energy functional that describes the segmentation problem. We model the problem as a combination of a region-based energy $E_{\text{cv, mod.}}(\Phi_{\text{loc}}(\vec{\Psi}))$ and an edge-based energy $E_{\text{gac}}(\Phi_{\text{loc}}(\vec{\Psi}))$:

$$E(\Phi_{\text{loc}}(\vec{\Psi})) = E_{\text{gac}}(\Phi_{\text{loc}}(\vec{\Psi})) + E_{\text{cv, mod.}}(\Phi_{\text{loc}}(\vec{\Psi})) \,. \tag{6.1}$$

The region-based energy $E_{\text{cv, mod.}}(\Phi_{\text{loc}}(\vec{\Psi}))$ is a slightly modified version of the energy by Chan and Vese that has been given in Eq. (5.93). We define it as

$$E_{\text{cv, mod.}}(\Phi_{\text{loc}}(\vec{\Psi})) = \lambda_1 \int_{\Omega} H(I - c_1) H(-\Phi)\, d\vec{x} + \lambda_2 \int_{\Omega} [1 - H(I - c_2)] H(\Phi)\, d\vec{x} \,, \tag{6.2}$$

with $\lambda_1 = \lambda_2 = 0.5$, $c_1 = 200$, and $c_2 = -200$. I denotes the considered CT slice and H is the Heaviside step function. This energy reaches its minimum when bony regions ($I > 200$) are located outside the segmented region, and air-filled regions ($I < -200$) are located inside the segmented region, respectively.

With the help of Eq. (5.106) and the explanations for the energy by Chan and Vese (Sect. 5.4.5.3), we obtain the following functional derivative:

$$\frac{\partial E_{\text{cv, mod.}}(\Phi_{\text{loc}}(\vec{\Psi}))}{\partial \Psi_i(\vec{x})} = \delta_{\Phi_{\text{loc}}(\vec{\Psi}(\vec{x}))}\big[-\lambda_1 H(I-c_1)+\lambda_2[1-H(I-c_2)]\big]\tilde{\Phi}_i(\vec{x}) \,. \quad (6.3)$$

When we evolve the level set according to the negative functional gradient, a constant factor λ_1 is added to the zero level set when the moving contour is located in a bony region so that the moving contour is forced inwards, and a constant factor λ_2 is subtracted from the zero level set when the moving contour is located in an air-filled region so that the moving contour is forced outwards, respectively. So, the effect of this energy on the moving contour is roughly equivalent to the offset in the weight update target from our former heuristic approach (Eq. (4.18)).

Replacing the level set function Φ in the edge-based *Geodesic Active Contours* energy (Eq. (1.53)) by our LDSSM $\Phi_{\text{loc}}(\vec{\Psi})$ yields

$$E_{\text{gac}}(\Phi_{\text{loc}}(\vec{\Psi})) = \int_{\Omega} \delta_{\Phi_{\text{loc}}(\vec{\Psi})}\, g\, |\nabla \Phi_{\text{loc}}|\, d\vec{x} \,, \quad (6.4)$$

where g is an edge-indicator function as defined in Eq. (1.46) and $\delta_{\Phi_{\text{loc}}(\vec{\Psi})}$ is the Dirac delta function that is nonzero only at the zero-crossings of $\Phi_{\text{loc}}(\vec{\Psi})$. This energy reaches its minimum when the moving contour is located in image regions with large gradient magnitudes. With the help of Eqs. (1.54) and (5.106), we obtain the functional differential as

$$\frac{\partial E_{\text{gac}}(\Phi_{\text{loc}}(\vec{\Psi}))}{\partial \Psi_i(\vec{x})} = \delta_{\Phi_{\text{loc}}(\vec{\Psi}(\vec{x}))}\left[-\mu\left\langle \nabla g, \frac{\nabla \Phi_{\text{loc}}}{|\nabla \Phi_{\text{loc}}|}\right\rangle - \nu\, g\, \operatorname{div}\left(\frac{\nabla \Phi_{\text{loc}}}{|\nabla \Phi_{\text{loc}}|}\right)\right]\tilde{\Phi}_i(\vec{x}) \,, \quad (6.5)$$

where $\mu \in \mathbb{R}$ and $\nu \in \mathbb{R}$ represent scalar weighting factors that control the influence of the advection term and the curve shortening term, respectively (see also Eq. (1.57)). As we do not want to favor short curves over long curves, we set ν to zero. For μ we choose the value one so that we finally obtain

$$\frac{\partial E_{\text{gac}}(\Phi_{\text{loc}}(\vec{\Psi}))}{\partial \Psi_i(\vec{x})} = -\delta_{\Phi_{\text{loc}}(\vec{\Psi}(\vec{x}))}\left\langle \nabla g, \frac{\nabla \Phi_{\text{loc}}}{|\nabla \Phi_{\text{loc}}|}\right\rangle \tilde{\Phi}_i(\vec{x}) \quad (6.6)$$

as the functional derivative of our edge-based energy, and the functional derivative of our combined energy reads

$$\frac{\partial E(\Phi_{\text{loc}}(\vec{\Psi}))}{\partial \Psi_i(\vec{x})} = \delta_{\Phi_{\text{loc}}(\vec{\Psi}(\vec{x}))}\bigg[-\left\langle \nabla g, \frac{\nabla \Phi_{\text{loc}}}{|\nabla \Phi_{\text{loc}}|}\right\rangle - \lambda_1 H(I-c_1) + \lambda_2[1-H(I-c_2)]\bigg]\tilde{\Phi}_i(\vec{x}) \,. \quad (6.7)$$

Table 6.1 Stepsizes y^k for the two-dimensional global-to-local refinement, depending on the radius d of the neighborhood $\Theta_d(\vec{y})$

Radius d	512	256	128	64	32	16	8	4	2	1	0
Stepsize y^k	100	10	10	10	10	1	0.1	0.1	0.1	0.1	0.1

Now, having obtained the functional derivative of our energy, we can use Algorithm 5.2 (introduced in Sect. 5.5.3) in order to obtain our global-to-local segmentation result. In step 1 of Algorithm 5.2, the initial global weight vector $\vec{w}^{\text{init}}$ is set to $\vec{0}$ so that the weight fields Ψ_i are initialized to zero as well in step 2 of Algorithm 5.2 ($\Psi_i(\vec{y}) = 0, \forall \vec{y} \in \Omega, i = 1 \ldots m$). The initial value for the radius d is chosen to $d = 512$ (step 3 of Algorithm 5.2), and the Chebyshev distance (or chessboard distance) is used to define the neighborhood according to Eq. (5.126): The neighborhood $\Theta_d(\vec{y})$ contains all elements that are located in a square region with the side length $2d + 1$ centered around the currently considered point $\vec{y}$.

According to the explanations in Sect. 5.5.2, we obtain the global solution by Tsai et al. when iterating Eq. (5.130) until convergence for the initial radius $d = 512$ (step 4 of Algorithm 5.2). Further reducing the radius (step 5 of Algorithm 5.2) yields our global-to-local refinement of the initial global solution. The stepsize y^k is thereby adjusted according to the radius d of the neighborhood $\Theta_d(\vec{y})$. The larger the radius d, the more robust the steepest descent scheme. So, larger stepsizes can be used for large radii d. The used stepsizes are shown in Table 6.1.

The gradients in Eq. (6.7) are numerically approximated via central differences and the Laplace operator in Eq. (5.130) is approximated via second order central differences [75], respectively. The Dirac delta function is approximated via the following smooth function [78] with $\epsilon = 1.5$:

$$\delta_x\epsilon = \begin{cases} 0 & \text{, if } |x| > \epsilon \\ \frac{1}{2\epsilon}[1 + \cos(\frac{\pi x}{\epsilon})] & \text{, if } |x| \leq \epsilon \,. \end{cases} \tag{6.8}$$

At the initial radius of $d = 512$, we employ the difference between consecutive level sets as the termination criterion for our level set evolution in order to obtain good and stable results for the global initial solution. This means that we terminate the iteration of Eq. (5.130) when the following condition is fulfilled:

$$||\Phi_{\text{loc}}(\vec{\Psi}^k) - \Phi_{\text{loc}}(\vec{\Psi}^{k-1})|| < 0.04 \tag{6.9}$$

For the radii $d = 256$ down to $d = 1$ we use a fixed number of 50 iterations as termination criterion and for the radius $d = 0$, i.e. for the completely local adaptation, we use 100 iterations, respectively.

In order to ensure stability according to Eq. (5.107) (with $\Delta x = 1$), we choose a value of $\beta = 0.2/y^k$ as weighting factor for the smoothness term in Eq. (5.130), and in consistency with the original approach by Tsai et al., we use $\alpha = 0$ as weighting factor for the shape term in Eq. (5.130), respectively. The parameters for our former local heuristic approach are chosen exactly as in Sect. 4.1.3.2. The only difference is that we use the mean shape as initial solution which means that we set all elements of the initial field of weight vectors to zero as for the above mentioned approaches. This is done in order to remove the influence of our global initial solution from the segmentation result. The influence of our global initial solution will be separately evaluated in Sect. 6.1.2.

6.1.1.2 Results

Some exemplary segmentation results, obtained with the global approach by Tsai et al. [129] and our new global-to-local approach, respectively, are shown in Fig. 6.1. Segmentation results for all datasets are shown in Appendix B. The corresponding resulting errors, together with the errors obtained with our former heuristic local approach, are depicted in Figs. 6.2 and 6.3. Additional boxplots for all 4 errors are shown in Fig. 6.4. From the errors can be seen that our new global-to-local segmentation approach is able to deliver more accurate segmentation results than the global approach by Tsai et al. We were able to achieve a lower root-mean-square deviation in 44 datasets, a lower Hausdorff distance in 36 datasets, and lower Jaccard and Dice distances in 43 datasets.

The mean errors over all 49 datasets and the corresponding standard deviations are shown in Table 6.2. The table shows that our new global-to-local segmentation approach outperforms the global approach by Tsai et al. in all four error measures. The average root-mean-square deviation is 32.5% lower, the average Hausdorff distance is 27.1% lower, the average Jaccard distance is 24.2% lower, and the average Dice distance is 25.3% lower. Similar improvements can be observed for the median errors which are given in Fig. 6.4 and Table 6.3. Like in Sect. 4.1.3 the statistical significance of this finding has been confirmed by using the left-sided Wilcoxon signed-rank test [138].

The null hypothesis that the population median of the left samples (our global-to-local segmentation results) is greater or equal to the population median of the right samples (the global segmentation results by Tsai et al.) can be rejected for all four error measures at the commonly used significance level of 0.05 (see last row of Table 6.3).

The best three results of our new global-to-local approach, according to the root-mean-square deviation and the Hausdorff distance, are obtained for datasets 18, 35, and 29 (see Fig. 6.1). One can nicely see how the global-to-local result almost perfectly reproduces the hand-segmented reference. For the Jaccard and Dice distances also dataset 33 provides very good error values. One can see that the segmentation is also almost perfect except for a small segment in the lower right of the dataset. On the other hand, the worst results are achieved for datasets 20, 45, and 2 (root-mean-square deviation and Hausdorff distance) and datasets 49, 16, and 2 (Jaccard and Dice distances). Dataset 20 is atypical due to an isolated frontal sinuses area in the upper left and dataset 45 is strongly deformed. This deformation leads to a poor global solution with distinctive frontal sinuses from which our global-to-local

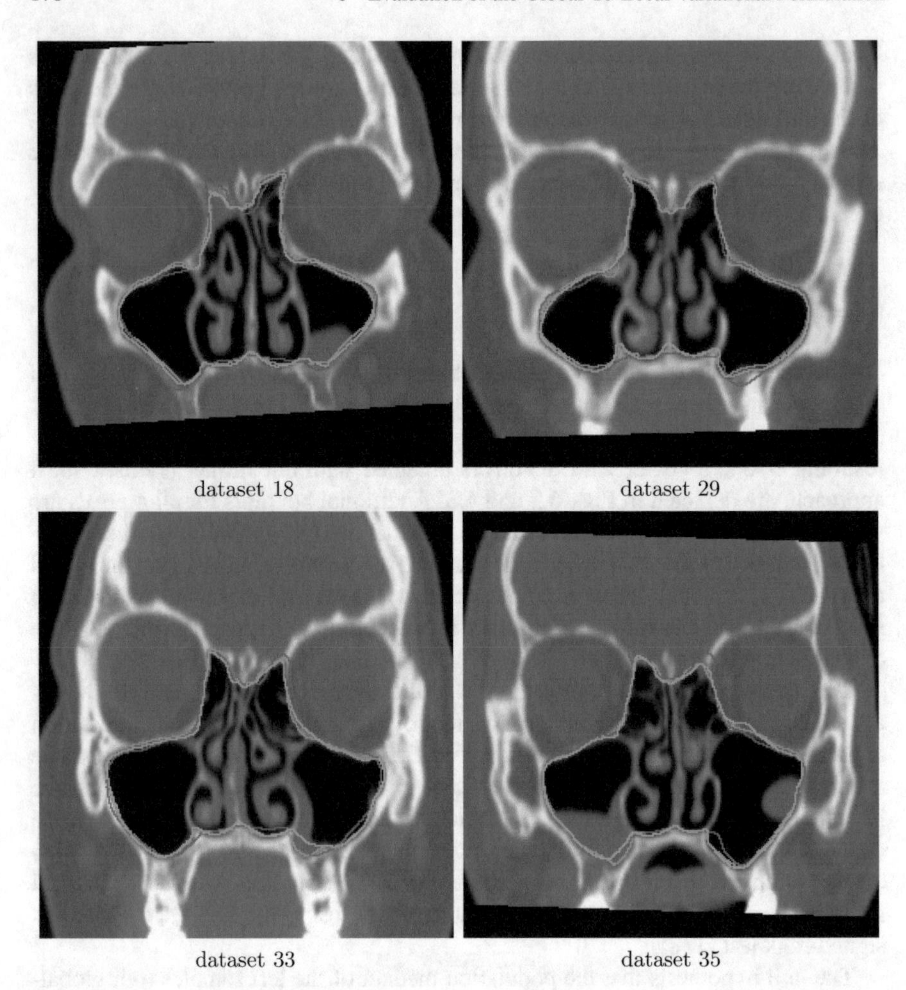

Fig. 6.1 Some exemplary results that show the advantage of our new global-to-local approach (*cyan line*) over the global approach by Tsai et al. (*magenta line*). The hand-segmented reference contour is shown as a *green line*

approach is not able to recover. For the other datasets (2, 16, and 49), it sticks out that the paranasal sinuses are severely inflamed. In the inflamed areas, no reliable image information is available over a wide range so that the global-to-local result closely resembles the global solution in these regions.

The largest improvements of our new global-to-local approach in comparison to the global approach by Tsai et al. can be seen in datasets 10, 43, and 48 for the root-mean-square deviation and for the Hausdorff distance, and in datasets 10, 5, and 3 for the Jaccard and Dice distances, respectively. In datasets 10, 43, and 48, the improvement is mostly due to the distinctive frontal sinuses which have been better

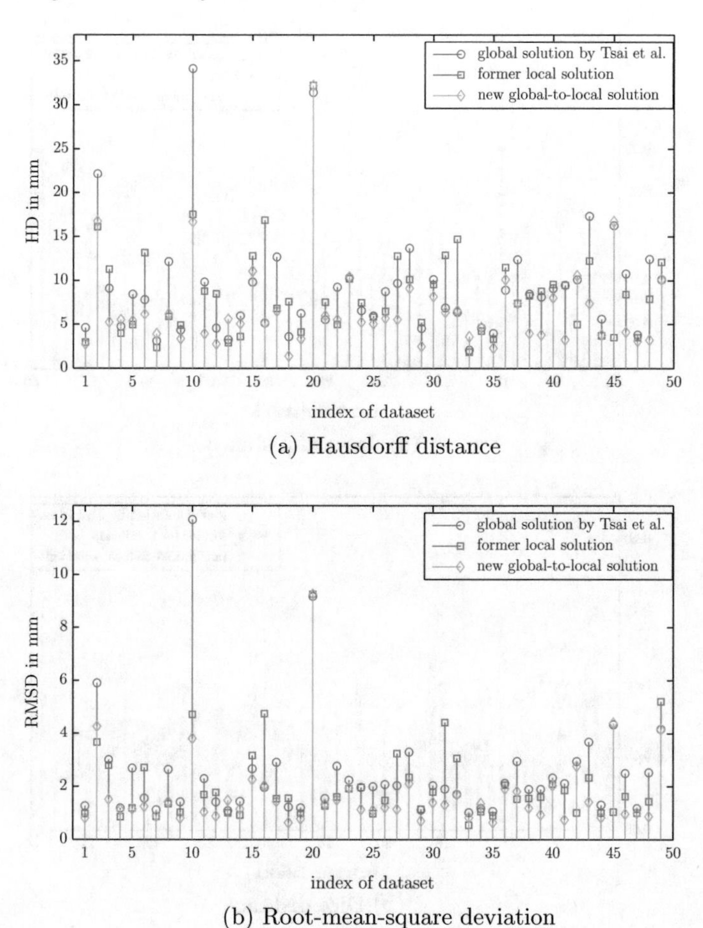

(a) Hausdorff distance

(b) Root-mean-square deviation

Fig. 6.2 Resulting Hausdorff distances (**a**) and root-mean-square deviations (**b**) obtained with the global approach by Tsai et al. (*magenta circles*), our former heuristic local approach (*red squares*), and our new global-to-local approach (*cyan diamonds*), respectively. The results are plotted against each dataset

approximated by our new global-to-local approach than by the global approach of Tsai et al.

Dataset 5 is not perfectly aligned which causes a big problem to the global solution. However, it can be seen that our global-to-local approach is even capable to deal with such difficult situations. For dataset 3, we have already found out in Sect. 4.1.3 that the overall shape seems to differ greatly from the other datasets so that no satisfactory solution can be obtained with a global model. Nevertheless, our new global-to-local approach handles also this problem very well.

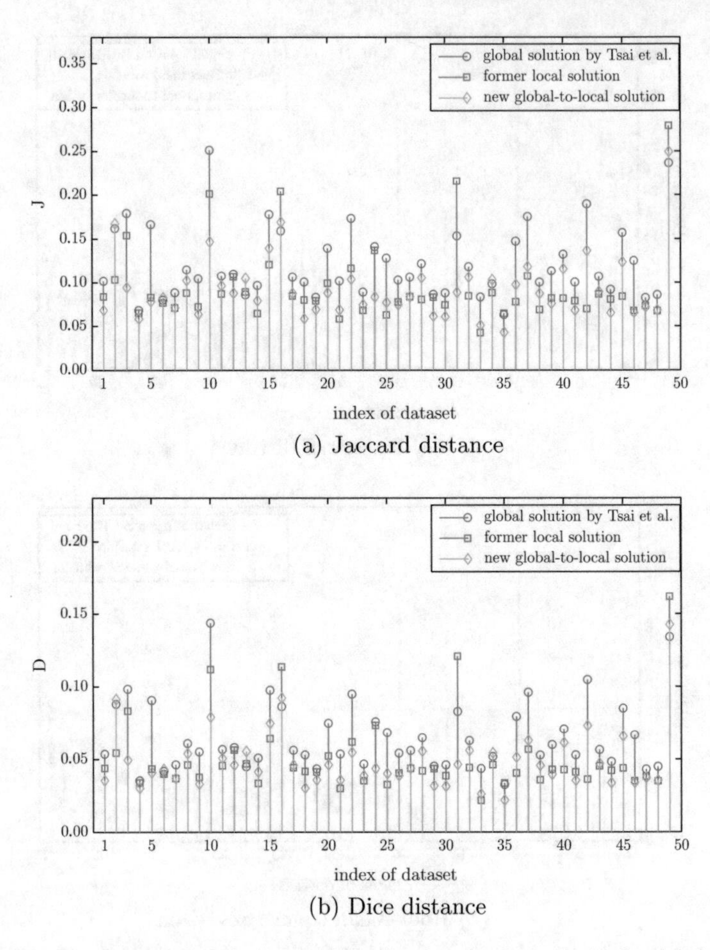

Fig. 6.3 Resulting Jaccard distances (**a**) and Dice distances (**b**) obtained with the global approach by Tsai et al. (*magenta circles*), our former heuristic local approach (*red squares*), and our new global-to-local approach (*cyan diamonds*), respectively. The results are plotted against each dataset

What can also be seen from the error plots in Figs. 6.2, 6.3, and 6.4 is that our new global-to-local segmentation approach is also able to improve the results of our former heuristic local approach. Table 6.2 shows that on average our new global-to-local approach segmentation approach outperforms our former heuristic local approach for the root-mean-square deviation and the Hausdorff distance. The average root-mean-square deviation is 17.5% lower and the average Hausdorff distance is 20.3% lower. Again, similar improvements can be observed for the median errors in Fig. 6.4 and Table 6.3. No significant improvements can be observed for the Jaccard distance and for the Dice distance. The mean distances are approximately equal for our new global-to-local approach and our former heuristic local approach. The statistical significance of these findings is again supported by the ρ-values of the

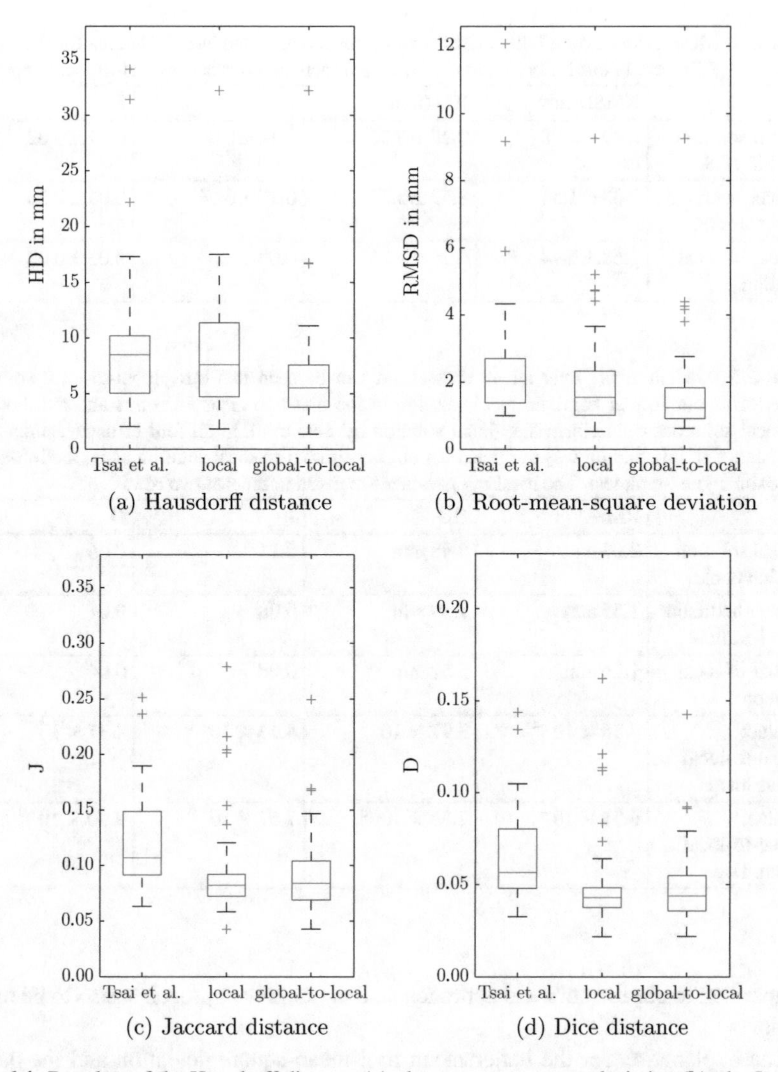

Fig. 6.4 Boxplots of the Hausdorff distance (**a**), the root-mean-square deviation (**b**), the Jaccard distance (**c**), and the Dice distance (**d**), respectively. The whiskers represent the highest/lowest datum within 1.5 times the interquartile range. All remaining data values are considered as outliers and given as *red crosses*. Subfigures **a**, **b** © [2014] IEEE. Reprinted, with permission, from [2]

left-sided Wilcoxon signed-rank test in the second to last row of Table 6.3: At a significance level of 0.05, the null hypothesis can be rejected for the root-mean-square deviation and the Hausdorff distance, but not for the Jaccard and Dice distances, respectively. However, what can be seen from the boxplots in Fig. 6.4c, d is that the number of outliers could be reduced for our new global-to-local approach in

Table 6.2 Mean errors over all 49 datasets and corresponding standard deviations for the global approach by Tsai et al., our former heuristic local approach, and our new global-to-local approach

	RMSD (mm)	HD (mm)	J	D
Global solution by Tsai et al.	2.49 ± 1.98	9.28 ± 6.30	0.12 ± 0.04	0.07 ± 0.02
Former heuristic local solution	2.03 ± 1.54	8.49 ± 5.31	0.10 ± 0.04	0.05 ± 0.03
Global-to-local solution	1.68 ± 1.44	6.76 ± 5.15	0.09 ± 0.04	0.05 ± 0.02

Table 6.3 Median errors over all 49 datasets. It can be seen that our global-to-local solution outperforms our former heuristic local solution in the first two error measures and that both of our local solutions outperform the global solution by Tsai et al. in all four error measures. The significance of this finding was evaluated for our new global-to-local solution using the left-sided Wilcoxon signed-rank test. The resulting ρ-values are given in the last two rows

	RMSD	HD	J	D
Global solution by Tsai et al.	2.03 mm	8.48 mm	0.11	0.06
Former heuristic local solution	1.55 mm	7.59 mm	0.08	0.04
Global-to-local solution	1.24 mm	5.52 mm	0.08	0.04
ρ-value: global-to-local versus local	7.36×10^{-4}	3.97×10^{-4}	4.13×10^{-1}	4.37×10^{-1}
ρ-value: global-to-local versus Tsai	6.35×10^{-9}	2.36×10^{-6}	7.57×10^{-9}	9.00×10^{-9}

comparison to our former local approach so that our new approach seems to be more stable.

The explanation for the better mean root-mean-square deviation and the better mean Hausdorff distance can be seen in the exemplary segmentation results that are depicted in Fig. 6.5. The examples show that our new global-to-local approach is superior to our former heuristic local approach especially in regions with weak or no gradient information and in regions with thin tubular structures. A region with weak gradient information is highlighted in Fig. 6.5 by a blue circle. While our former heuristic local approach overfits the data under consideration, our new global-to-local approach sticks closer to the correct solution in regions where only unsatisfactory image information is available. The reason for the overfitting of the local approach is that the contour is forced downwards by the strong gradients at both sides of the uncertain region inside the blue circle. The contour at both sides of the uncertain

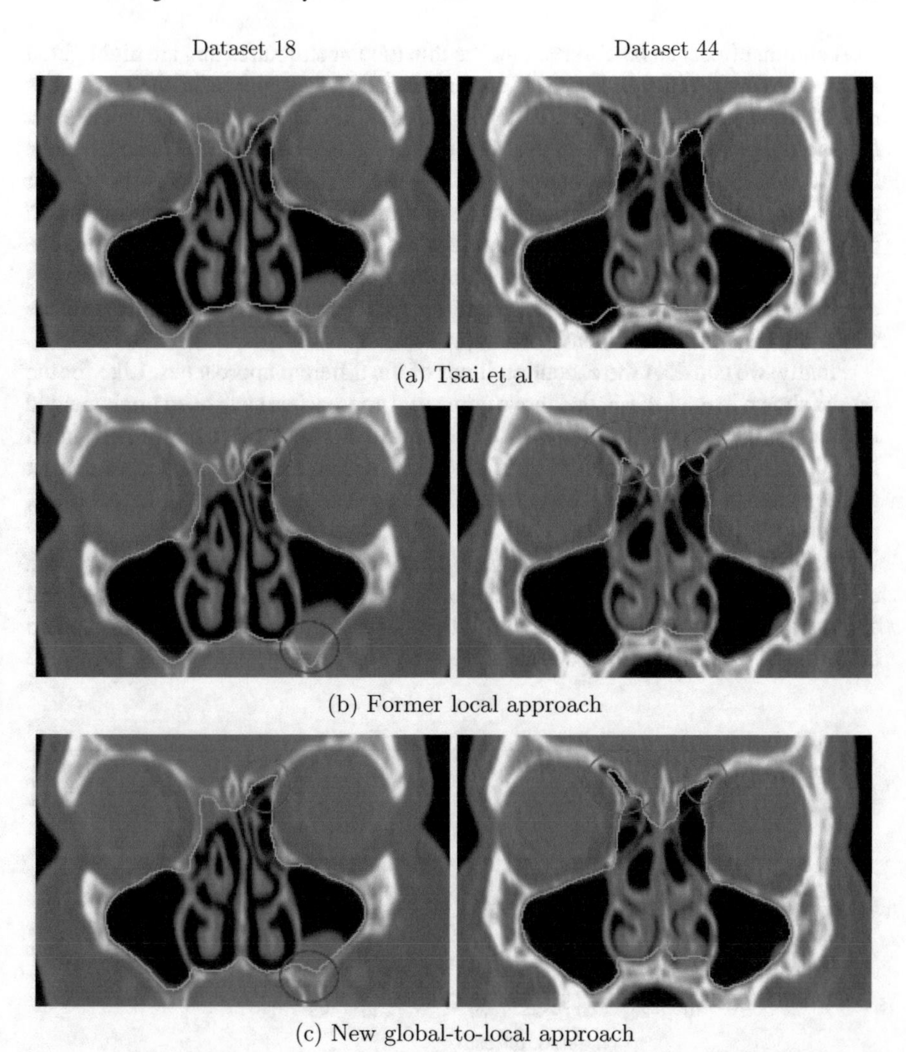

Fig. 6.5 Segmentation results for two exemplary datasets, obtained with the global approach by Tsai et al. (**a**), our former heuristic local approach (**b**), and our new global-to-local approach (**c**). The improvements are highlighted by *circles*. © [2014] IEEE. Reprinted, with permission, from [2]

region adapts to the strong vertical gradients and the remaining contour follows due to the smoothing of the resulting weight fields in line 9 of Algorithm 3.2. In contrast to this, when using our new formulation from Sect. 5.5, a mean descent direction is obtained within the currently considered window. When using this mean descent direction to update the weight fields, erroneous gradients have less influence on the direction of the contour evolution so that a better segmentation result is obtained.

A similar effect can be observed for the thin tubular structures that are highlighted by red circles in Fig. 6.5. For our former heuristic local approach, the contour is mostly attracted by the gradients at the edges of the thin structures which exert forces in opposite directions so that the subsequent smoothing step cancels out the local weight updates and the contour is not able to enter the frontal sinuses. In contrast to this, when using our new formulation, one obtains a mean descent direction which pulls the local contour inside the frontal sinuses. This small but important difference between our former heuristic local solution and our new global-to-local solution has a strong impact on the root-mean-square deviation and on the Hausdorff distance which explains the above mentioned improvements.

Finally, we consider the execution times of the different approaches. Like for the results shown in Sect. 4.1.3, the above compared approaches have been implemented in C++ (single-threaded program) and the evaluation has been performed on an Intel Core2Quad CPU with 2.83 GHz clock speed. The execution times per dataset are depicted in Fig. 6.6 and the average execution times are shown in Table 6.4, respectively.

The time difference between our new global-to-local approach and the approach by Tsai et al., i.e. the additional time that is needed by our new global-to-local approach in order to refine the global solution, is on average 169.38 s with a standard deviation of 4.49 s (dashed cyan line in Fig. 6.6).

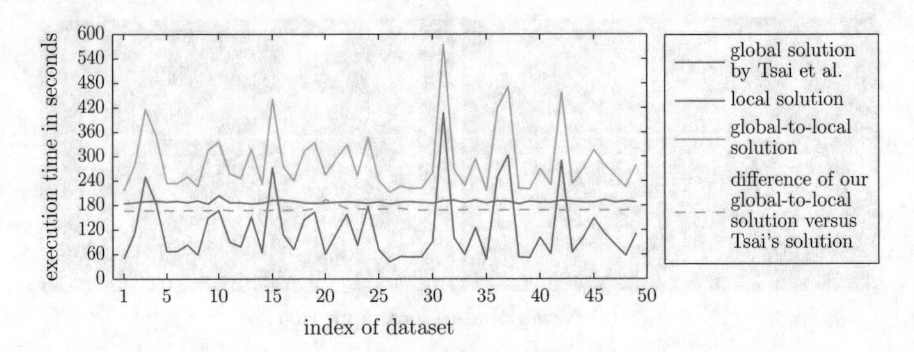

Fig. 6.6 Execution times per dataset for the global approach by Tsai et al. (*magenta line*), our local approach (*red line*), and our new global-to-local approach (*cyan line*). The additional cost of the global-to-local adaptation, in comparison to the initial global adaptation, is shown as a *dashed cyan line*

Table 6.4 Average execution time and standard deviation per dataset for the global approach by Tsai et al., our local approach, and our new global-to-local approach

Global solution by Tsai et al.	118.48 s ± 79.02 s
Former heuristic local solution	188.31 s ± 3.40 s
New global-to-local solution	287.86 s ± 78.69 s

What stands out in particular is that our new global-to-local approach is not slower than our former local approach when we neglect the global initial solution (dashed cyan line in Fig. 6.6, compare also with the results from Sect. 6.1.2.2). In fact, our new global-to-local approach is on average 18.93 s (approximately 10%) faster than our former local approach even though it has been iterated for 550 iterations in contrast to 500 iterations for the local approach (10% increase). This is most probably due to the fact that our new global-to-local approach circumvents the computationally intensive computation of the outer product in the Cauchy step size from Eq. (3.19) which is needed for our former local approach but not for our new global-to-local approach.

Once again, it should be mentioned that the runtime of our former local approach can be reduced significantly when the local weight update loop in lines 5–8 of Algorithm 3.2 would be parallelized. The same is true for our new global-to-local approach as the functional gradient from Eq. (5.106) can be computed element-wise before it is combined to the functional differential by the integral in Eq. (5.129). Furthermore, for the considered task, an increase in accuracy has more value than a short execution time.

6.1.2 New Global-To-Local Approach Versus Former Segmentation Framework

In a second setup, we want to evaluate the influence of our global initial solution from Sect. 4.1.2 on the results that we can obtain with our new global-to-local approach. For this purpose, we compare our new global-to-local approach against our former local segmentation framework.

6.1.2.1 Experimental Setup

For this experiment, the experimental setup for our former heuristic local approach is almost identical to the setup that has been described in Sect. 6.1.1.1. However, as mentioned above, we want to evaluate the influence of our global initial solution from Sect. 4.1.2 on the segmentation results. So, the results for our former global approach and our former local heuristic approach are obtained exactly as described in Sect. 4.1.3.2. This means in particular that we use our former global solution in order to initialize our former heuristic local approach.

The results for our new global-to-local approach are obtained as explained in Sect. 6.1.1.1 but as well with one important difference: We use the outcome of our former global approach not only in order to initialize our former heuristic local approach but also our new global-to-local approach. This means that we set the initial global weight vector $\vec{w}^{\text{init}}$ in step 1 of Algorithm 5.2 to the outcome of our former global approach. Consequently, we additionally reduce the initial radius d

that defines the neighborhood $\Theta_d(\vec{y})$ to $d = 128$ so that in total 500 iterations are needed for each dataset in order to obtain the solution. All other parameters of our global-to-local approach remain unchanged.

6.1.2.2 Results

The segmentation errors obtained with our former heuristic local approach and our new global-to-local approach are depicted in Figs. 6.7 and 6.8. The corresponding segmentation results are shown in Appendix C. Additional boxplots for all 4 error

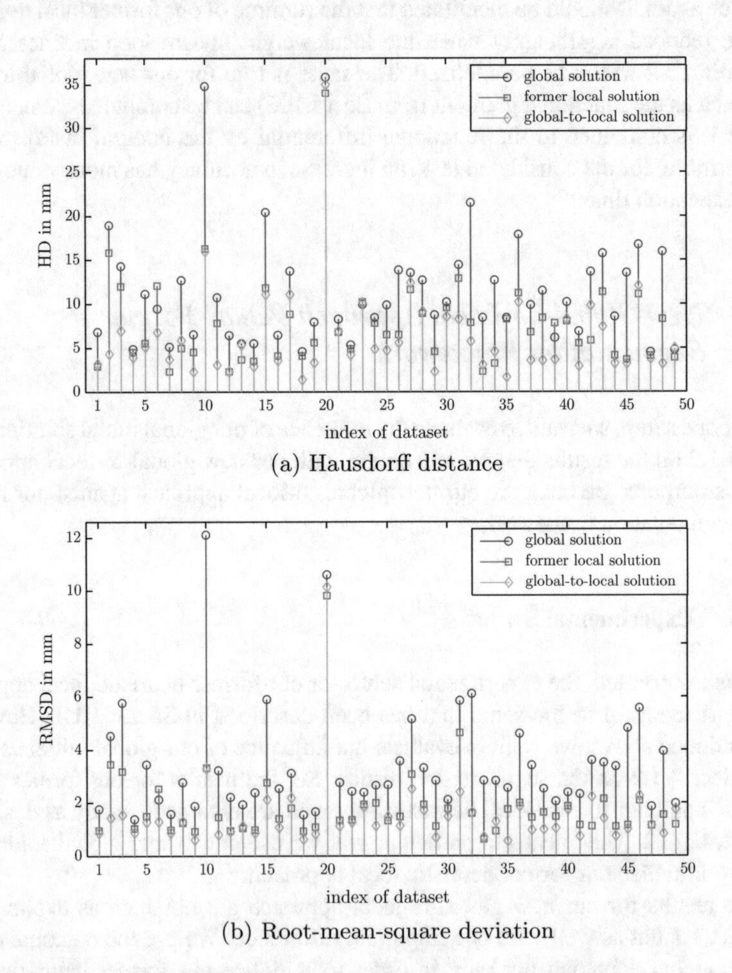

Fig. 6.7 Resulting Hausdorff distances (**a**) and root-mean-square deviations (**b**) obtained with our former global approach (*blue circles*), our former heuristic local approach (*red squares*), and our new global-to-local approach (*cyan diamonds*), respectively. The results are plotted against each dataset

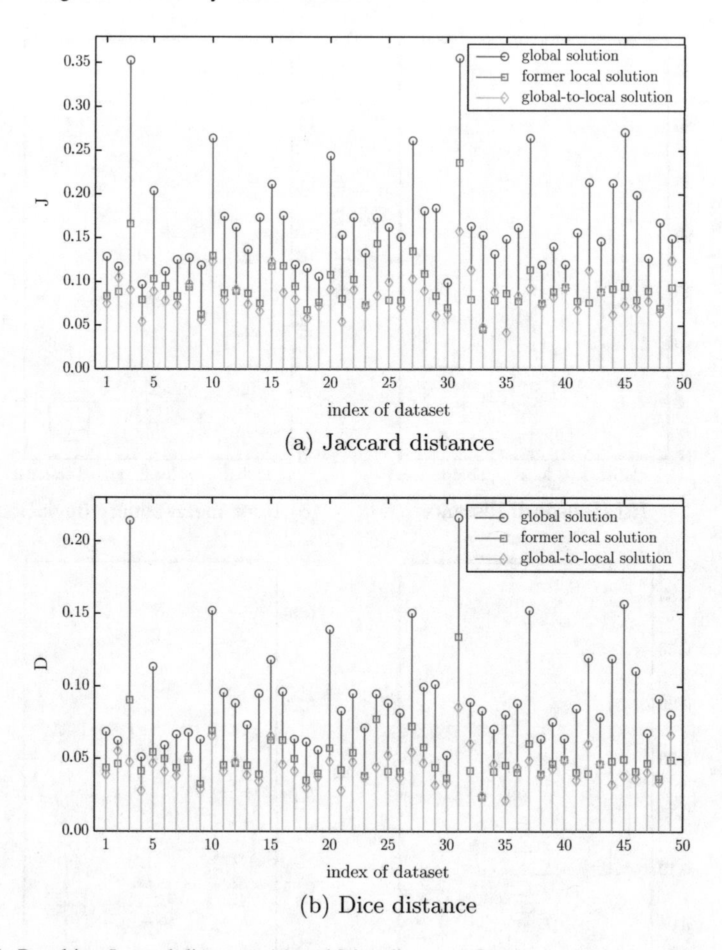

Fig. 6.8 Resulting Jaccard distances (**a**) and Dice distances (**b**) obtained with our former global approach (*blue circles*), our former heuristic local approach (*red squares*), and our new global-to-local approach (*cyan diamonds*), respectively. The results are plotted against each dataset

measures are shown in Fig. 6.9. For comparison, we included the results of our initial global solution in the quantitative evaluation.

When we compare our former heuristic local approach and our new global-to-local approach, we can see that even when we use the outcome of our former global approach as initial solution for both approaches, our new global-to-local approach is able to deliver more accurate segmentation results than our former local approach: We obtained a lower Hausdorff distance in 33 datasets. A lower root-mean-square

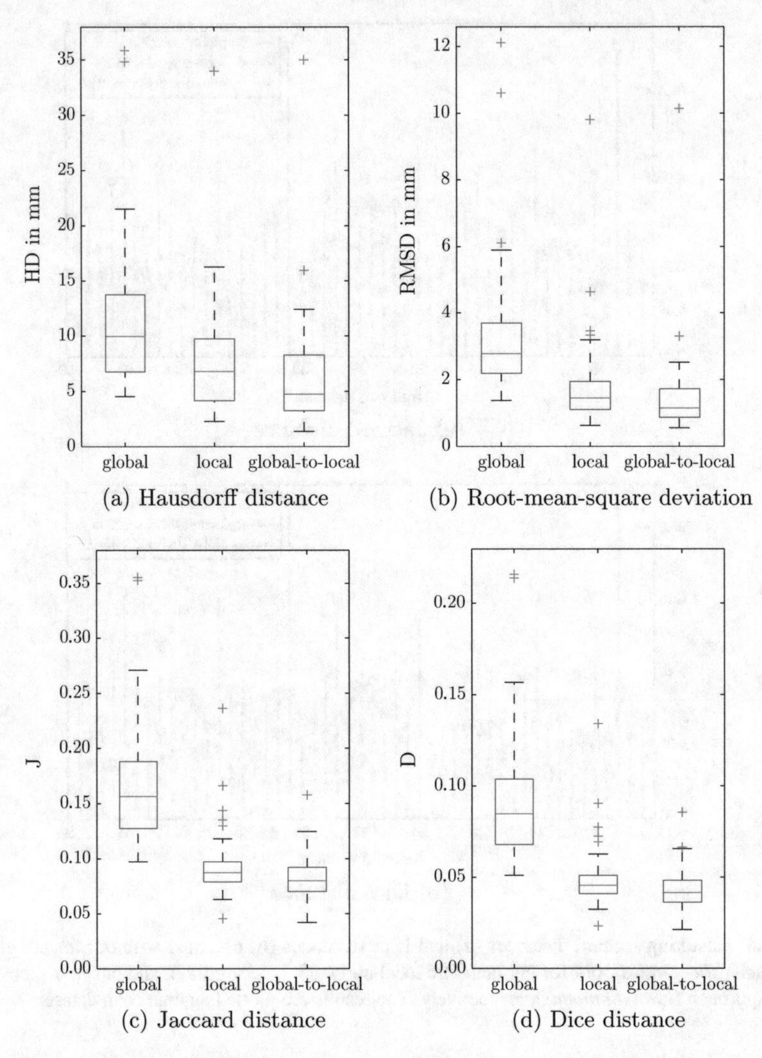

Fig. 6.9 Boxplots of the Hausdorff distance (**a**), the root-mean-square deviation (**b**), the Jaccard distance (**c**), and the Dice distance (**d**), respectively. The whiskers represent the highest/lowest datum within 1.5 times the interquartile range. All remaining data values are considered as outliers and given as *red crosses*

deviation and lower Jaccard and Dice distances have been achieved in 37 datasets. The mean errors over all 49 datasets and the corresponding standard deviations are shown in Table 6.5. It can be seen that our new global-to-local segmentation approach outperforms our former heuristic local approach in all four error measures. The average root-mean-square deviation is 17.4% lower, the average Hausdorff distance is 18.3% lower, the average Jaccard distance is 11.5% lower, and the average Dice

Table 6.5 Mean errors over all 49 datasets and corresponding standard deviations for our former global approach, our former heuristic local approach, and our new global-to-local approach

	RMSD (mm)	HD (mm)	J	D
Former global solution	3.44 ± 2.08	11.52 ± 6.59	0.17 ± 0.06	0.09 ± 0.04
Former heuristic local solution	1.82 ± 1.41	7.74 ± 5.18	0.09 ± 0.03	0.05 ± 0.02
Global-to-local solution	1.50 ± 1.39	6.32 ± 5.30	0.08 ± 0.02	0.04 ± 0.01

Table 6.6 Median errors over all 49 datasets. It can be seen that our new global-to-local solution outperforms our former heuristic local solution and our former global solution in all four error measures. The significance of this finding was evaluated using the left-sided Wilcoxon signed-rank test. The resulting ρ-values are given in the last two rows

	RMSD	HD	J	D
Former global solution	2.78 mm	9.95 mm	0.16	0.09
Former heuristic local solution	1.45 mm	6.79 mm	0.09	0.05
Global-to-local solution	1.15 mm	4.74 mm	0.08	0.04
ρ-value: glob.-to-local versus local	5.45×10^{-5}	6.51×10^{-4}	1.85×10^{-4}	1.88×10^{-4}
ρ-value: glob.-to-local versus global	5.72×10^{-10}	6.90×10^{-10}	5.72×10^{-10}	5.72×10^{-10}

distance is 11.2% lower. Similar improvements can be observed for the median errors which are given in Fig. 6.9 and Table 6.6.

The statistical significance of these findings has again been confirmed by using the left-sided Wilcoxon signed-rank test [138]. The ρ-values are given in the second to last row of Table 6.6. The null hypothesis that the population median of the left samples (our new global-to-local results) is greater or equal to the population median of the right samples (our former local segmentation results) can be rejected for all four error measures at the commonly used significance level of 0.05.

Now that we have compared our individual approaches with each other, we want to evaluate how the results have changed in comparison to the previous results from Sect. 6.1.1. In summary, we can say that our global approach is inferior to the global approach by Tsai et al. For example, Tables 6.2 and 6.5 show that the average root-mean square deviation is 38.4% worse, the average Hausdorff distance is 24.2% worse, the average Jaccard distance is 39.4% worse, and the average Dice distance is 43.8% worse for our global approach in comparison to the global approach by Tsai et al. This is most likely due to the fact that our global approach makes no use of the property that air-filled regions are most probably located inside the paranasal sinuses.

However, what can also be seen is that using our global solution as initial solution for our former heuristic local approach as well as our new global-to-local approach, improves the results for both approaches. The combination of our former heuristic local approach with our former global approach as initial solution improves the results from Sect. 6.1.1 by 10.5% for the average root-mean-square deviation and 8.8% for the average Hausdorff distance. The average Jaccard and Dice distances improve only marginally by 1.2%, respectively 1.9%, each. The combination of our new global-to-local approach with our former global approach as initial solution improves the average results from Sect. 6.1.1 by 10.4% for the root-mean-square deviation, 6.6% for the Hausdorff distance, 10.1% for the Jaccard distance, and 10.9% for the Dice distance, respectively.

We can see that the improvements in the root-mean-square deviation and the Hausdorff distance are approximately equal for our former heuristic local approach and our new global-to-local approach. However, it is striking that the results for the Jaccard and Dice distances improve much more for our new global-to-local approach in comparison to our former heuristic local approach when we use the outcome of our former global approach as initial solution instead of the mean shape (about 10% improvement versus virtually no improvement). The most improvement in the results for the Jaccard and Dice distances can be observed in datasets 49, 16, 2, 45, 13, and 37. All these datasets have in common that the global approach by Tsai et al. resulted in a global solution where large parts of the contour are located in regions where no reliable image information is available because of severely inflamed paranasal sinuses, or where a strong deviation between the global solution and the hand-segmented reference exists (dataset 45).

So, the most likely explanation for the improved performance of our new global-to-local approach with regard to the Jaccard and Dice distances is that our former global solution is designed in such a way that it reaches its minimum only when large parts of the contour that determines the segmentation result are located on the edges of the input data (c.f. Sect. 4.1.2). Thus, large parts of the contour are located in areas with reliable image information. This favors our global-to-local adaptation, as reliable mean descent directions can be determined based on the gradients of adjacent edges.

Like in the previous experiments, the approaches have all been implemented in C++ and the evaluation has been performed on an Intel Core2Quad CPU with 2.83 GHz clock speed.

The execution times for each dataset are depicted in Fig. 6.10 and the average execution times are shown in Table 6.7, respectively. When comparing the execution times from Table 6.7 with the execution times from Table 6.4, it can be seen that our former global approach is approximately 13 times (one order of magnitude) faster than the global approach by Tsai et al.

The runtime of our former heuristic local approach increases only insignificantly when our former global approach is used to provide the initial solution (2.54% execution time increase in comparison to Sect. 6.1.1.2). This is because our former

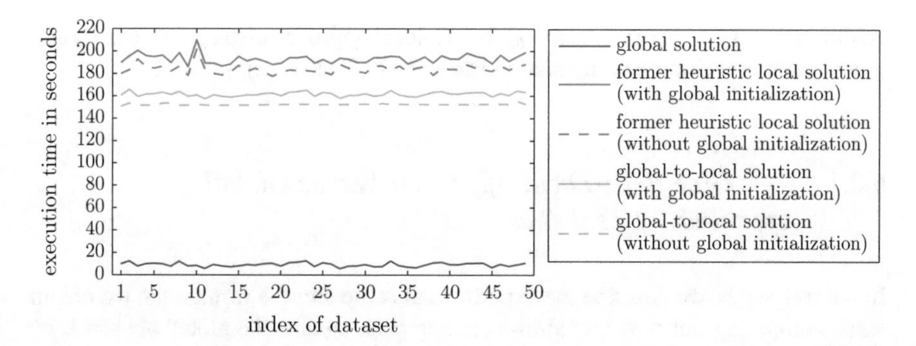

Fig. 6.10 Execution times per dataset for our former global approach (*blue line*), our former heuristic local approach (*red line*), and our new global-to-local approach (*cyan line*). The additional cost of the local adaptation or the global-to-local adaptation, in addition to the initial global solution, is shown as a *dashed red line* or a *dashed cyan line*, respectively

Table 6.7 Average execution time and standard deviation per dataset for our global approach, our former local approach, and our new global-to-local approach

Former global solution	$9.16\,\text{s} \pm 1.75\,\text{s}$
Former heuristic local solution	$193.08\,\text{s} \pm 4.19\,\text{s}$
New global-to-local solution	$161.59\,\text{s} \pm 1.92\,\text{s}$

global approach is also about one order of magnitude faster than our former heuristic local approach (c.f. Sect. 4.1.3.3).

The execution time of our new global-to-local approach decreases when the outcome of our former global approach is used as initial solution (43.41% execution time reduction in comparison to Sect. 6.1.1.2). This is because in our global-to-local solution from Sect. 6.1.1.2, most iterations are used to obtain an as good as possible initial global solution. When our former global solution is used for initialization, significantly fewer iterations are needed.

6.2 Extracting the Nasal Cavity and the Paranasal Sinuses from Three-Dimensional CT Data

Now that we have shown the potential of our new global-to-local variational formulation in a two-dimensional segmentation task, we go one step further and extract the combined outer bony boundary of the nasal cavity and the paranasal sinuses from three-dimensional CT data as well.

For this purpose, we compare the results of our new global-to-local approach against the results obtained with the global approach by Tsai et al. (Sect. 6.2.1).

Additionally, we compare our new global-to-local approach also against our former global paranasal sinuses segmentation framework (Sect. 6.2.2).

6.2.1 New Global-To-Local Approach Versus Global Approach by Tsai et al.

In the following, we describe the experimental setup and the results that we obtain when comparing our new global-to-local approach against the global approach by Tsai et al. for the above-mentioned three-dimensional segmentation task.

6.2.1.1 Experimental Setup

Identical to the experimental setups explained in Sects. 6.1 and 4.1, we are interested in extracting the combined outer bony boundary of the nasal cavity and the paranasal sinuses. However, this time we do not extract two-dimensional CT slices from the database that has been introduced in Sect. 4.1.1. Instead, we extract a three-dimensional volume of interest from all datasets which incorporates the full three-dimensional paranasal sinuses information. For this purpose, we again keep the first CT dataset fixed and rigidly align the remaining CT datasets to the first dataset with regard to rotation, translation, and scale by using a similarity transformation.

From the now rigidly aligned datasets, we identify a common volume of interest with the help of the bounding box that encloses the hand-segmented paranasal sinuses from all datasets. Additionally, a safety margin of 20 voxels on each side of the bounding box is included in the volume of interest so that the segmentations are located in a sufficient distance from the borders of the volume of interest. The thus defined volume of interest has a resolution of $256 \times 242 \times 312$ voxels with a uniform voxel spacing of 0.46 mm. The voxels of the individual datasets are transformed to this volume of interest via trilinear interpolation. Those voxels in the volume of interests that have no counterpart in the original data volume are replaced by air (Hounsfield value -1000). After these steps, we obtain 48 volumes with identical resolution that contain the CT data of the rigidly-aligned paranasal sinuses and another 48 volumes that contain the corresponding hand-segmentations, respectively.[1]

The protocol of the experimental evaluation is equivalent to the experiments in Sects. 6.1 and 4.1: In a leave-one-out approach, we always use the information from $n = 47$ hand-segmentations to generate the shape models that are used in the segmentation of the remaining dataset. Like for the two-dimensional evaluation, we utilize the $m = 10$ most important eigenshapes in order to define the SSM subspace. The resulting eigenshapes are also used by our LDSSM (Chap. 3).

---

[1] We use 48 datasets in contrast to 49 datasets that have been used in the previous experiments as one dataset contains truncated frontal sinuses. This dataset is not considered in the three-dimensional evaluation.

Table 6.8 Stepsizes y^k for the three-dimensional global-to-local refinement, depending on the radius d of the neighborhood $\Theta_d(\vec{y})$

Radius d	512	256	128	64	32	16	8	4	2	1	0
Stepsize y^k	100	10	10	10	5	5	1	1	0.5	0.5	0.1

Since the segmentation problem is the same as in Sect. 6.1 – apart from the higher dimension – we use the same energy functional in order to describe the three-dimensional segmentation problem that has already been defined in Eq. (6.1) to describe the two-dimensional segmentation problem. The functional derivative of this energy has been given in Eq. (6.7). We also choose the same parameters $\lambda_1 = \lambda_2 = 0.5$, $c_1 = 200$, $c_2 = -200$, and $\alpha = 0$. Only the weighting factor of the smoothness term is modified to $\beta = 0.1$ in order to ensure stability according to Eq. (5.107) and to account for the increased complexity of the three-dimensional segmentation problem in comparison to the two-dimensional segmentation problem.

Now, with a mathematical description of the segmentation problem at hand, we use our global-to-local segmentation approach (Algorithm 5.2) in order to obtain the segmentation result. Like in Sect. 6.1, the initial global weight vector $\vec{w}^{\text{init}}$ is chosen to $\vec{0}$ so that all weight fields Ψ_i are initialized to zero as well, and we use the Chebyshev distance (or chessboard distance) in order to define the neighborhood $\Theta_d(\vec{y})$ according to Eq. (5.126). The initial value for the radius d is again chosen to $d = 512$ and we also choose the stepsize y^k of the steepest descent scheme depending on the value of the radius. The incorporated stepsizes are shown in Table 6.8. The gradients in Eq. (6.7) are numerically approximated via central differences, the Laplace operator in Eq. (5.130) is approximated via second order central differences, and the Dirac delta function is approximated via Eq. (6.8) (with $\epsilon = 1.5$), respectively. At the initial radius of $d = 512$, we terminate the level set evolution when the difference between consecutive level sets is less than 0.04 (Eq. (6.9)), for the radii $d = 256$ down to $d = 1$ we use a fixed number of 50 iterations, and for the radius $d = 0$ we use 100 iterations, respectively.

6.2.1.2 Results

The segmentation errors, obtained with the global approach by Tsai et al. and our new global-to-local approach, are shown in Figs. 6.11 and 6.12. Additional boxplots for all four errors are shown in Fig. 6.13. Similar to the two-dimensional evaluation, it can be seen that also in 3D our new global-to-local segmentation approach is able to deliver more accurate segmentation results than the global approach by Tsai et al. A lower root-mean-square deviation could be achieved in 46 datasets, a lower Hausdorff distance in 28 datasets, and lower Jaccard and Dice distances in all 48 datasets, respectively.

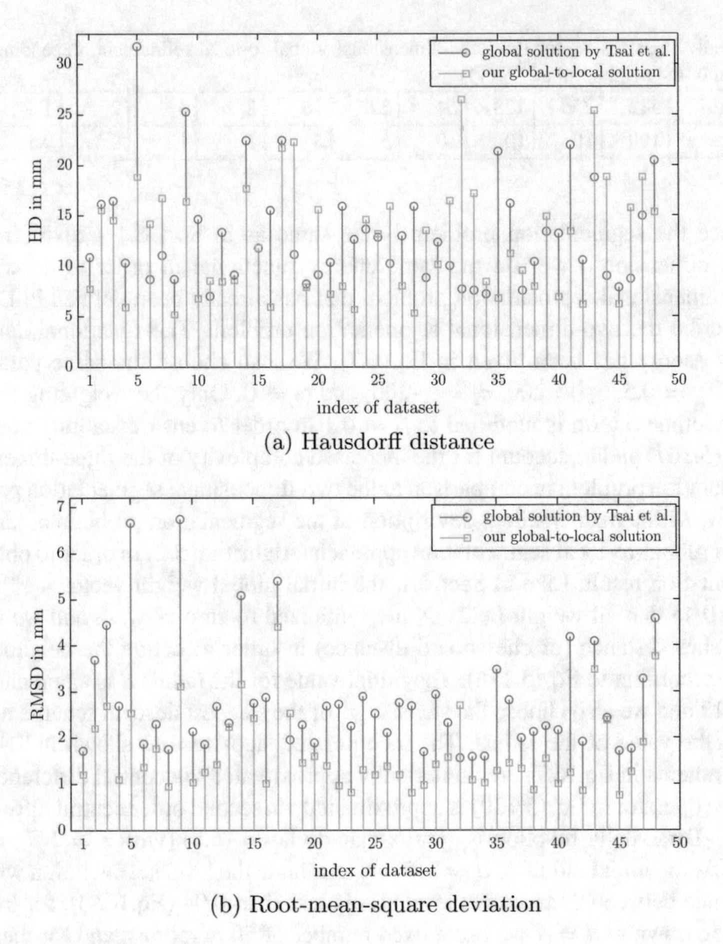

Fig. 6.11 Resulting Hausdorff distances (**a**) and root-mean-square deviations (**b**) obtained with the global approach by Tsai et al. (*magenta circles*) and our new global-to-local approach (*cyan squares*), respectively. The results are plotted against each dataset

Some exemplary segmentation results for our new global-to-local approach are shown in Fig. 6.14. The mean errors over all 48 datasets and the corresponding standard deviations are shown in Table 6.9. It can be seen that the average root-mean-square deviation of our new global-to-local approach is 40.1% lower than the average root-mean-square deviation of the global approach by Tsai et al. The average Jaccard distance is 51.2% lower and the average Dice distance is 54% lower, respectively. Similar improvements can be observed for the median errors which are given in Fig. 6.13 and Table 6.10.

For the average Hausdorff distance, only a slight improvement of 6.1% can be observed, and the median Hausdorff distance is even 0.42 mm (3.5%) higher for our new global-to-local approach in comparison to the global approach by Tsai et al.

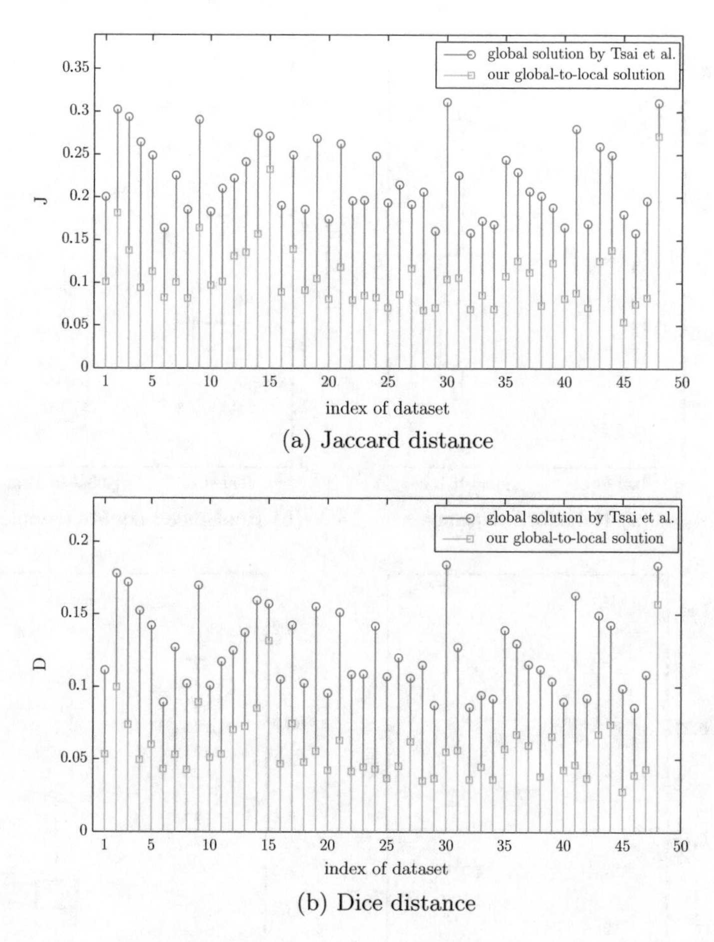

Fig. 6.12 Resulting Jaccard distances (**a**) and Dice distances (**b**) obtained with the global approach by Tsai et al. (*magenta circles*) and our new global-to-local approach (*cyan squares*), respectively. The results are plotted against each dataset

So, our new global-to-local approach and the global approach by Tsai et al. perform approximately equally well with regard to the Hausdorff distance. This is due to the fact that, because of the increased complexity of the three-dimensional segmentation problem, we have allowed our new global-to-local solution to become more local in comparison to the two-dimensional analysis from Sect. 6.1.1 (we use a fixed weighting factor of $\beta = 0.1$ instead of the adaptive weighting factor $\beta = 0.2/y^k$ which we used for the 2D problem). We have observed already for the two-dimensional problem that an increased locality leads to a worse Hausdorff distance (see the explanations regarding the results of our former local approach and our new global-to-local approach in Sect. 6.1.1).

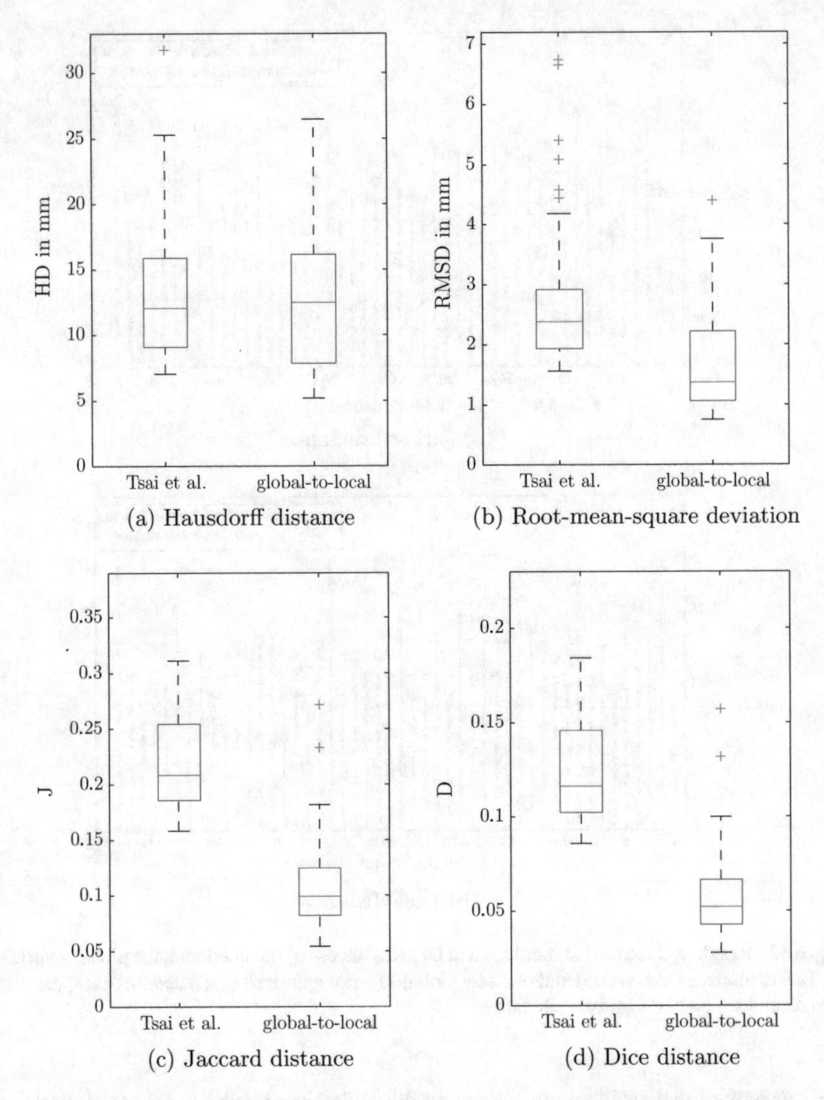

Fig. 6.13 Boxplots of the Hausdorff distance (**a**), the root-mean-square deviation (**b**), the Jaccard distance (**c**), and the Dice distance (**d**). The whiskers represent the highest/lowest datum within 1.5 times the interquartile range. All remaining data values are considered as outliers and given as *red crosses*

The statistical significance of the improvements in the median errors has again been investigated with the help of the left-sided Wilcoxon signed-rank test. The corresponding ρ-values are given in the last row of Table 6.10. The null hypothesis that the population median of the left samples (our new global-to-local segmentation results) is greater or equal to the population median of the right samples (the global segmentation results by Tsai et al.) can be rejected for the root-mean-square deviation

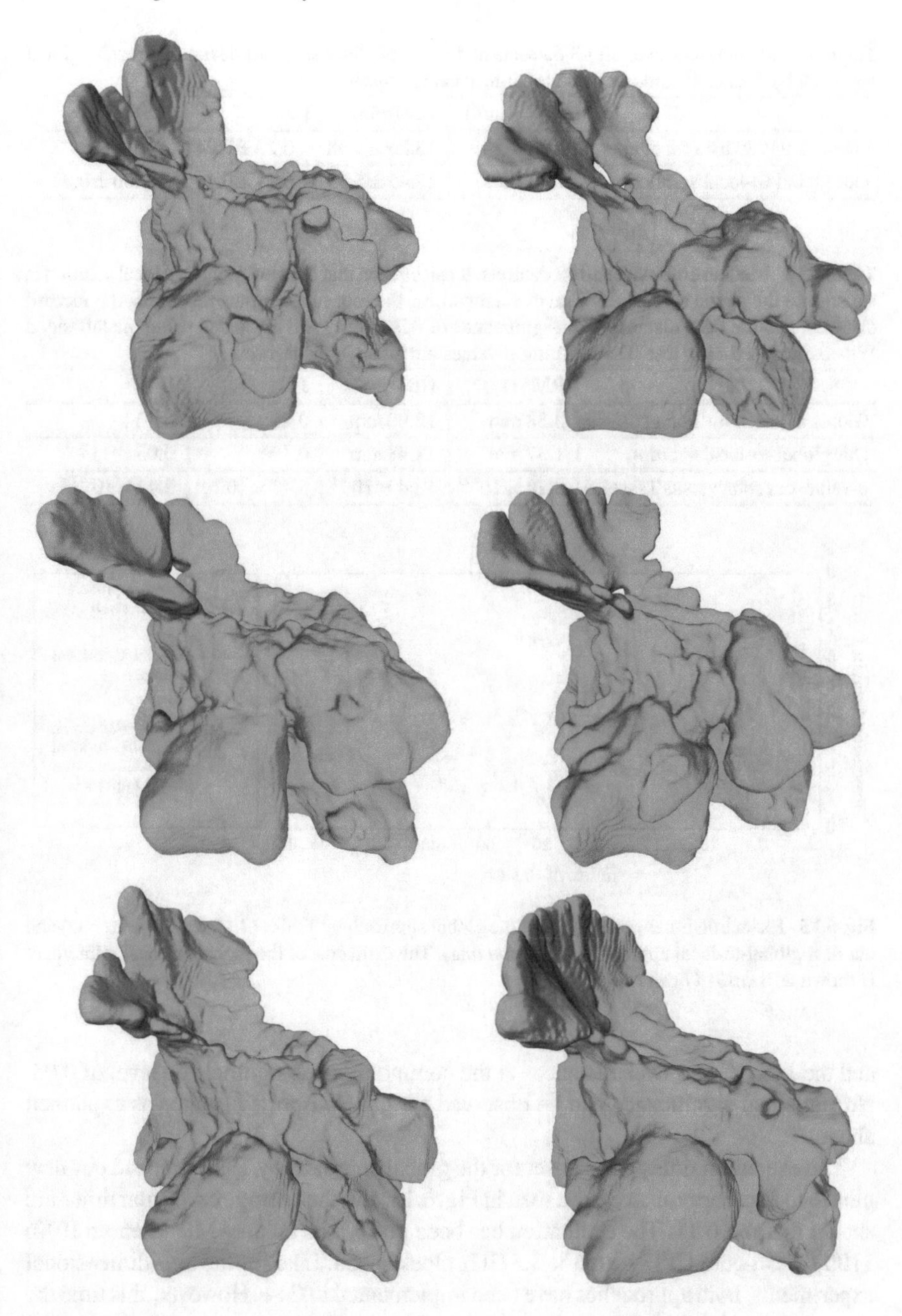

Fig. 6.14 Exemplary segmentation results for our new global-to-local approach

Table 6.9 Mean errors over all 48 datasets and corresponding standard deviations for the global approach by Tsai et al. and our new global-to-local approach

	RMSD (mm)	HD (mm)	J	D
Global solution by Tsai et al.	2.80 ± 1.24	13.26 ± 5.38	0.22 ± 0.04	0.12 ± 0.03
Our global-to-local solution	1.67 ± 0.85	12.45 ± 5.55	0.11 ± 0.04	0.06 ± 0.02

Table 6.10 Median errors over all 48 datasets. It can be seen that our new global-to-local solution is superior to the global solution by Tsai et al., regarding the root-mean-square deviation, the Jaccard distance, and the Dice distance. The significance of this finding was evaluated using the left-sided Wilcoxon signed-rank test. The resulting ρ-values are given in the last row

	RMSD	HD	J	D
Global solution by Tsai et al.	2.38 mm	12.06 mm	0.21	0.12
Our global-to-local solution	1.37 mm	12.48 mm	0.10	0.05
ρ-value: our soln. versus Tsai et al.	8.20×10^{-9}	1.64×10^{-1}	8.42×10^{-10}	8.42×10^{-10}

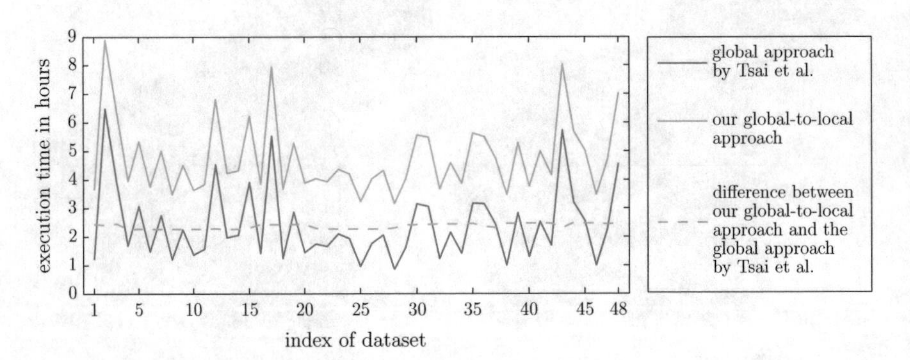

Fig. 6.15 Execution times per dataset for the global approach by Tsai et al. (*solid magenta line*) and our new global-to-local approach (*solid cyan line*). The extra cost of the global-to-local refinement is shown as a *dashed cyan line*

and the Jaccard and Dice distances at the commonly used significance level of 0.05. No statistical significance can be observed for the Hausdorff distance as explained above.

The execution times per dataset for the global approach by Tsai et al. and our new global-to-local approach can be seen in Fig. 6.15, and the average execution times are shown in Table 6.11. The evaluation has been performed on an AMD Phenom II X6 1100 T hexa-core CPU with 6×3.3 GHz clock speed. Like for the two-dimensional experiments, both approaches have been implemented in C++. However, this time the update of the weight fields Ψ_i (step 4 of Algorithm 5.2) is performed in parallel for several weight fields. This has been realized by using *OpenMP* [125]. The additional time needed by our new global-to-local approach in order to refine the global initial solution by Tsai et al. is on average 2 h and 22 min. So, on average, our new global-

Table 6.11 Average execution time and standard deviation per dataset for the global approach by Tsai et al. and our new global-to-local approach

Global solution by Tsai et al.	2 h 23 m 51 s ± 78 m
Our global-to-local solution	4 h 45 m 51 s ± 78 m 44 s

to-local refinement approximately doubles the execution time of the global approach by Tsai et al., with the result that the root-mean-square deviation and the Jaccard and Dice distances are approximately halved.

6.2.2 New Global-To-Local Approach Versus Former Segmentation Framework

At next, we discuss the experimental setup and the results that we obtain when comparing our new global-to-local segmentation approach against our former global paranasal sinuses segmentation approach (Sect. 4.1.2) for the above-mentioned three-dimensional segmentation problem.

6.2.2.1 Experimental Setup

The results of our former global paranasal sinuses segmentation approach are obtained as described in Sect. 4.1.2. The processing chain of the approach is depicted in Fig. 4.6 and the required parameters are chosen as mentioned in Sect. 4.1.3.2: We choose a standard deviation of $\sigma_{\text{gauss}} = 1.0$ for the Gaussian smoothing kernel K_{gauss}, the binarization threshold t for the intensity weighted gradient magnitude E is chosen to 0.35, and the *Nelder-Mead simplex method* is terminated when the absolute difference between the normalized function values of the objective function is less than $1e^{-10}$ (Eq. (4.16)).

The results of our new global-to-local approach are obtained as explained in Sect. 6.2.1 with one important difference: Instead of initializing the weight fields to zero, we set the initial global weight vector $\vec{w}^{\text{init}}$ to the outcome of our former global approach so that the weight fields are initialized to the global solution. Additionally, we reduce the initial radius d that defines the neighborhood $\Theta_d(\vec{y})$ to $d = 128$. So, 500 iterations are needed for each dataset in order to obtain the global-to-local refined solution when we start from the global initial solution.

6.2.2.2 Results

Some exemplary segmentation results for our new global-to-local approach are shown in Fig. 6.16. The outcomes of the error measures for each dataset are shown in Figs. 6.17 and 6.18. Additional boxplots for all errors are depicted in Fig. 6.19.

The mean errors and corresponding standard deviations, averaged over all 48 datasets, are shown in Table 6.12. It can be seen that for the Jaccard and Dice distances our new global-to-local approach is superior to our former global approach in all 48 datasets. The average Jaccard distance is 0.13 (56.5%) better and the average Dice distance is 0.08 (59.9%) better, respectively. For the root-mean-square deviation, our new global-to-local approach is superior to our former global approach in 47 datasets and for the Hausdorff distance it is superior in 40 datasets. Hence, on average our new global-to-local approach outperforms the global approach for these two error measures as well.

Similar improvements can also be observed for the medians of all four error measures that are given in Fig. 6.19 and in Table 6.13. Like in all previous experiments, the statistical significance of these improvements has been confirmed by a left-sided Wilcoxon signed-rank test. The ρ-values are given in the last row of Table 6.13. We can reject the null hypothesis that the population median of the left samples (the new global-to-local results) is greater or equal to the population median of the right samples (the global results) for all four error measures at the commonly used significance level of 0.05.

When we compare the results from this section against the results from the previous section, it can be seen that, similar to the two-dimensional evaluation, our former global approach is inferior to the global approach by Tsai et al. for the three-dimensional evaluation as well (see e.g. the boxplots in Figs. 6.19 and 6.13). The statistical significance of this statement can also be verified by a Wilcoxon signed-rank test. More specifically, the average root-mean square deviation is 27.8% worse, the average Hausdorff distance is 26.8% worse, the average Jaccard distance is 4.3% worse, and the average Dice distance is 6.5% worse for our global approach in comparison to the global approach by Tsai et al.

However, it can also be seen once again that using our former global approach in order to provide the initial solution for our new global-to-local approach improves the results of our new global-to-local approach. In more detail, the combination of our new global-to-local approach and our former global approach improves the results from Sect. 6.2.1 by 4.7% for the average root-mean-square deviation, 6.4% for the average Hausdorff distance, 7.0% for the average Jaccard distance, and 7.2% for the average Dice distance. The improvements in the median errors are even greater. They amount 17.8, 16.1, 13.6, and 14.2%, respectively.

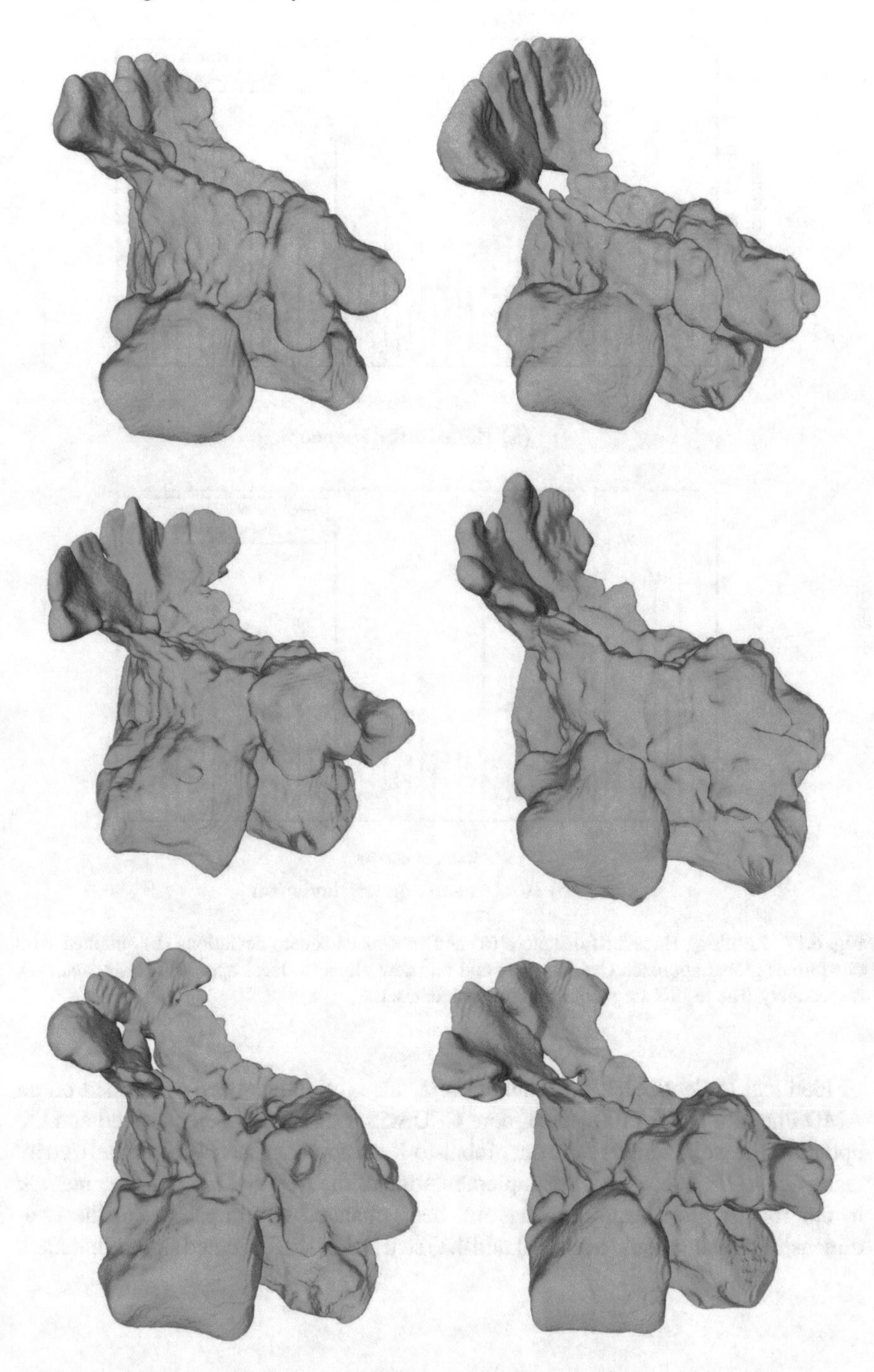

Fig. 6.16 Exemplary segmentation results for our new global-to-local approach

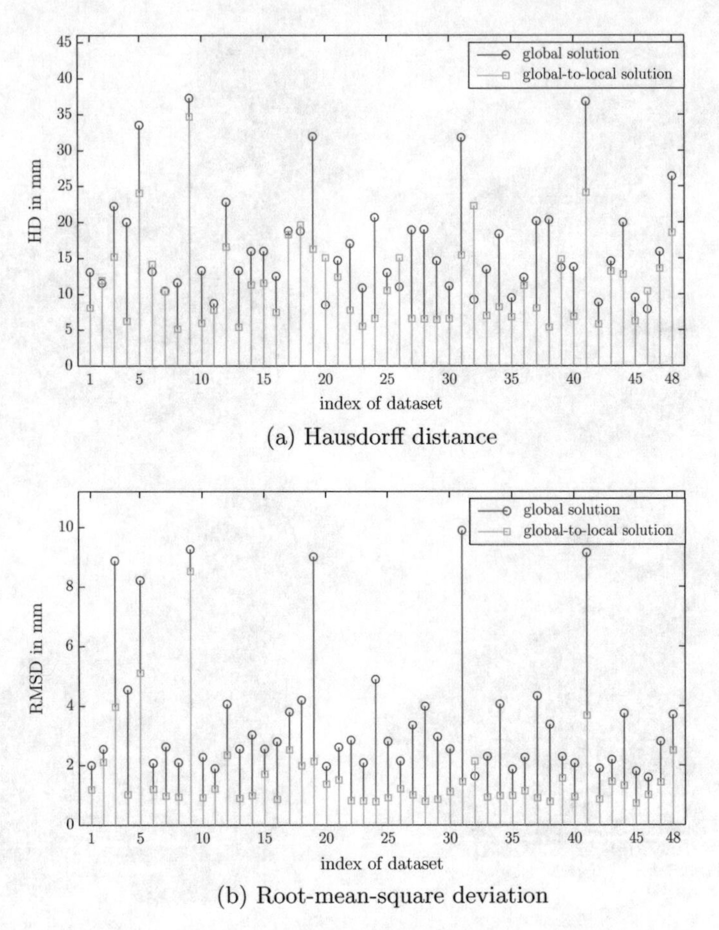

Fig. 6.17 Resulting Hausdorff distances (**a**) and root-mean-square deviations (**b**) obtained with our former global approach (*blue circles*) and our new global-to-local approach (*cyan squares*), respectively. The results are plotted against each dataset

Identical to Sect. 6.2.1, the evaluation in this section has been performed on an AMD Phenom II X6 1100 T hexa-core CPU with 6 × 3.3 GHz clock speed and the update of the weight fields Ψ_i in our global-to-local approach has been parallelized by using *OpenMP*. In contrast, the implementation of the *Nelder-Mead simplex method* in our former global approach has not been changed in comparison to the two-dimensional evaluations (Sects. 6.1 and 4.1) so that it is still executed single-threaded.

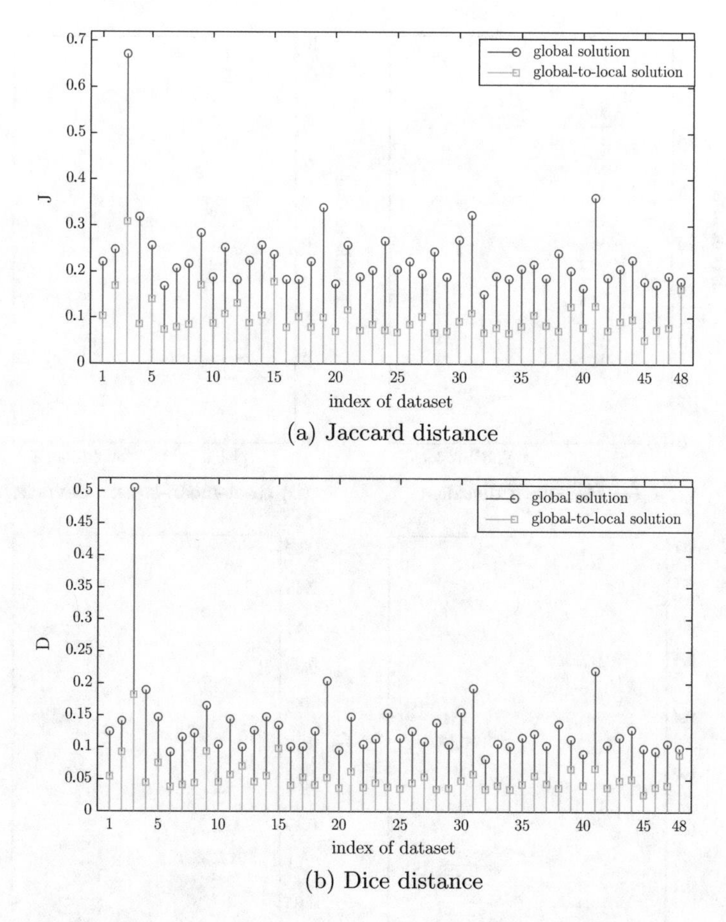

Fig. 6.18 Resulting Jaccard distances (**a**) and Dice distances (**b**) obtained with our former global approach (*blue circles*) and our new global-to-local approach (*cyan squares*), respectively. The results are plotted against each dataset

The execution times per dataset of our former global approach and our new global-to-local approach are depicted in Fig. 6.20 and the mean execution times are shown in Table 6.14. It can be seen that the additional time needed by our new global-to-local approach in order to refine our global initial solution is on average 2 h, 10 min, and 22 s. This approximately corresponds to an execution time increase by a factor of 22.

However, the extra execution time required for our new global-to-local solution is well spent since the results that we obtain with our new global-to-local approach are much better than the results from our former global approach. So, it is more fair to compare the execution time of our new global-to-local approach with our global initial solution from this section against the execution time of our new global-to-local approach with Tsai et al.'s global initial solution (i.e. the weight fields are initialized to zero) from Sect. 6.2.1. By doing this, we can see that another benefit of using our

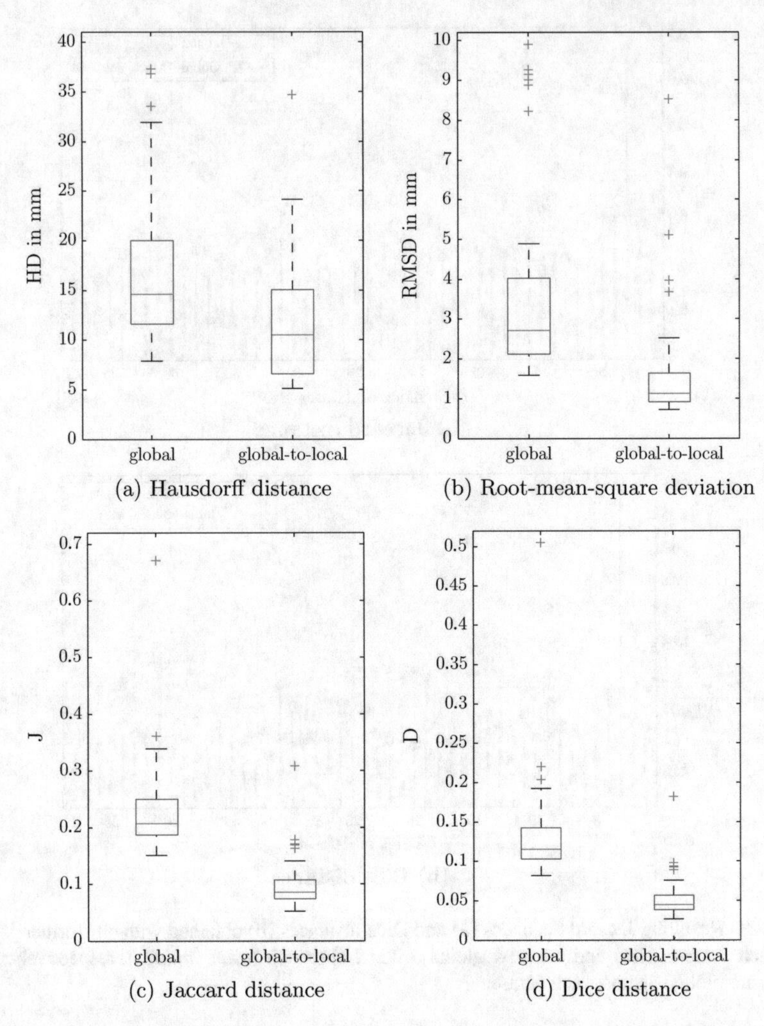

Fig. 6.19 Boxplots of the Hausdorff distance (**a**), the root-mean-square deviation (**b**), the Jaccard distance (**c**), and the Dice distance (**d**). The whiskers represent the highest/lowest datum within 1.5 times the interquartile range. All remaining data values are considered as outliers and given as *red crosses*

Table 6.12 Mean errors over all 48 datasets and corresponding standard deviations for our former global approach and our new global-to-local approach

	RMSD (mm)	HD (mm)	J	D
Former global solution	3.57 ± 2.26	16.82 ± 7.41	0.23 ± 0.08	0.13 ± 0.06
Global-to-local solution	1.60 ± 1.35	11.66 ± 6.20	0.10 ± 0.04	0.05 ± 0.03

Table 6.13 Median errors over all 48 datasets. It can be seen that our new global-to-local solution outperforms our former global solution in all four error measures. The significance of this finding was evaluated using the left-sided Wilcoxon signed-rank test. The resulting ρ-values are given in the last row

	RMSD	HD	J	D
Former global solution	2.71 mm	14.61 mm	0.21	0.12
Global-to-local solution	1.13 mm	10.48 mm	0.09	0.04
ρ-value: glob.-to-local versus global	9.55×10^{-10}	9.01×10^{-7}	8.42×10^{-10}	8.42×10^{-10}

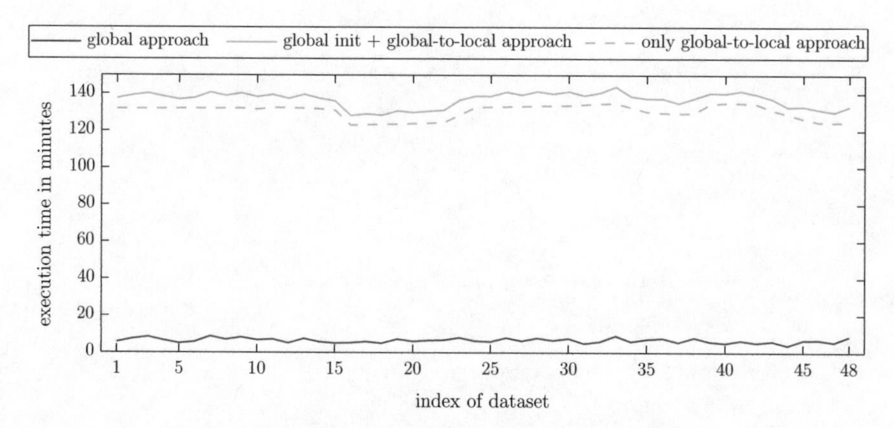

Fig. 6.20 Execution times per dataset for our former global approach (*solid blue line*) and our new global-to-local approach (*solid cyan line*). The extra cost of the global-to-local refinement, in addition to the initial global solution, is shown as a *dashed cyan line*

Table 6.14 Average execution time and standard deviation per dataset for our former global approach and our new global-to-local approach

Former global solution	6 m 30 s ± 1 m 11 s
New global-to-local solution	2 h 16 m 52 s ± 4 m 1 s

former global approach as initial solution for our new global-to-local approach, in addition to the better results, is a greatly reduced average execution time (2 h, 16 min, and 52 s vs. 4 h, 45 min, and 51 s).

Chapter 7
Conclusion and Outlook

In this thesis, we successfully addressed one major problem of *Statistical Shape Models* (SSMs), namely that the number of training samples often does not suffice to model the intra-class variability of complex shapes. For this purpose, we introduced a *Locally Deformable Statistical Shape Model* (LDSSM) that enhances conventional global SSMs by allowing a different model approximation in each point of the underlying data domain. Our LDSSM is an extension of a well-known global SSM in which a single weight vector is used to control the modeled shapes. In our LDSSM, this weight vector is replaced by a *smooth* vector-valued weight function. The modeled shape in each point is thus uniquely determined by the value of the weight function in this point.

The smoothness constraint, imposed on the weight function, ensures that the modeled shape remains continuous and that it is consistent with the global SSM in local areas of the data domain. By relaxing the smoothness constraint, one can seamlessly adapt the degree of locality. So, one is able to continuously adapt our LDSSM from modeling shapes that are restricted to the subspace of feasible shapes from the global SSM to modeling shapes that are not restricted at all. Our approach neither introduces artificial variations that cannot be explained through the training shapes nor does it need any arbitrarily predefined segments, neither in the spatial nor in the frequency domain.

The main application for our LDSSM is to act as a high-level shape prior for solving image segmentation problems. So, we additionally introduced an image segmentation framework that iteratively adapts the weight function of our LDSSM until the modeled shape approximates the desired structure in the image as good as possible. In the spirit of the most prominent SSM-based image segmentation approaches, like e.g. the *Active Shape Model* approach by Cootes et al. [32], the weight adaptation is achieved by iteratively computing a shape update in the vicinity of the currently modeled shape and to restrict this shape update back to the extended space of feasible shapes from our LDSSM.

C. Last, *From Global to Local Statistical Shape Priors*, Studies in Systems, Decision and Control 98, DOI 10.1007/978-3-319-53508-1_7

We have demonstrated the great performance of our LDSSM-based iterative segmentation framework by automatically extracting the outer bony border of the nasal cavity and the paranasal sinuses from computed tomography slices as well as by automatically segmenting the bones in the human knee. With only 48 or rather 5 training shapes at hand, we were able to obtain segmentation results with an average root mean square error smaller than two and one millimeters, respectively. To the best of our knowledge, our approach is the first that can provide a fully-automatic segmentation of the complex endonasal structures with sufficient accuracy. We have also shown that a global SSM-constraint segmentation approach is not able to deliver such accurate segmentation results as we can obtain with the help of our LDSSM. This is due to the high inter-patient variability of the paranasal sinuses and due to the very limited amount of training shapes for the bones in the human knee.

Another possible application for our LDSSM is to fill the holes of incomplete range scans. For this purpose, we have proposed an extension of our iterative framework that makes use of the RANSAM (*RANdom SAmple Matching*) method in order to automatically estimate the rigid transformation parameters (translation, rotation, and scale). This modification allows us to fit our LDSSM to incomplete range scans without any user interaction. Also for this application, we have shown that a local adaptation of the model parameters substantially improves the fitting results. For incomplete range scans of faces, we were able to obtain a natural-looking approximation of the complete face scans with only as few as 30 training shapes at hand.

Despite the good performance of our iterative image segmentation framework, its heuristic motivation leads to two considerable drawbacks:

1. The shape update step is solely based on the image gradient.
2. Statistical shape information is used only *after* the shape update step.

We addressed both issues by showing that the shape update step in our framework can be motivated with the help of the famous level set framework introduced by Osher and Sethian in [93]. Based on this observation, we subsequently presented an elegant variational formulation which integrates the shape update and the model adaptation step into one combined energy functional that has to be minimized in order to obtain the desired segmentation result. Our proposed energy functional depends directly on the vector-valued weight function of our LDSSM so that the evolving shape is at any time bound to the extended subspace of feasible shapes from our LDSSM. This general formulation of a local shape-constraint segmentation problem enables the use of many different data terms, e.g. the famous edge-based *Geodesic Active Contours* energy by Caselles et al. [25] (later extended by Li et al. [78]) but also the famous region-based *Active Contours Without Edges* energy by Chan and Vese [26].

The improvements of our new variational formulation in comparison to our heuristic formulation have been shown once again by extracting the outer bony border of the nasal cavity and the paranasal sinuses from computed tomography data. In addition, it has also been shown that our new approach outperforms the well-established approach by Tsai et al. [129], which uses a global shape prior, when extracting the paranasal sinuses from two-dimensional CT slices as well as from the whole three-dimensional CT volume.

A lot of work has already been done in order to integrate statistical shape knowledge into variational image segmentation approaches. However, to the best of our knowledge, we were the first to integrate the concept of partitioning the model, known-from statistical shape models based on point correspondences, into the variational image segmentation framework. All existing variational approaches that allow deviations from the trained subspace of feasible shapes are based on the concept of a relaxed global (or completely local) shape prior.

So far, we have only considered statistical shape models based on implicit shape representations where the training shapes are given as the zero level set of a higher-dimensional signed distance function. We have decided for this shape representation, because we think that it provides a more elegant and flexible way to describe shapes compared to explicit shape representations based on point correspondences while yielding similar or even better shape approximation results. This has been demonstrated in the present thesis by approximating range scans of faces. However, the idea of our LDSSM is easily transferable also to other representations, e.g. point-based shape models like the famous *Point Distribution Model* by Cootes et al. [32] or models based on dense correspondences like the *Statistical Deformation Model* by Rueckert et al. [110].

7.1 Open Topics and Future Work

One drawback of variational image segmentation approaches is that the solutions obtained by functional gradient descent are not required to be global optima of the corresponding energy functionals, since in general the energy functionals are not convex. A possible solution to this problem would be to consider only convex energy functionals. One approach to define convex energy functionals for variational image segmentation problems with a linear global shape prior has been proposed by Cremers et al. in [41].

Another way to obtain globally optimal solutions would be to employ other image segmentation methods that are known to produce globally optimal results, like e.g. graph cut methods [21]. First approaches for image segmentation and tracking with the help of a single, elastically deformable template shape using graphical methods have been presented e.g. by Felzenszwalb [51] or Schoenemann and Cremers [113]. However, these approaches work only with a single shape prior and they make no use of the statistical distribution inside a class of training shapes.

Other approaches try to address this issue by iteratively projecting the shape-constrained graph cut segmentation result back into the subspace of feasible shapes and using the obtained model-constraint shape as a new prior in the next graph cut iteration (e.g. [54, 85]). However, like the iterative variational approaches, these iterative graph cut approaches are no more required to produce a globally optimal segmentation result. An approach that is guaranteed to deliver a globally optimal segmentation result for global, or even local, statistical shape priors is still an open topic. So, working in the direction of such an approach would be an interesting area for future research.

Appendix A
Results from Sect. 4.1.3

The images in this chapter depict all paranasal sinus segmentation results obtained with our global segmentation approach from Sect. 4.1.2 and our local segmentation approach from Sect. 3.3.4. The experimental setup has been explained in Sect. 4.1.3 where you can also find the quantitative evaluation.

In Fig. A.1, The hand-segmented reference contours are always shown as a green line, the global segmentation results are always shown as a blue line, and the local results are always shown as a red line, respectively.

C. Last, *From Global to Local Statistical Shape Priors*, Studies in Systems, Decision and Control 98, DOI 10.1007/978-3-319-53508-1

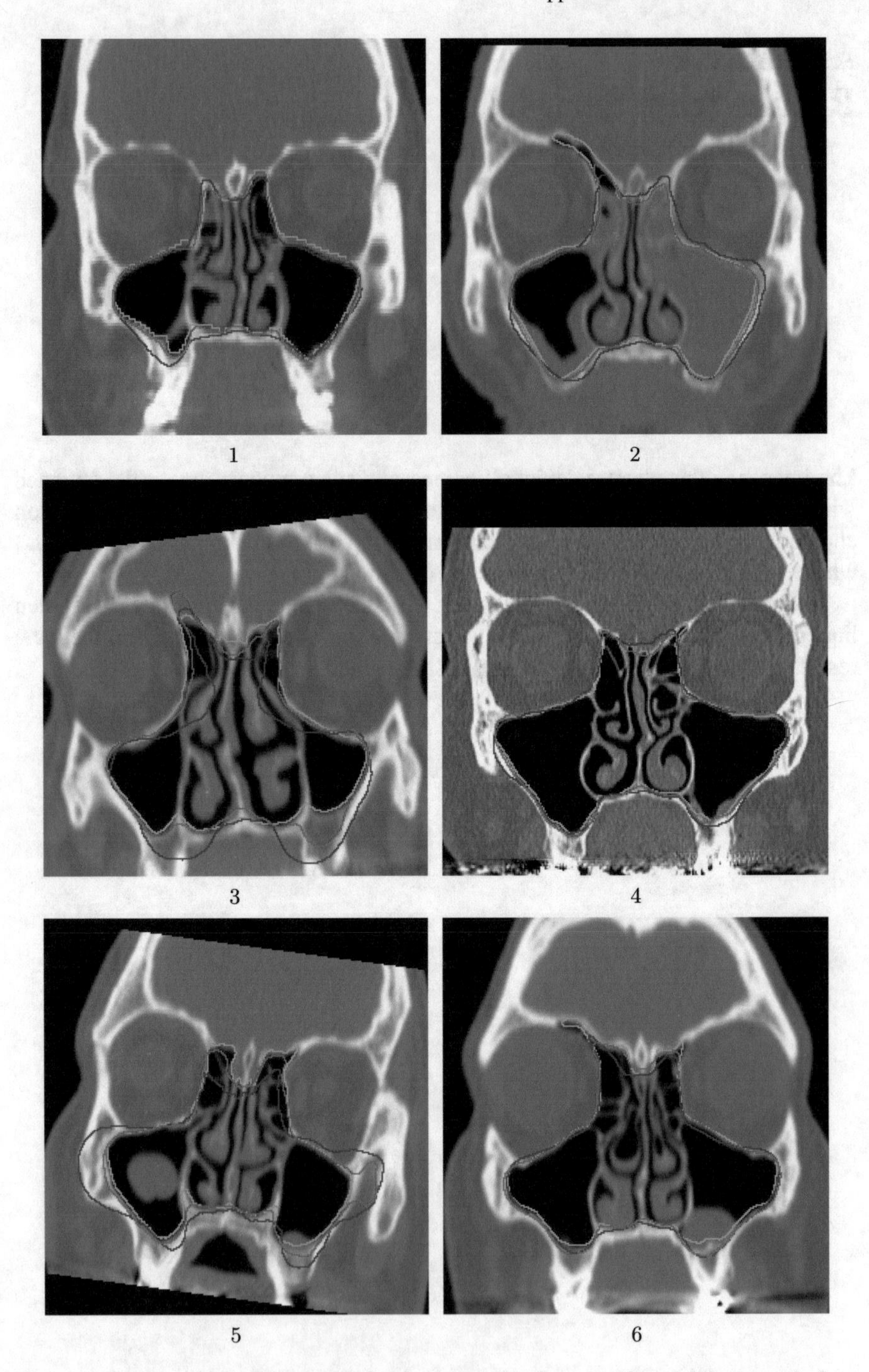

Fig. A.1 Segmentation results: *green* hand-segmented reference; *blue* global approach from Sect. 4.1.2; *red* local approach from Sect. 3.3.4

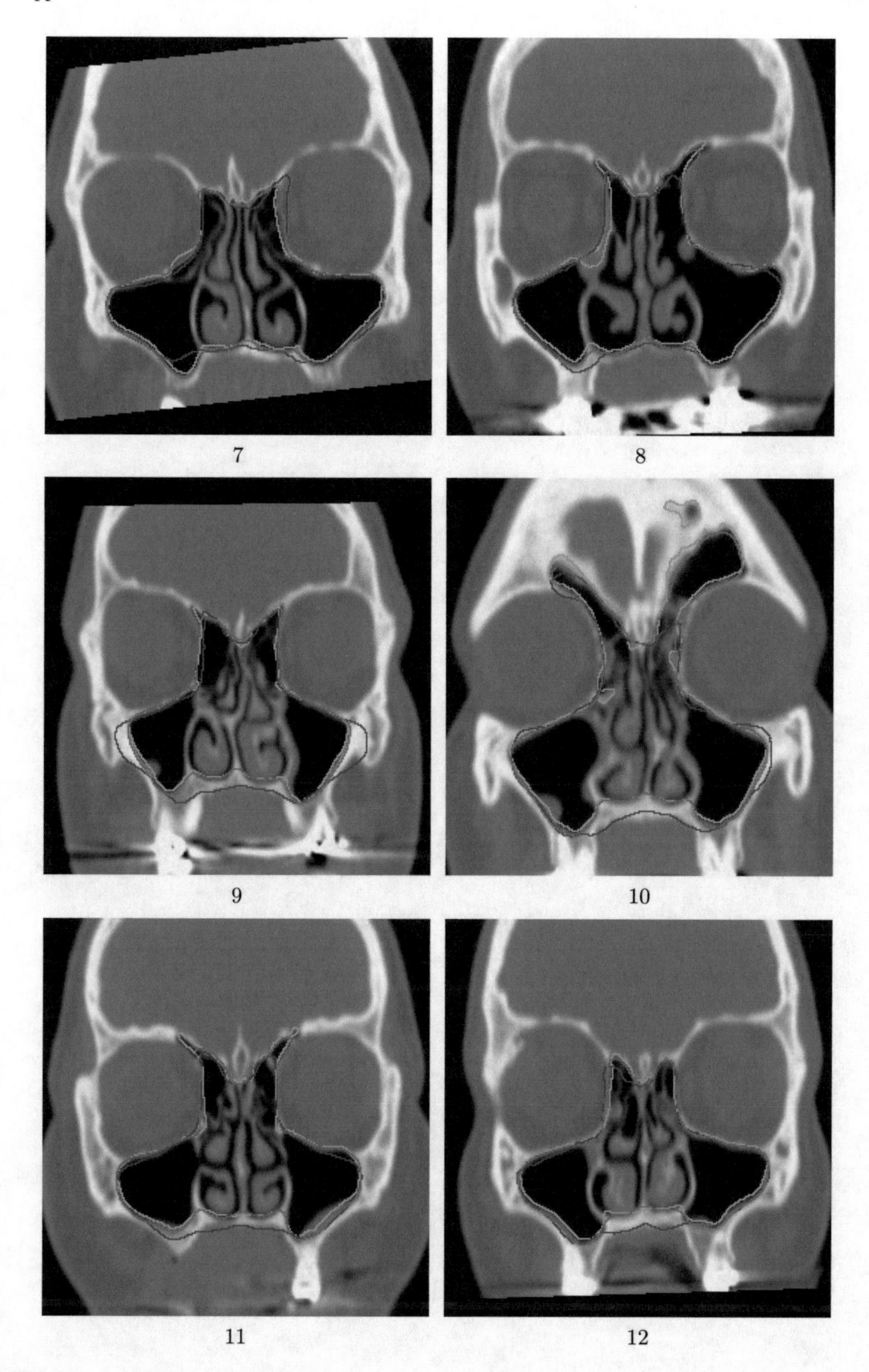

Fig. A.1 (continued)

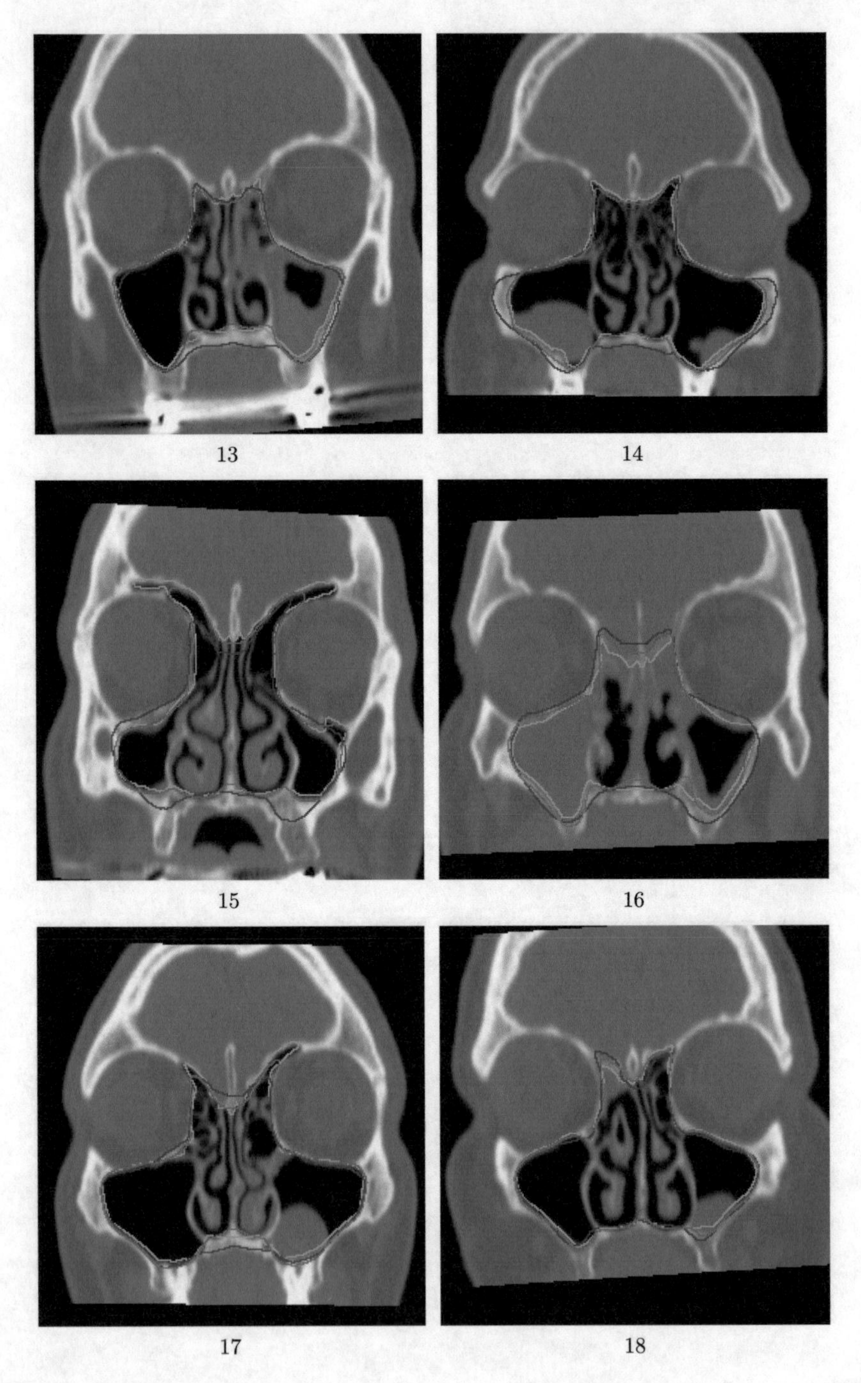

Fig. A.1 (continued)

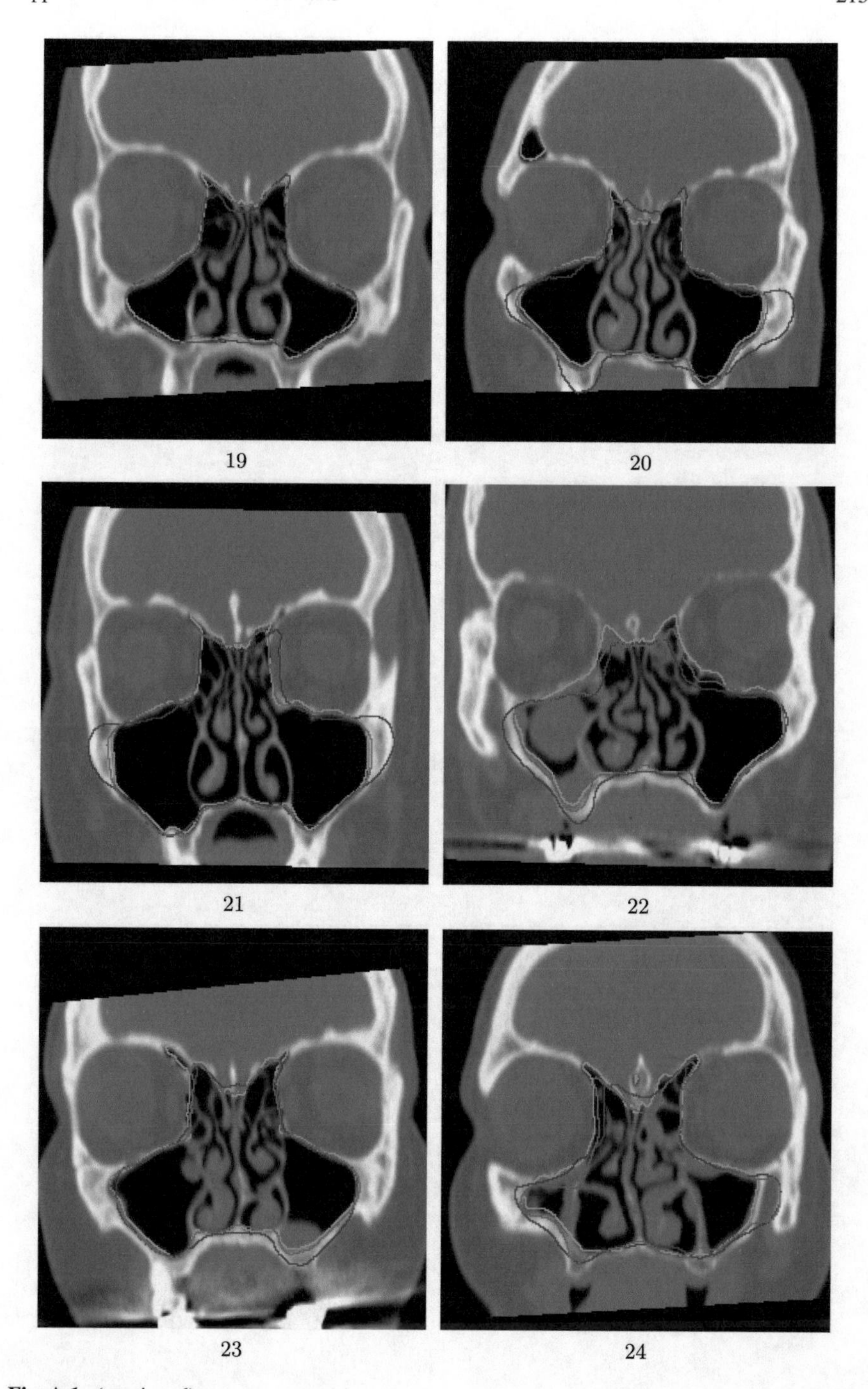

Fig. A.1 (continued)

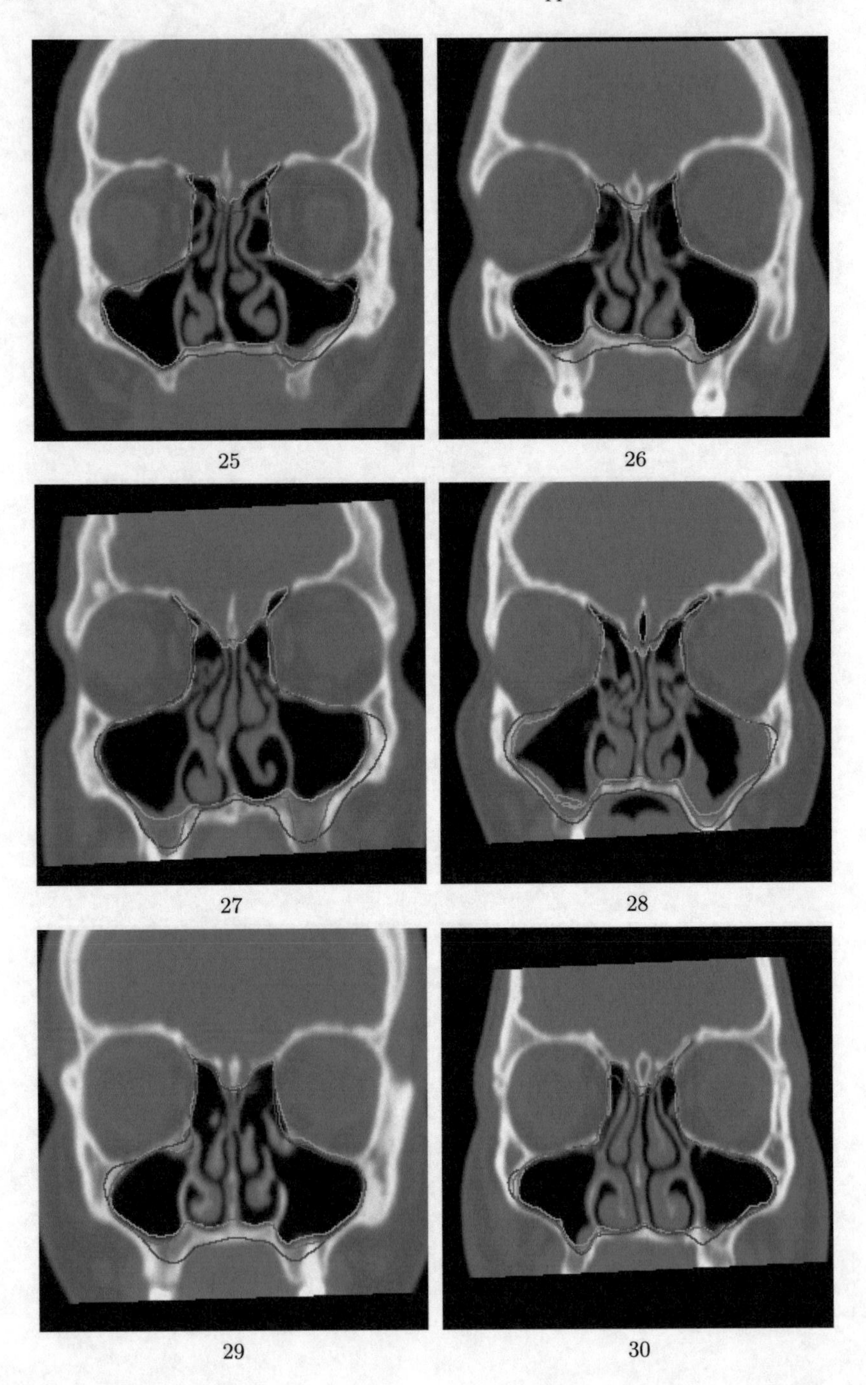

Fig. A.1 (continued)

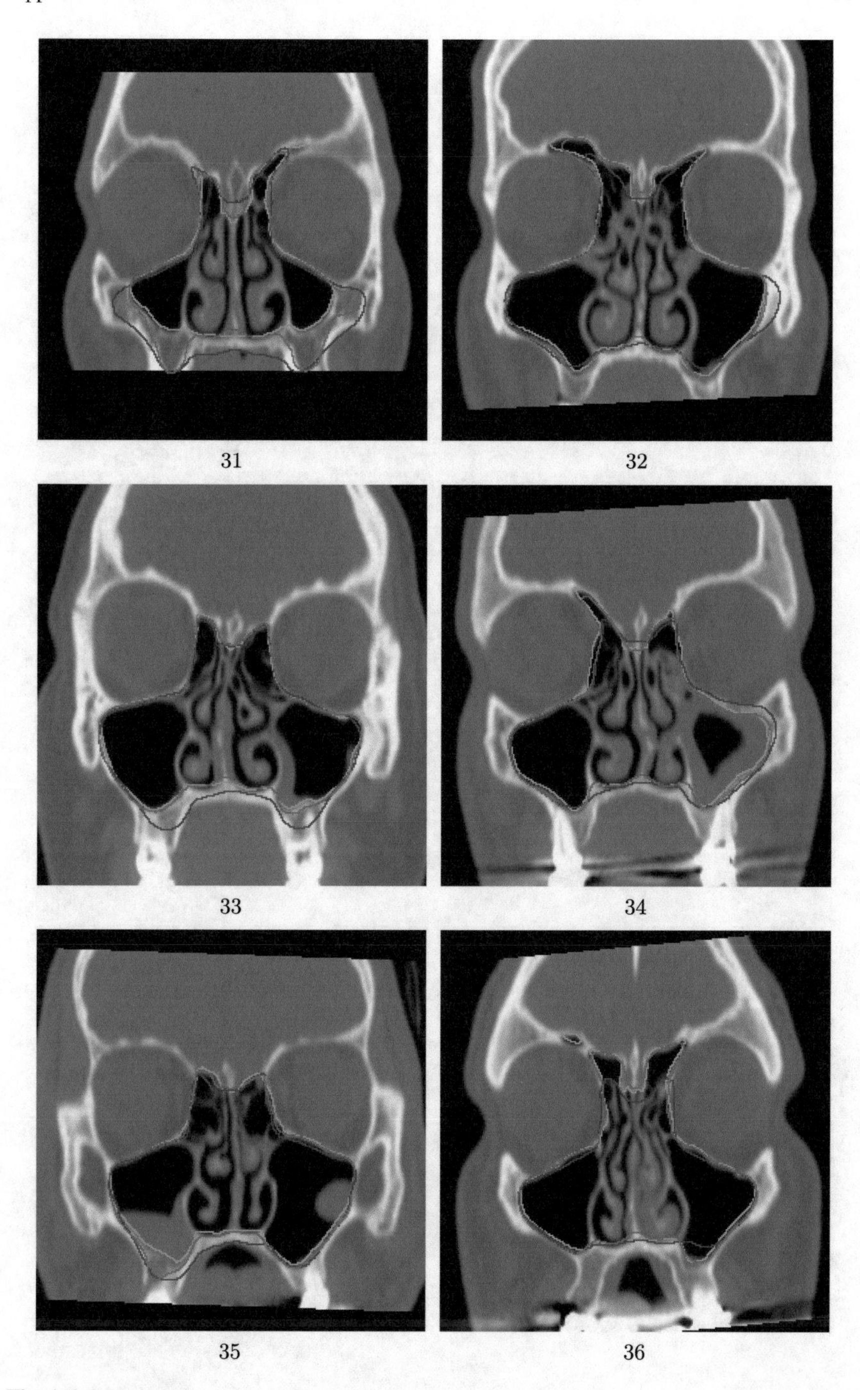

Fig. A.1 (continued)

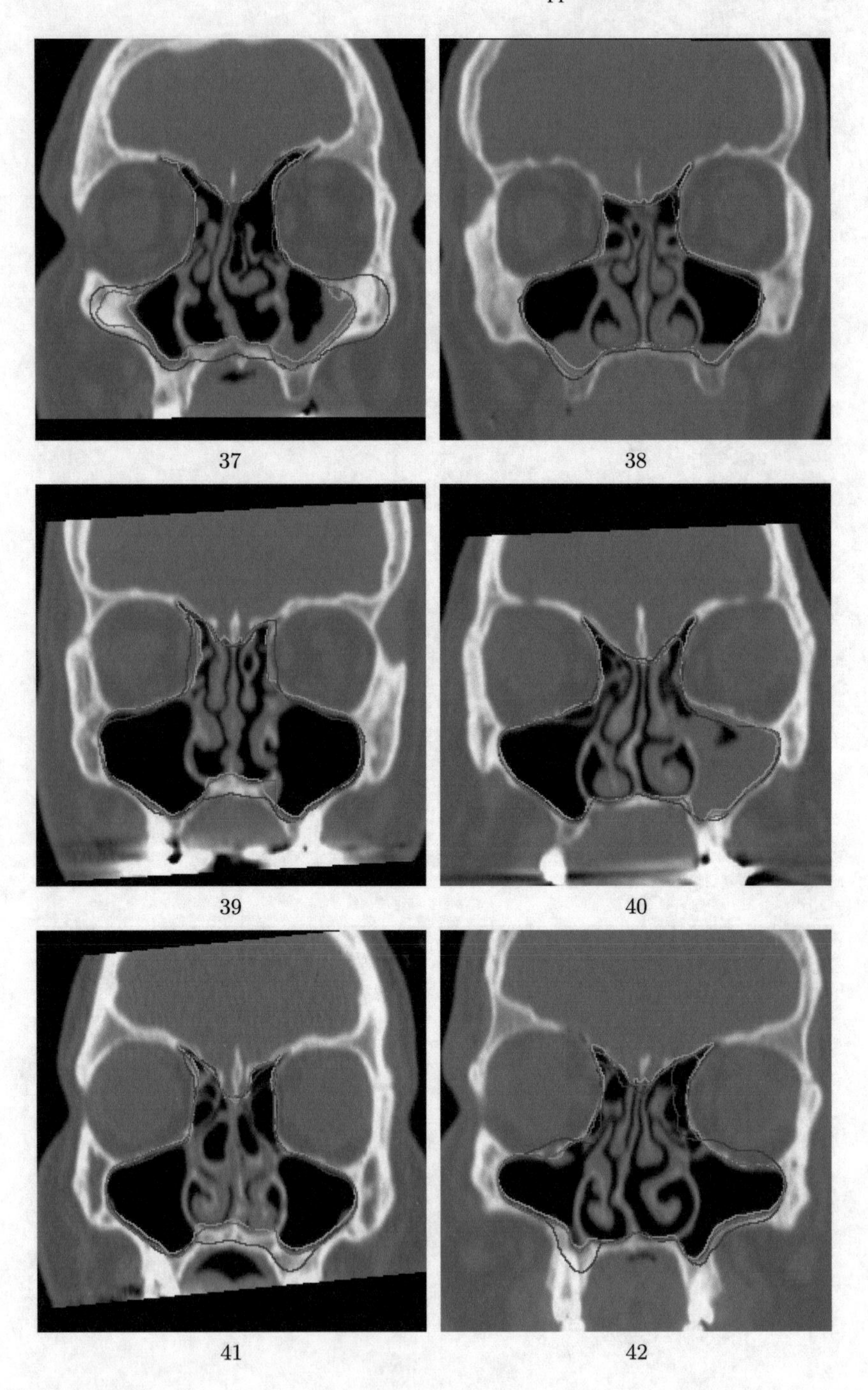

Fig. A.1 (continued)

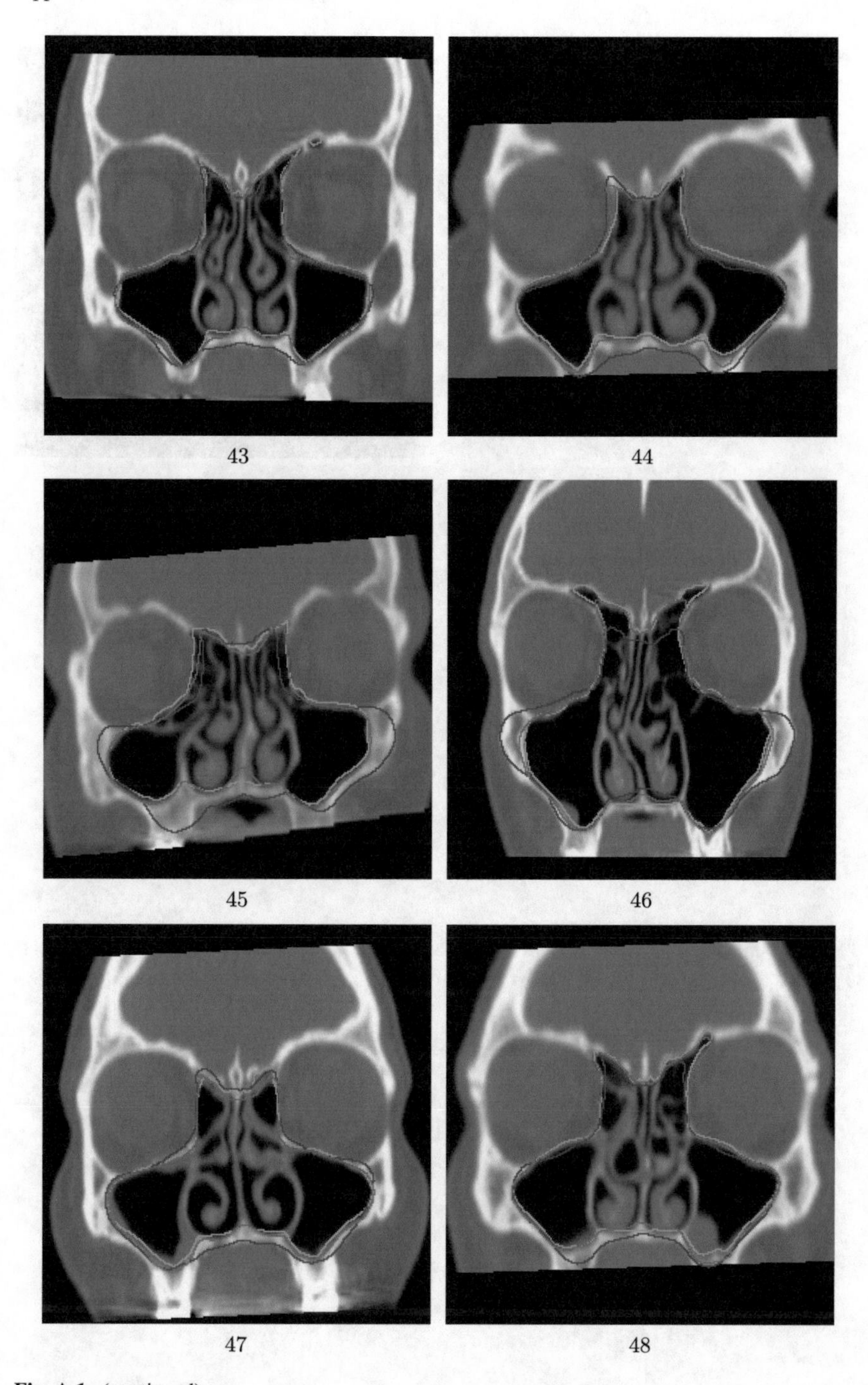

Fig. A.1 (continued)

Fig. A.1 (continued)

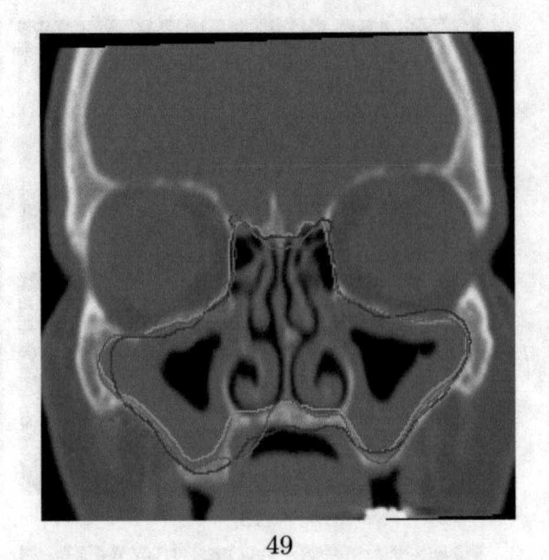

49

Appendix B
Results from Sect. 6.1.1

The images in this chapter depict all paranasal sinus segmentation results obtained with our global-to-local approach from Sect. 5.5 compared to the segmentation results obtained with the global approach by Tsai et al. [130] that has been introduced in Sect. 5.3. The experimental setup has been explained in Sect. 6.1.1 where you can also find the quantitative evaluation.

In Fig. B.1, the hand-segmented reference contours are always shown as a green line, the global segmentation results by Tsai et al. are always shown as a magenta line, and our global-to-local results are always shown as a cyan line, respectively.

C. Last, *From Global to Local Statistical Shape Priors*, Studies in Systems, Decision and Control 98, DOI 10.1007/978-3-319-53508-1

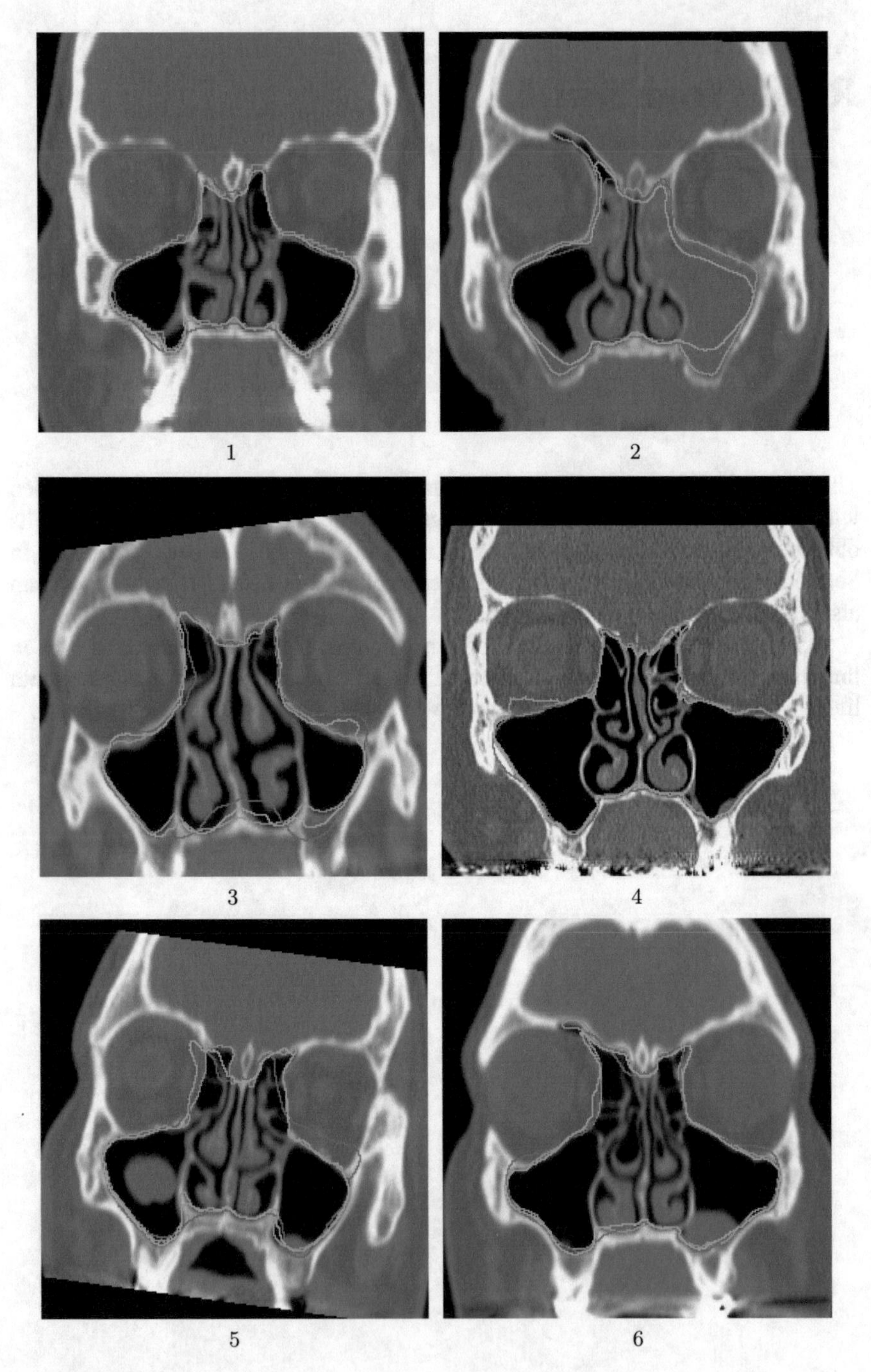

Fig. B.1 Segmentation results: *green* hand-segmented reference; *magenta* global approach from [130]; *cyan* global-to-local approach from Sect. 5.5

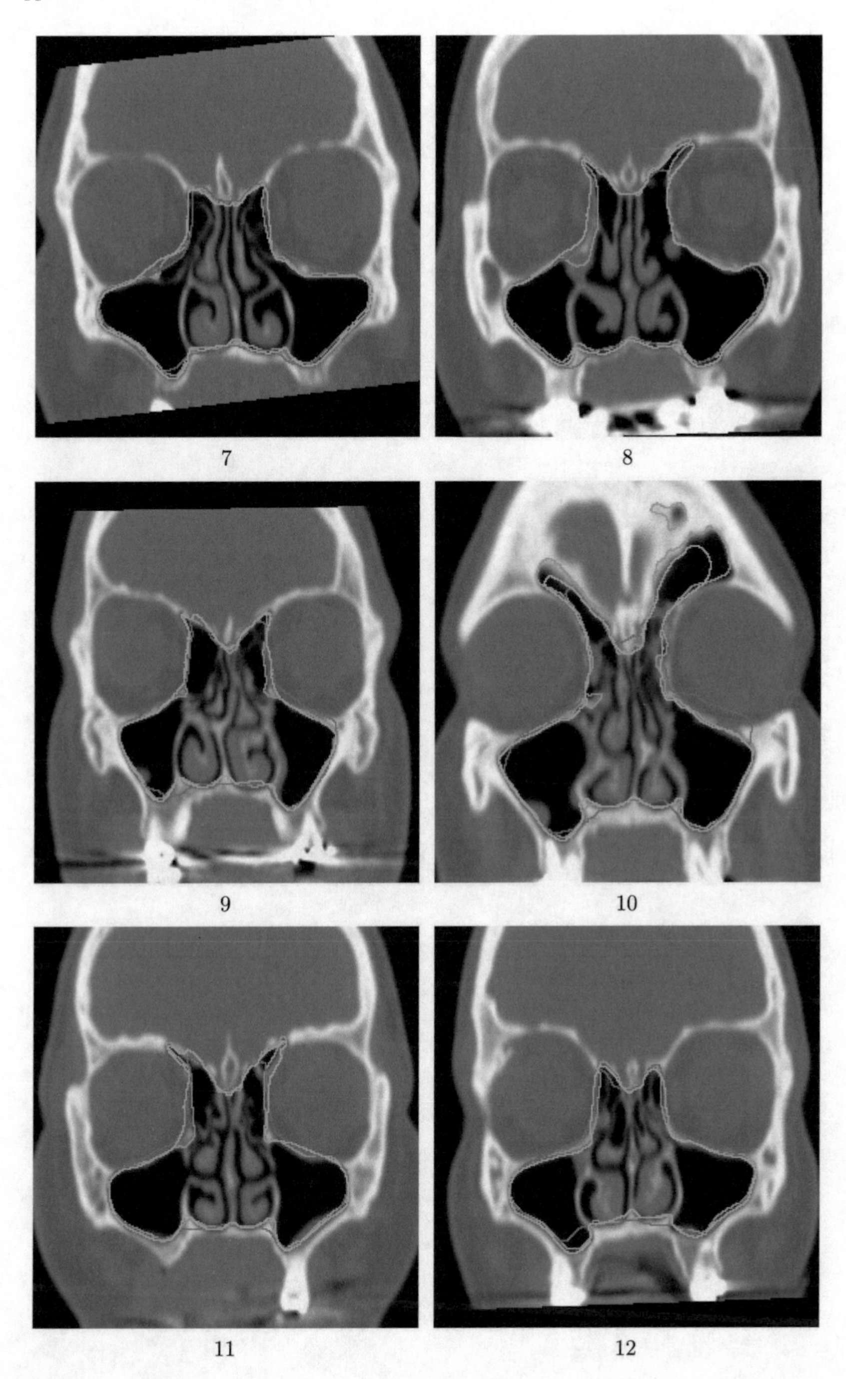

Fig. B.1 (continued)

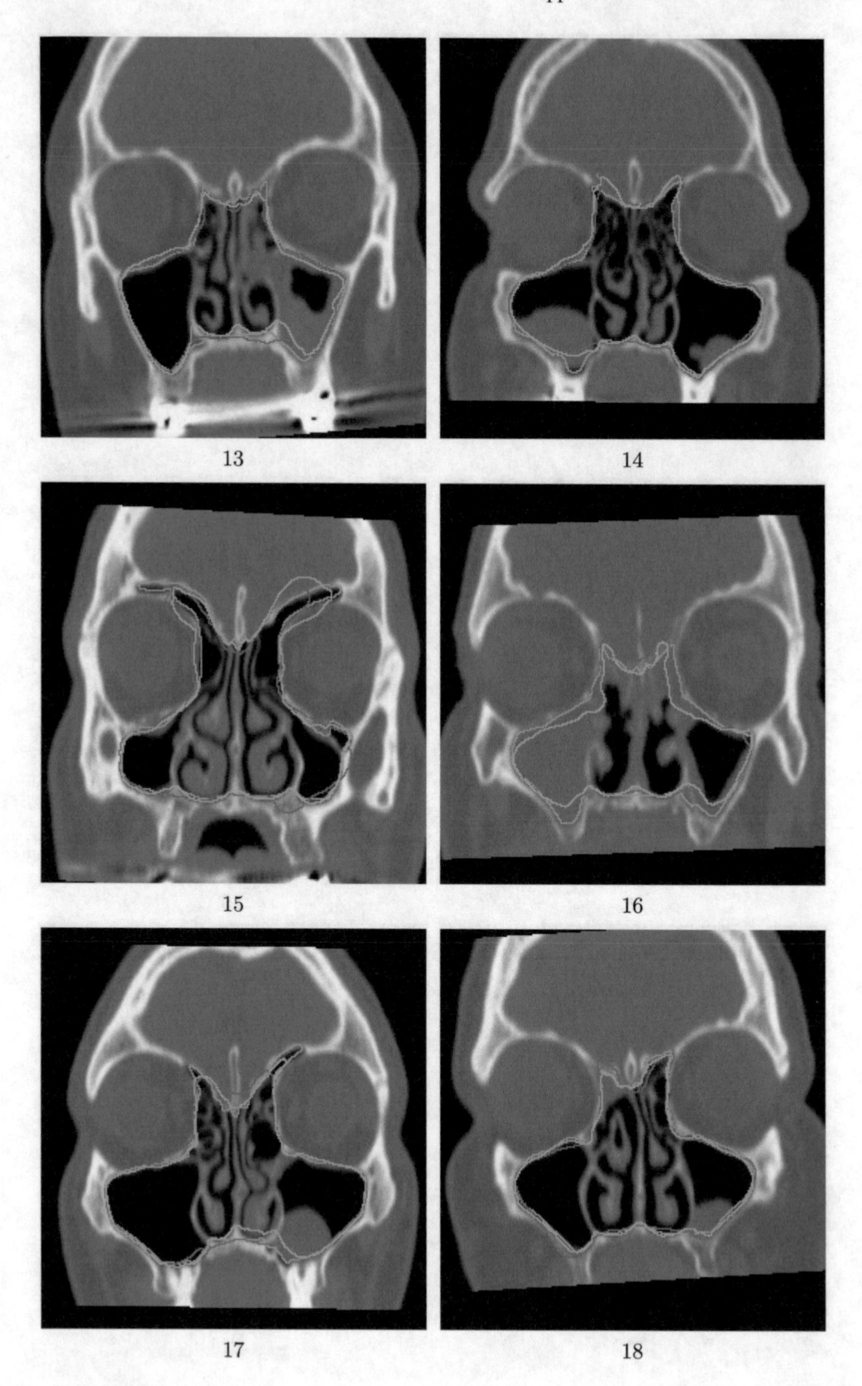

Fig. B.1 (continued)

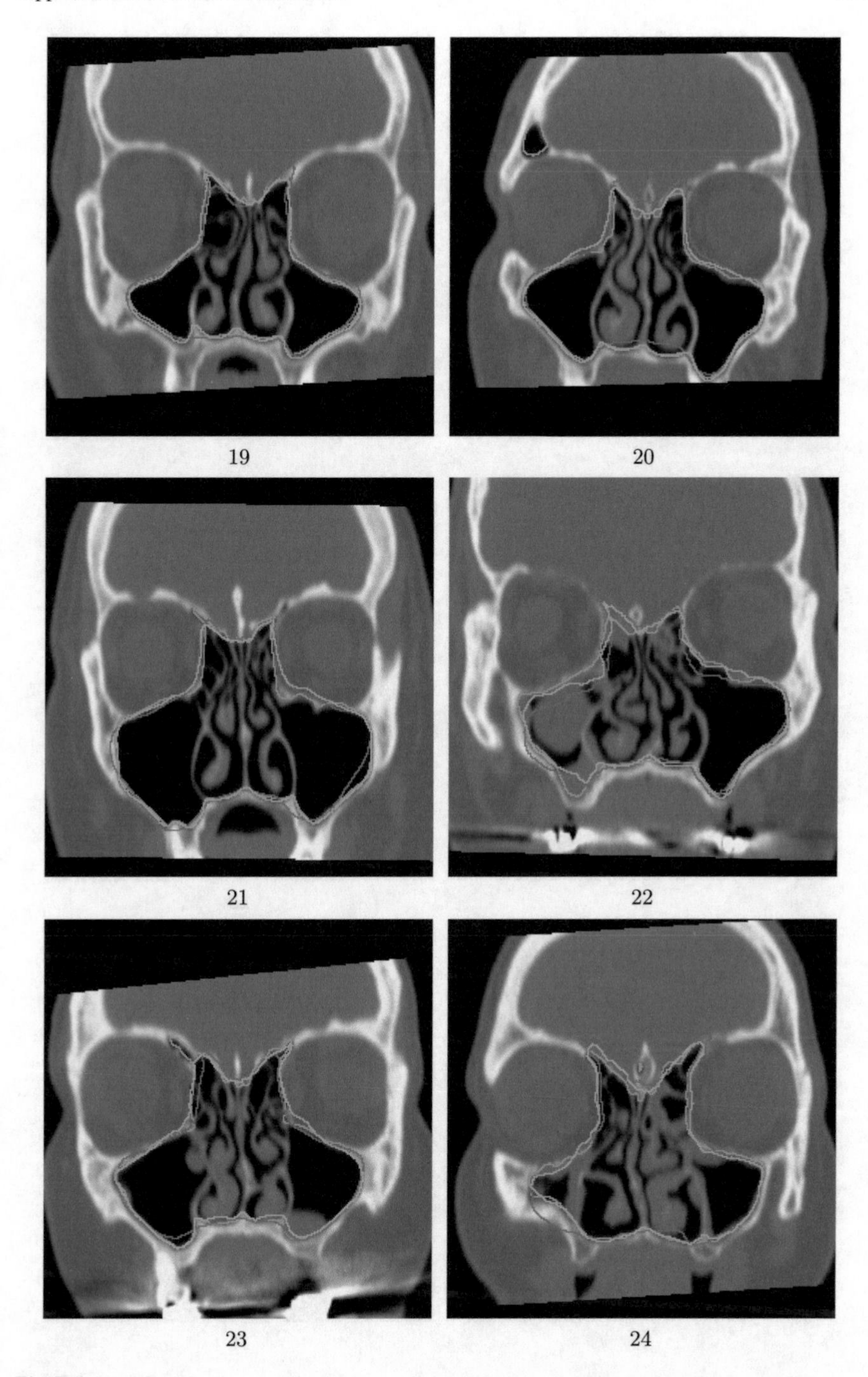

Fig. B.1 (continued)

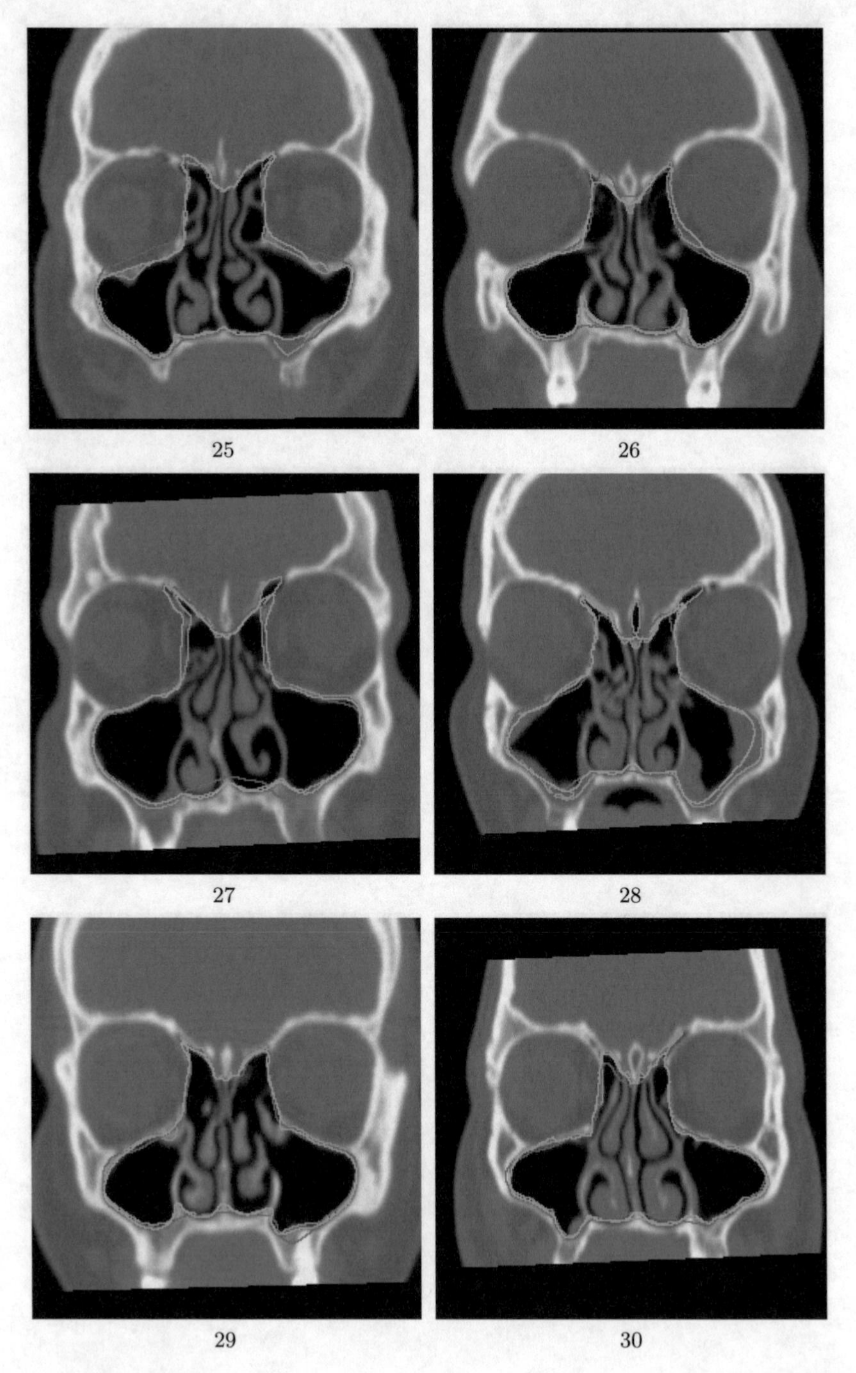

Fig. B.1 (continued)

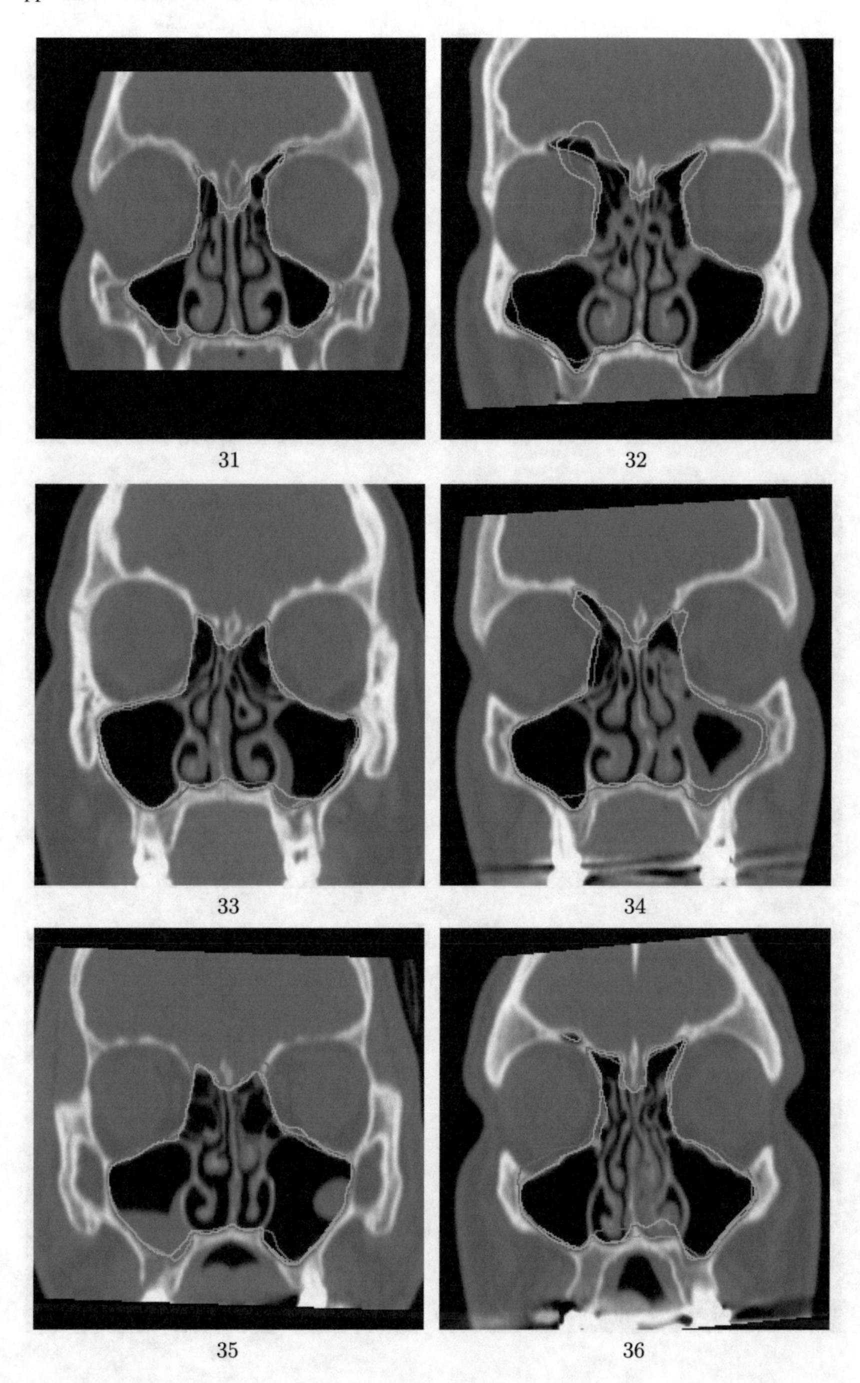

Fig. B.1 (continued)

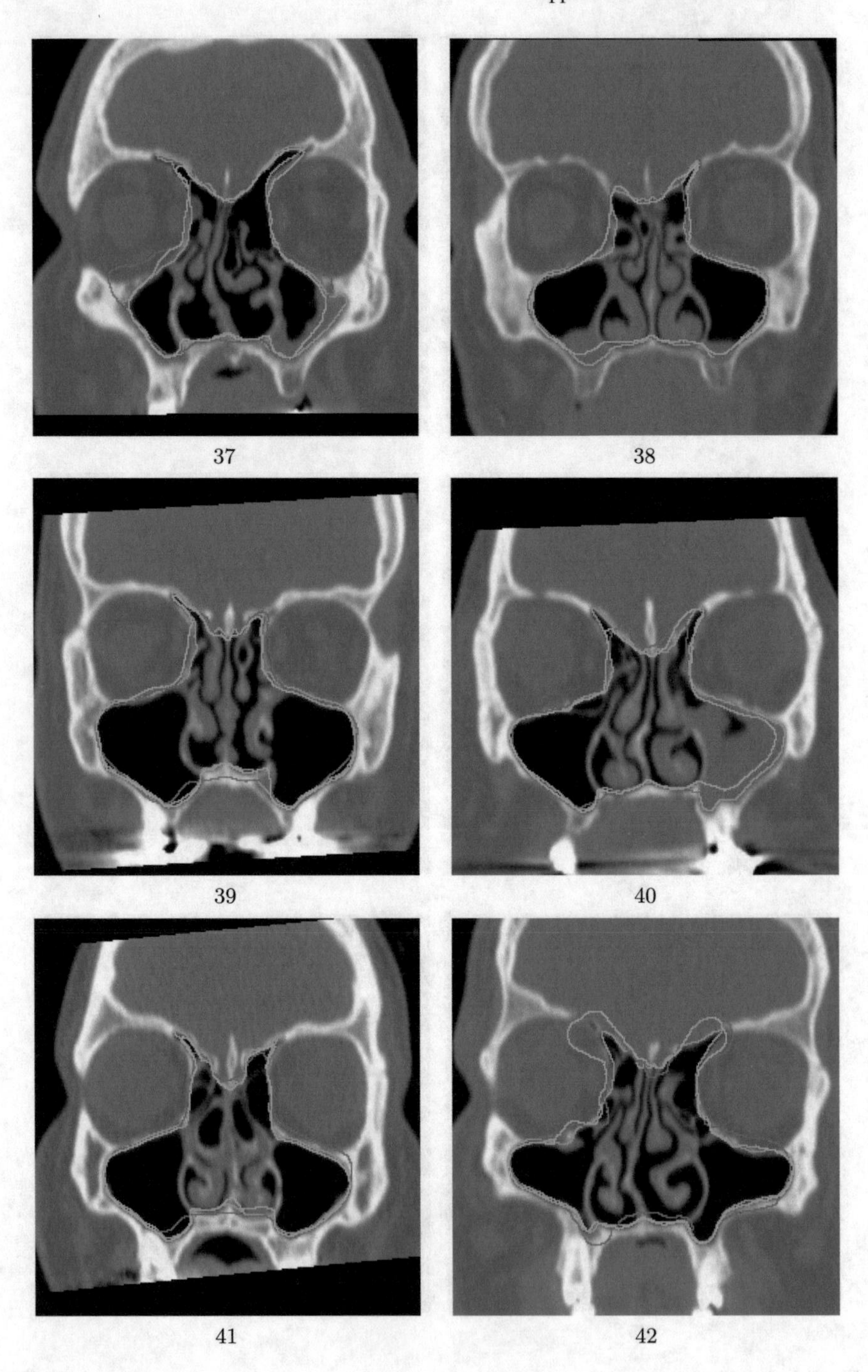

Fig. B.1 (continued)

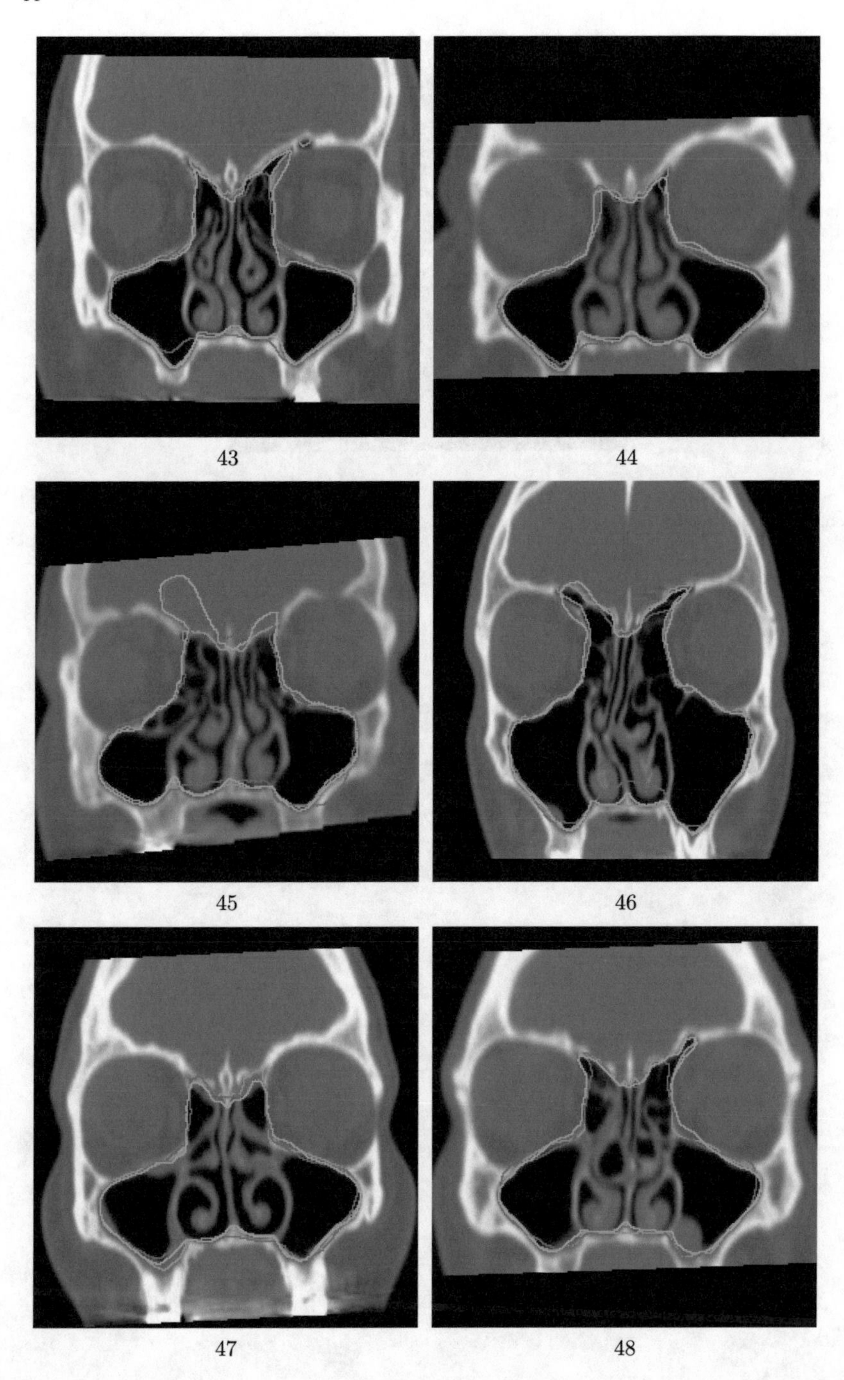

Fig. B.1 (continued)

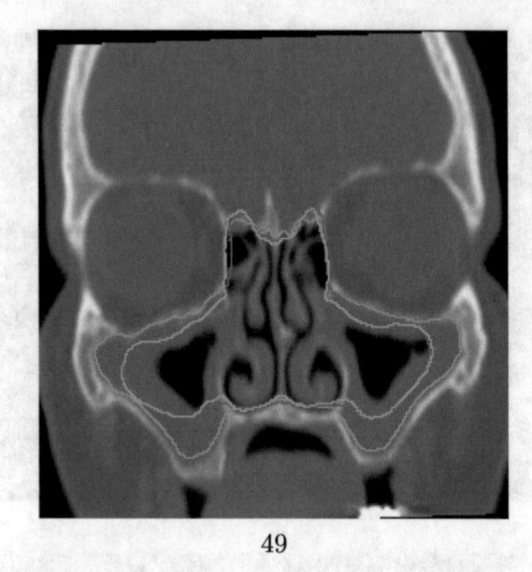

49

Fig. B.1 (continued)

Appendix C
Results from Sect. 6.1.2

The images in this chapter depict all paranasal sinus segmentation results obtained with our global-to-local approach from Sect. 5.5 compared to our former heuristic local approach from Sect. 3.3.4. The experimental setup has been explained in Sect. 6.1.2 where you can also find the quantitative evaluation.

In Fig. C.1, the hand-segmented reference contours are always shown as a green line, our former local results are always shown as a red line, and our new global-to-local results are always shown as a cyan line, respectively.

C. Last, *From Global to Local Statistical Shape Priors*, Studies in Systems, Decision and Control 98, DOI 10.1007/978-3-319-53508-1

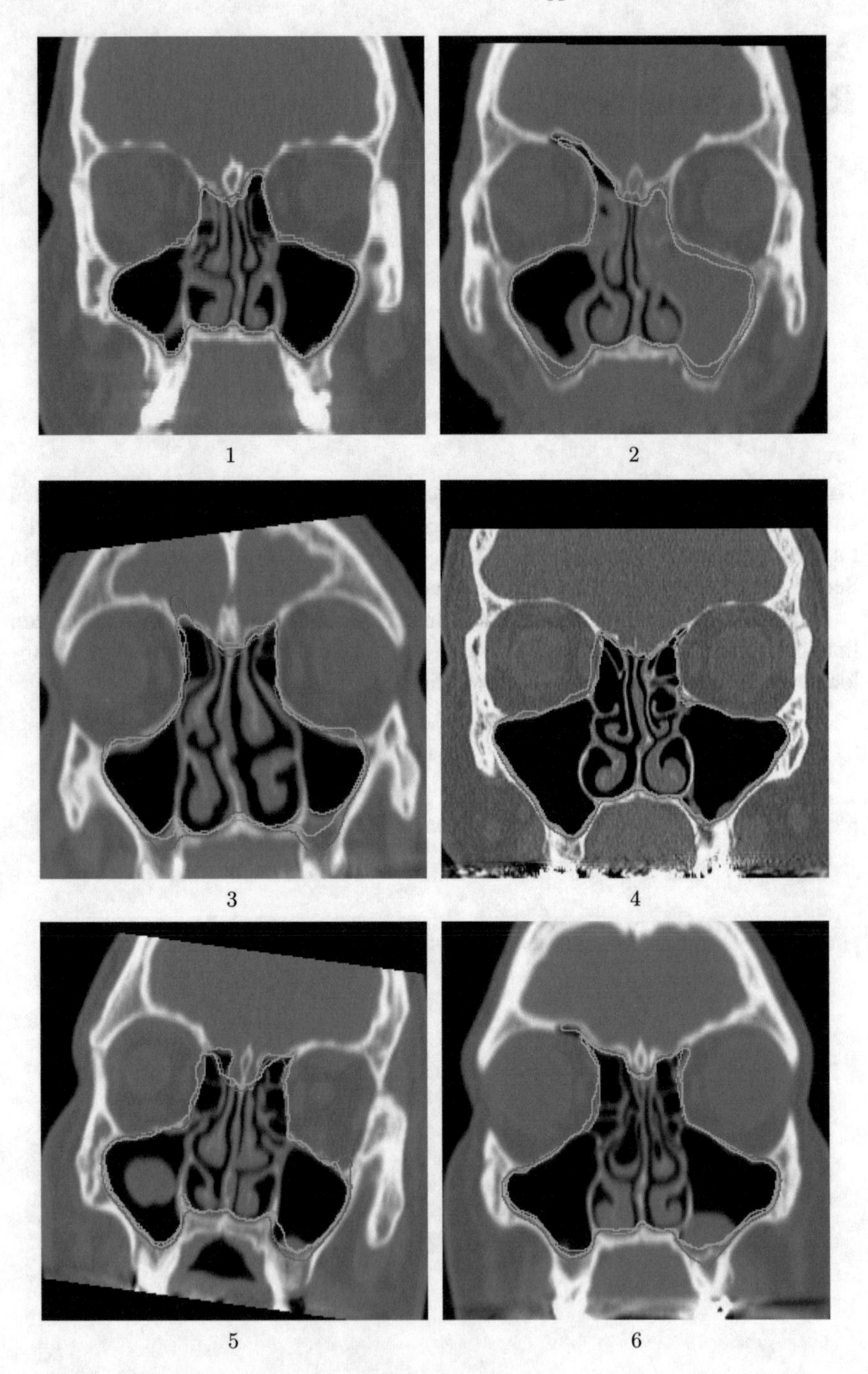

Fig. C.1 Segmentation results: *green* hand-segmented reference; *red* local approach from Sect. 3.3.4; *cyan* global-to-local approach from Sect. 5.5

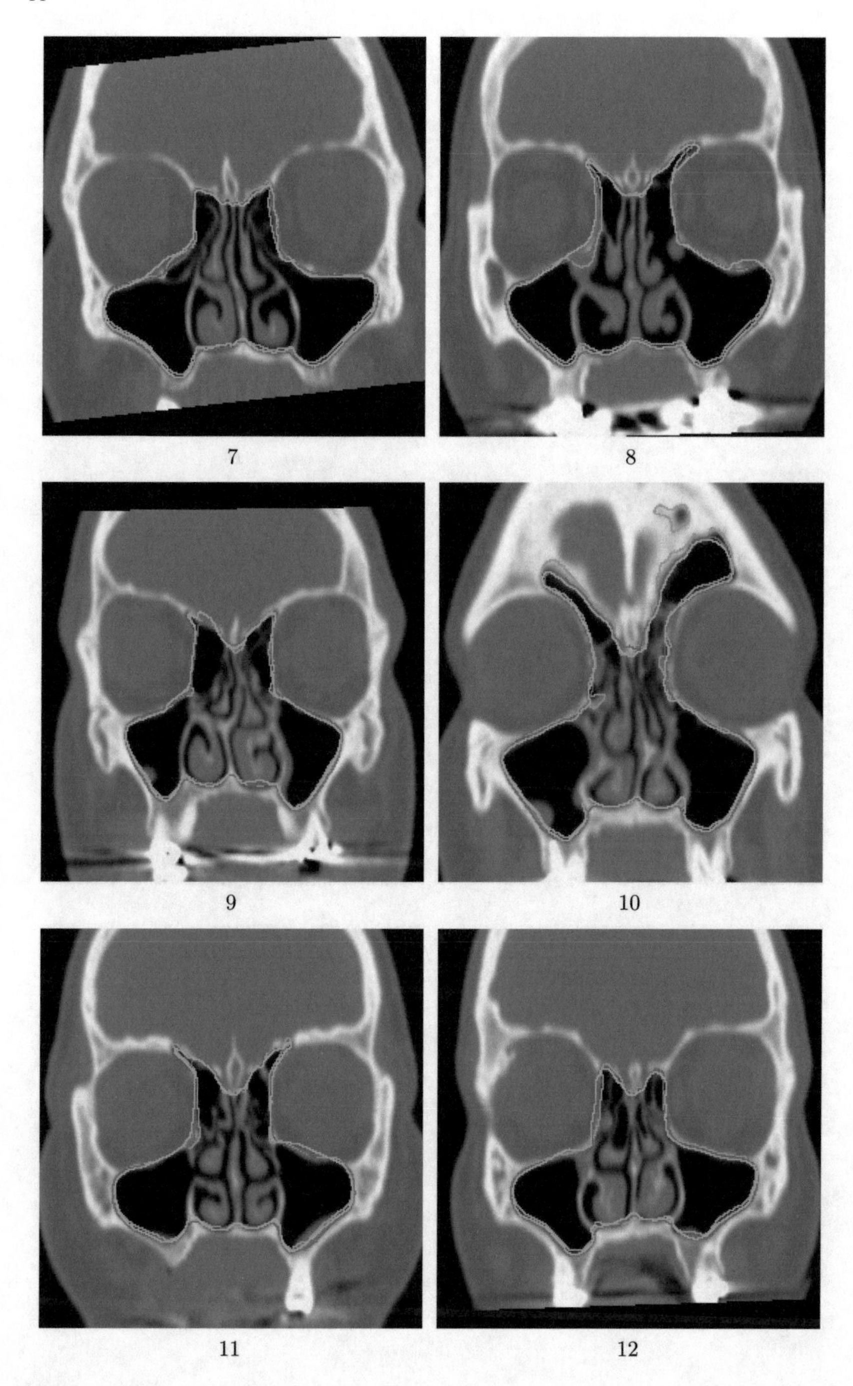

Fig. C.1 (continued)

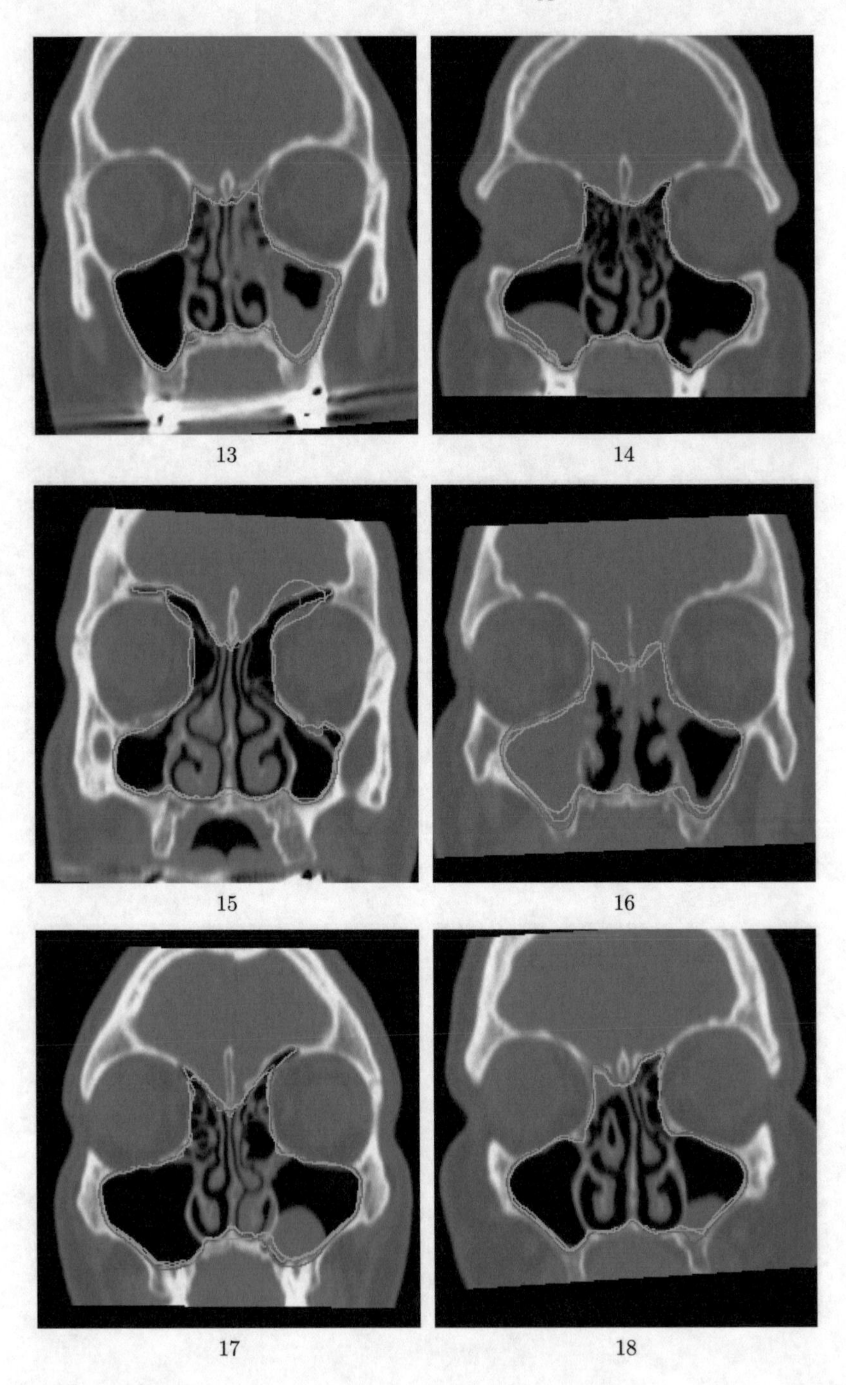

Fig. C.1 (continued)

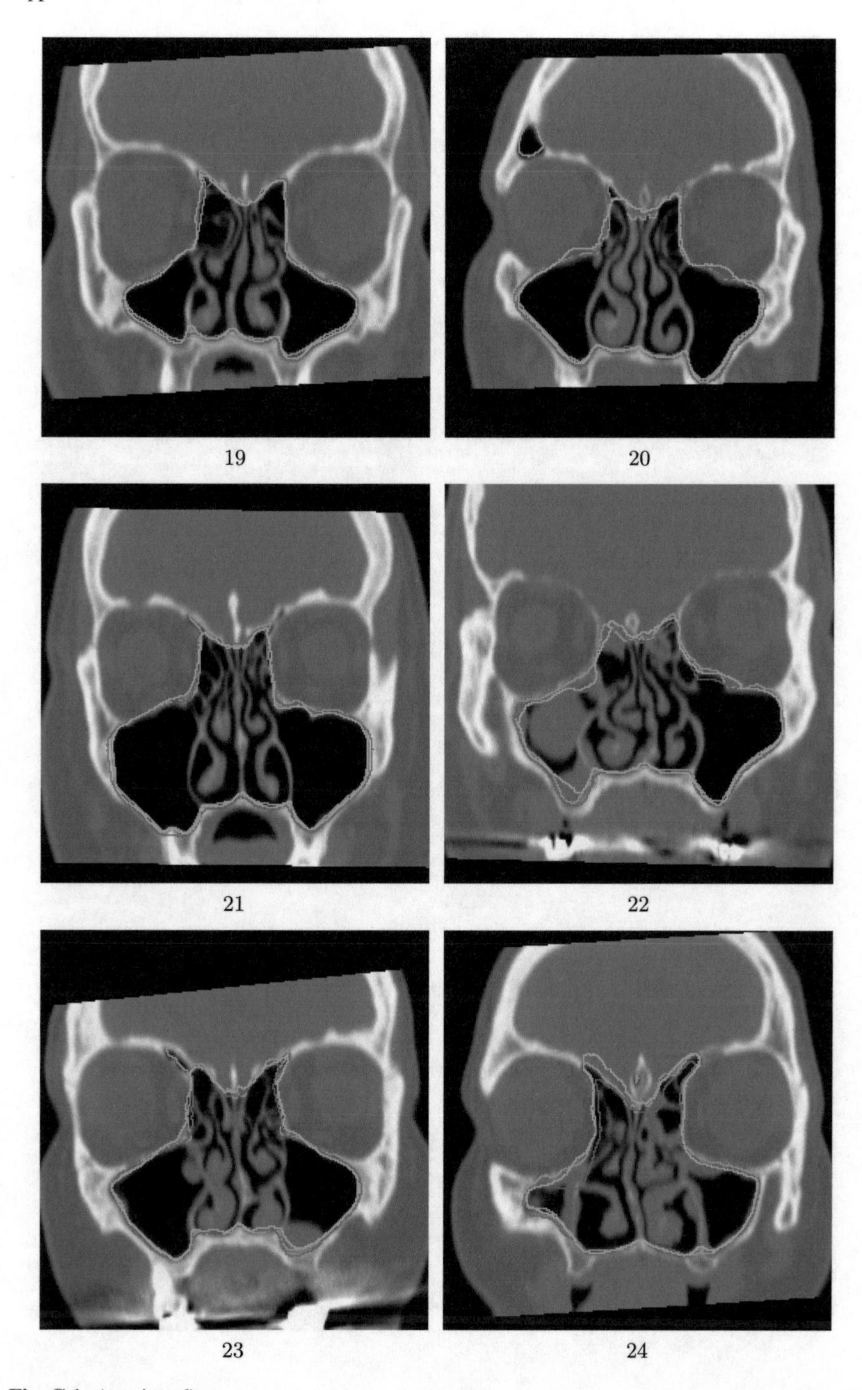

Fig. C.1 (continued)

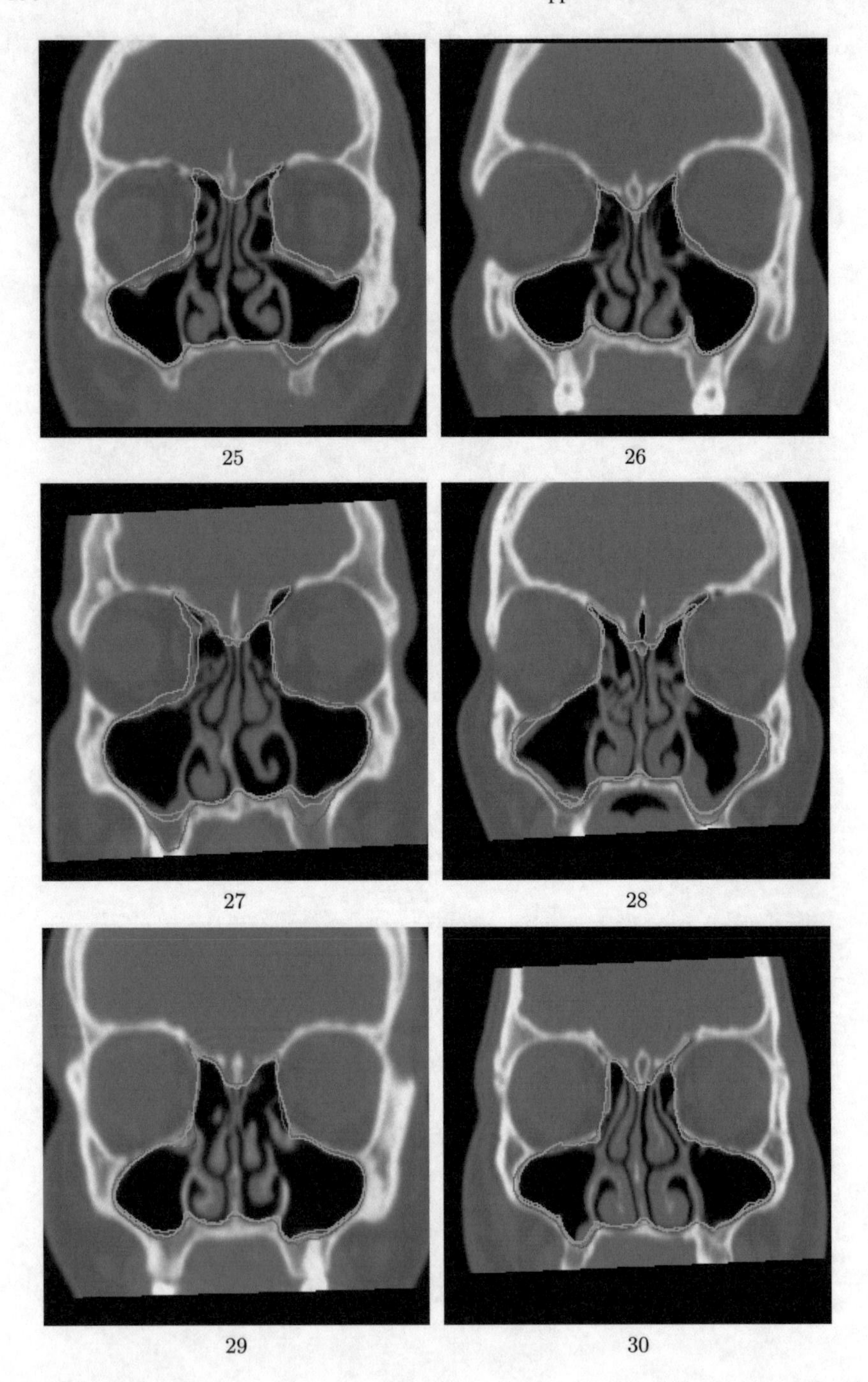

Fig. C.1 (continued)

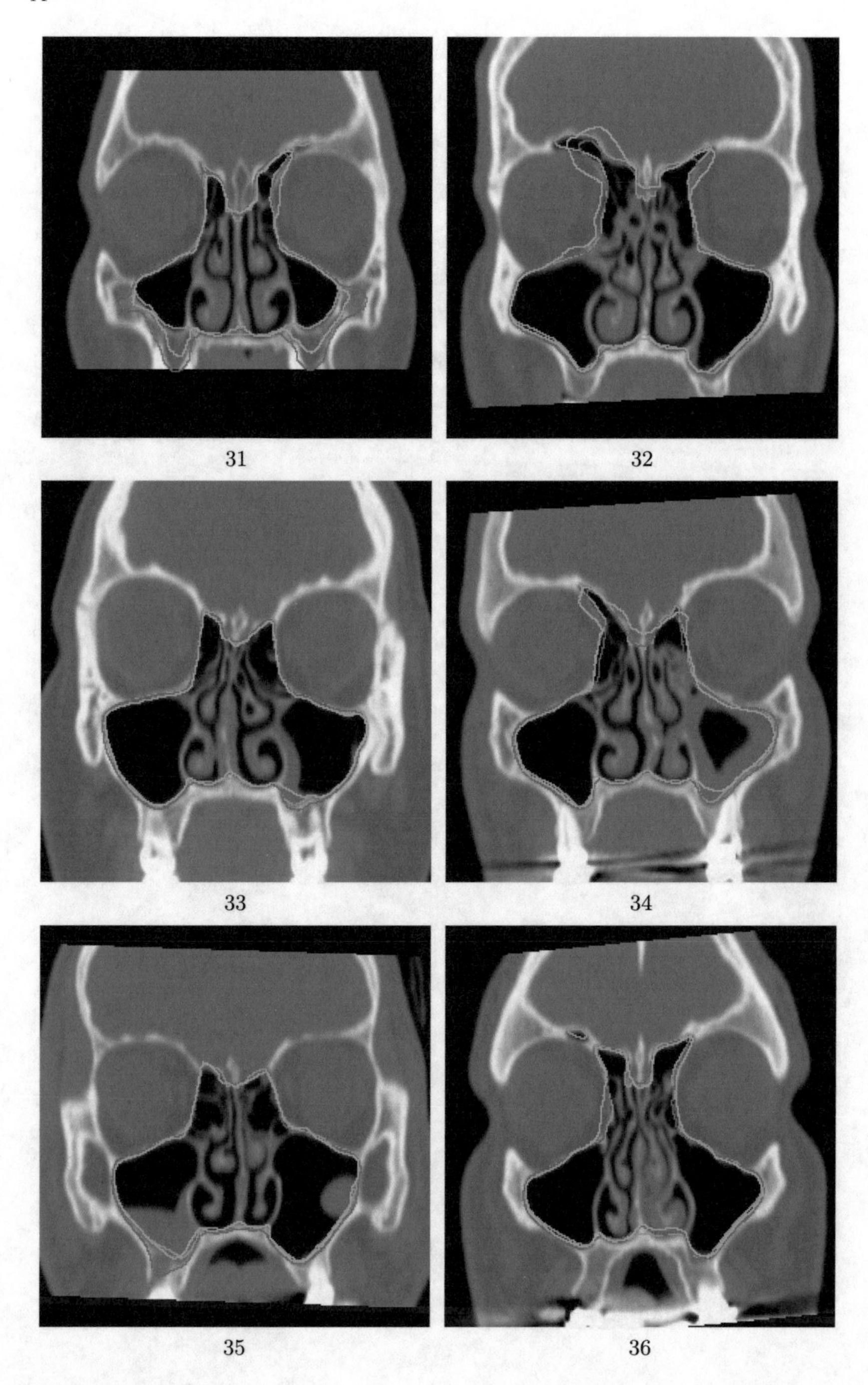

Fig. C.1 (continued)

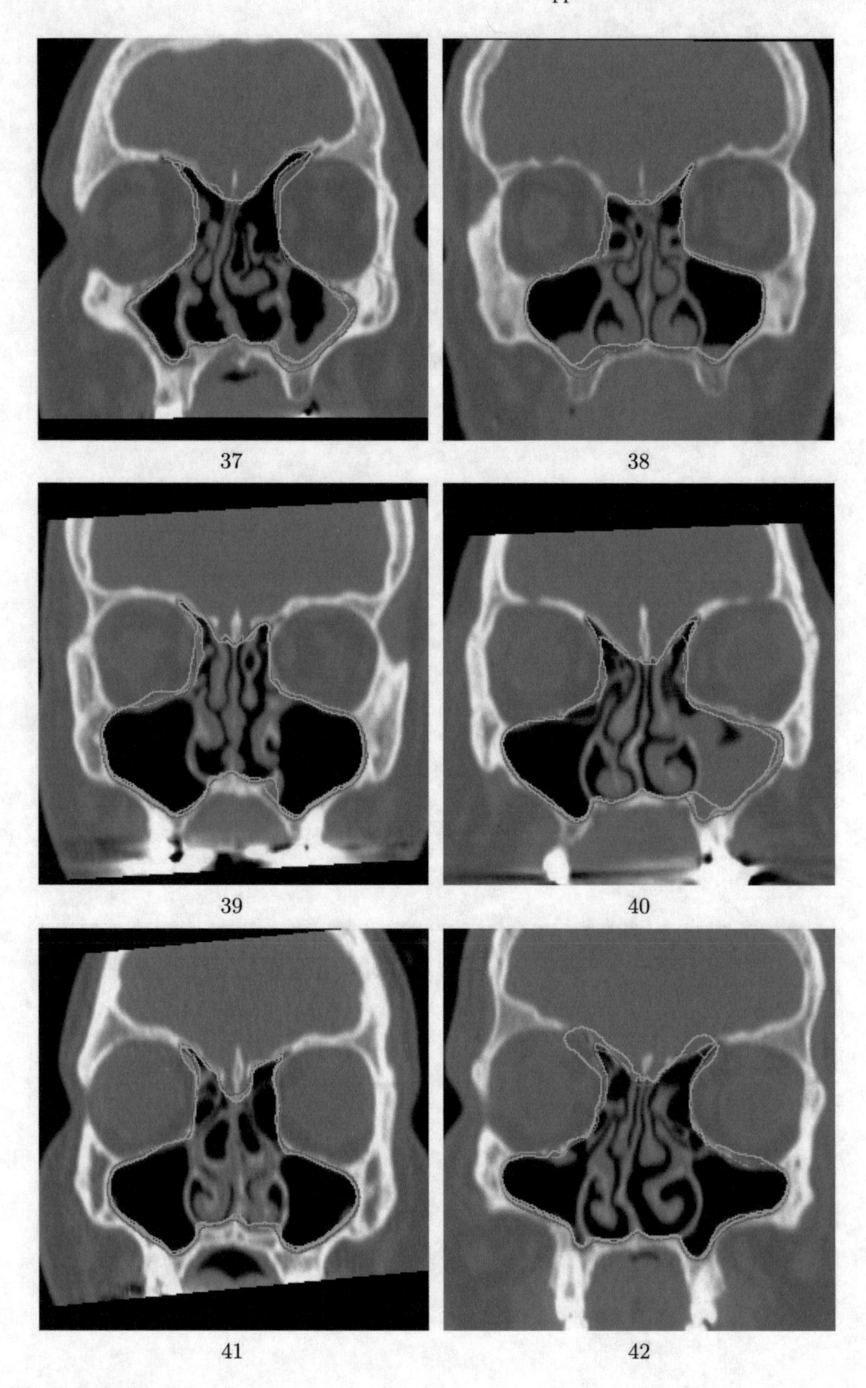

Fig. C.1 (continued)

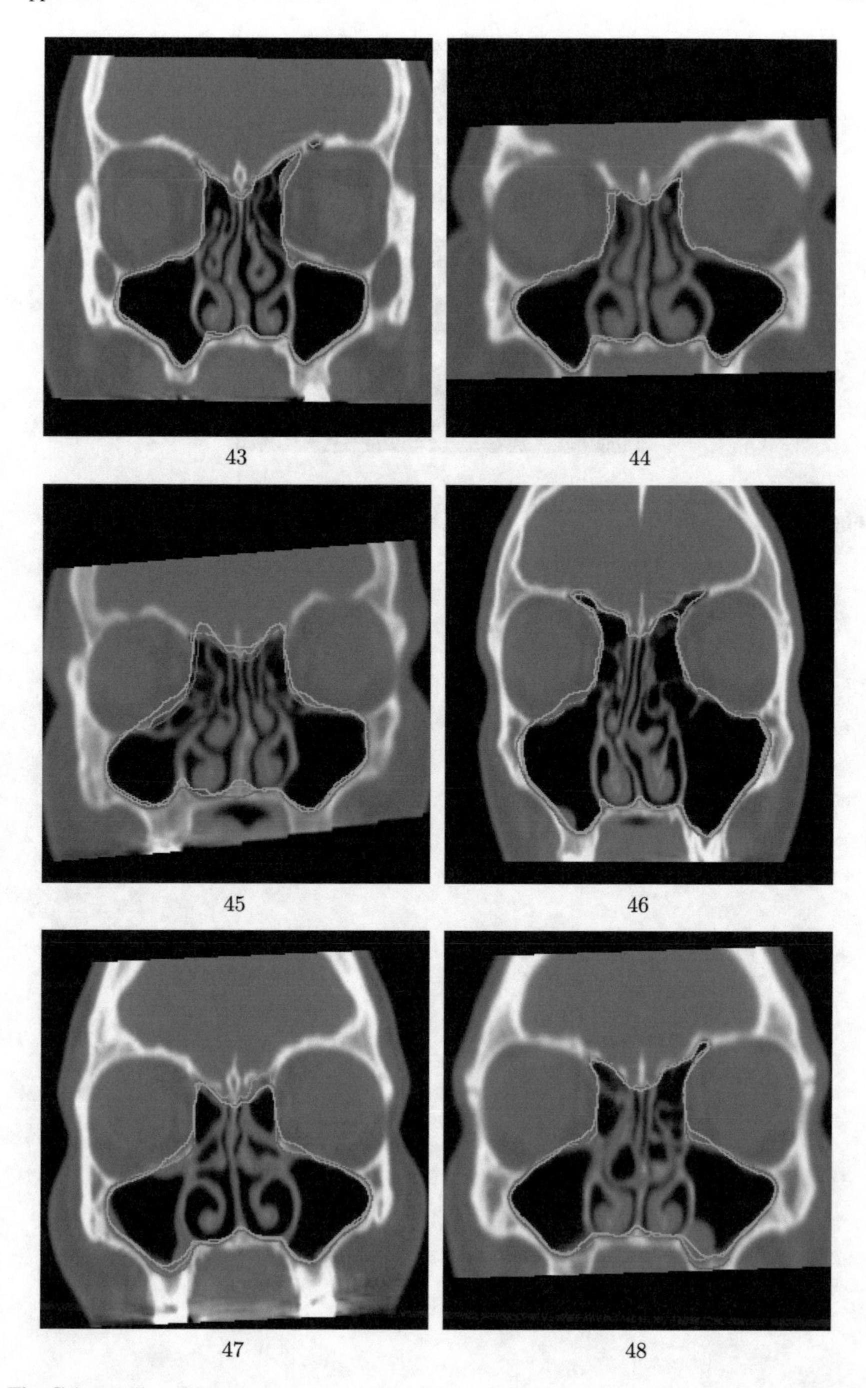

Fig. C.1 (continued)

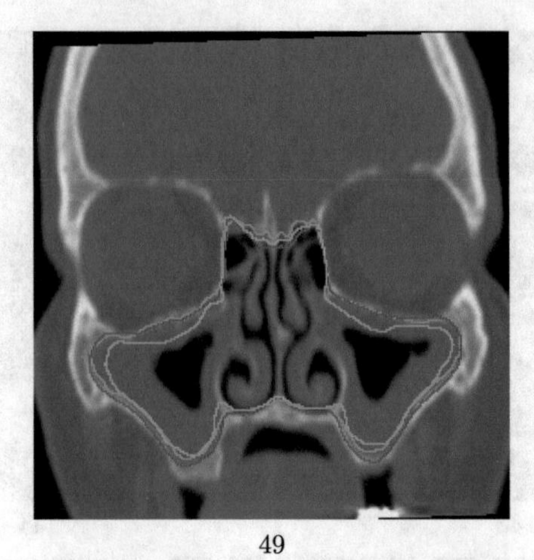

49

Fig. C.1 (continued)

Appendix D
The Sample Covariance Matrix

Before we derive the sample covariance matrix, we first discuss the covariance matrix of a random vector: Given an u-dimensional random vector

$$\vec{x} = (x_1, \dots, x_u)\,, \tag{D.1}$$

i.e. each element x_i of the vector $\vec{x}$ is a real-valued random variable, we search for the covariance matrix $\mathbf{\Sigma}$ that represents all pairwise covariances between two elements x_i and x_j of the vector $\vec{x}$:

$$\mathbf{\Sigma} = \mathrm{Cov}(\vec{x}) = \begin{pmatrix} \mathrm{Cov}(x_1, x_1) & \dots & \mathrm{Cov}(x_u, x_1) \\ \vdots & \ddots & \vdots \\ \mathrm{Cov}(x_1, x_u) & \dots & \mathrm{Cov}(x_u, x_u) \end{pmatrix}. \tag{D.2}$$

The covariance between two random variables is defined as [23, Eq. (16.150)]

$$\mathrm{Cov}(x, y) = E([x - E(x)][y - E(y)])\,, \tag{D.3}$$

where $E(x)$ denotes the expected value of the random variable x. So, Eq. (D.2) can be written as

$$\mathbf{\Sigma} = \begin{pmatrix} E([x_1 - E(x_1)][x_1 - E(x_1)]) & \dots & E([x_u - E(x_u)][x_1 - E(x_1)]) \\ \vdots & \ddots & \vdots \\ E([x_1 - E(x_1)][x_u - E(x_u)]) & \dots & E([x_u - E(x_u)][x_u - E(x_u)]) \end{pmatrix}. \tag{D.4}$$

Now, when we denote the expected value of a matrix as the expected values of its entries, Eq. (D.4) simplifies to

C. Last, *From Global to Local Statistical Shape Priors*, Studies in Systems, Decision and Control 98, DOI 10.1007/978-3-319-53508-1

$$\boldsymbol{\Sigma} = E\left[\begin{pmatrix} [x_1 - E(x_1)][x_1 - E(x_1)) & \dots & [x_u - E(x_u)][x_1 - E(x_1)] \\ \vdots & \ddots & \vdots \\ [x_1 - E(x_1)][x_u - E(x_u)) & \dots & [x_u - E(x_u)][x_u - E(x_u)] \end{pmatrix}\right]. \tag{D.5}$$

With $E(\vec{x}) = (E(x_1), \dots, E(x_u))$, this can be further simplified to

$$\boldsymbol{\Sigma} = E([\vec{x} - E(\vec{x})]^T [\vec{x} - E(\vec{x})]) . \tag{D.6}$$

By denoting the expected value of the random vector $\vec{x}$ as $\vec{\mu} = E(\vec{x})$ and by making use of an algebraic formula for the variance: $E([x - E(x)]^2) = E(x^2) - E(x)^2$ [23, Eq. (16.53)], we finally obtain

$$\boldsymbol{\Sigma} = E([\vec{x} - \vec{\mu}]^T [\vec{x} - \vec{\mu}]) \tag{D.7}$$

$$= E(\vec{x}^T \vec{x}) - \vec{\mu}^T \vec{\mu} \tag{D.8}$$

as the covariance matrix of the random vector $\vec{x}$.

Now, for a sample of the random vector $\vec{x}$ that consists of n vectors $\vec{x}_i$, one commonly uses the sample mean $\hat{\vec{\mu}}$ as an unbiased estimate of the real expected value $\vec{\mu}$:

$$\hat{\vec{\mu}} = \frac{1}{n} \sum_{i=1}^{n} \vec{x}_i , \tag{D.9}$$

so that

$$\hat{E}(\vec{x}^T \vec{x}) = \frac{1}{n} \sum_{i=1}^{n} \vec{x}_i^{\,T} \vec{x}_i \tag{D.10}$$

$$= \frac{1}{n} \left(\vec{x}_1^T, \dots, \vec{x}_n^T\right) \begin{pmatrix} \vec{x}_1 \\ \vdots \\ \vec{x}_n \end{pmatrix}. \tag{D.11}$$

When we stack all row vectors $\vec{x}_i$ on top of each other and denote the thus obtained matrix as $\mathbf{A}$, Eq. (D.11) can be written as

$$\hat{E}(\vec{x}^T \vec{x}) = \frac{1}{n} \mathbf{A}^T \mathbf{A}, \quad \text{with } \mathbf{A} = \begin{pmatrix} \vec{x}_1 \\ \vdots \\ \vec{x}_n \end{pmatrix} \in \mathbb{R}^{n \times u} . \tag{D.12}$$

Now, when we insert Eqs. (D.12) and (D.9) into Eq. (D.8), we obtain an estimate for the sample covariance matrix as

$$\hat{\boldsymbol{\Sigma}}_{\text{bias}} = \frac{1}{n} \mathbf{A}^T \mathbf{A} - \hat{\vec{\mu}}^T \hat{\vec{\mu}} . \tag{D.13}$$

However, it can be shown that the expectation value of $\hat{\mathbf{\Sigma}}_{\text{bias}}$ is biased to the expectation value of the population covariance by a factor of $\frac{n-1}{n}$ [23, Chap. 16.3.1.2].

In order to address this bias, one commonly uses the corrected sample covariance matrix $\hat{\mathbf{\Sigma}}$ instead which is defined as

$$\hat{\mathbf{\Sigma}} = \frac{n}{n-1}\hat{\mathbf{\Sigma}}_{\text{bias}} \tag{D.14}$$

$$= \frac{1}{n-1}\mathbf{A}^T\mathbf{A} - \frac{n}{n-1}\hat{\vec{\mu}}^T\hat{\vec{\mu}}. \tag{D.15}$$

When the sample has zero mean (i.e. $\hat{\vec{\mu}} = \vec{0}$), Eq. (D.15) further reduces to

$$\hat{\mathbf{\Sigma}} = \frac{1}{n-1}\mathbf{A}^T\mathbf{A}. \tag{D.16}$$

Appendix E
More Examples for the Global Variational Formulation

In Sect. 5.3.2, we derived the gradient for one of the three segmentation problems that have been considered by Tsai et al. in [130]. In this chapter, we will give two more examples in order to clarify the idea of their approach.

The two other segmentation problems that have been considered by Tsai et al. are the so-called *Binary Mean Model* [140] and the *Binary Variance Model* [140]. The goal of the *Binary Mean Model* is to partition an image into two regions based on differing mean intensities. It is given as [140]

$$E_{\text{binary}} = -\frac{1}{2}(\mu - \nu)^2 , \tag{E.1}$$

where μ and ν denote the mean intensities inside and outside the modeled shape, respectively, so that

$$\mu = \frac{\int_\Omega I H(-\Phi_{\text{glob}}(\vec{w}))\, d\vec{x}}{\int_\Omega H(-\Phi_{\text{glob}}(\vec{w}))\, d\vec{x}} \tag{E.2}$$

and

$$\nu = \frac{\int_\Omega I H(\Phi_{\text{glob}}(\vec{w}))\, d\vec{x}}{\int_\Omega H(\Phi_{\text{glob}}(\vec{w}))\, d\vec{x}} . \tag{E.3}$$

The goal of the *Binary Variance Model* is to segment an image into two regions with differing mean intensity variances. It reads as [140]

$$E_{\text{variance}} = -\frac{1}{2}(\sigma_\mu^2 - \sigma_\nu^2)^2 , \tag{E.4}$$

where σ_μ^2 and σ_ν^2 denote the intensity variances inside and outside the modeled shape, respectively, so that

$$\sigma_\mu^2 = \frac{\int_\Omega I^2 H(-\Phi_{\text{glob}}(\vec{w}))\, d\vec{x}}{\int_\Omega H(-\Phi_{\text{glob}}(\vec{w}))\, d\vec{x}} - \mu^2 \tag{E.5}$$

C. Last, *From Global to Local Statistical Shape Priors*, Studies in Systems, Decision and Control 98, DOI 10.1007/978-3-319-53508-1

and

$$\sigma_\nu^2 = \frac{\int_\Omega I^2 H(\Phi_{\text{glob}}(\vec{w}))\, d\vec{x}}{\int_\Omega H(\Phi_{\text{glob}}(\vec{w}))\, d\vec{x}} - \nu^2. \tag{E.6}$$

Similar to the example in Sect. 5.3.2, the gradient of the *Binary Mean Model* can be obtained with the help of the chain rule as

$$\begin{aligned}\frac{\partial E_{\text{binary}}}{\partial w_i} = (\nu - \mu)\Bigg[& \Bigg(\frac{\frac{\partial}{\partial w_i}\int_\Omega I H(-\Phi_{\text{glob}}(\vec{w}))\, d\vec{x}}{\int_\Omega H(-\Phi_{\text{glob}}(\vec{w}))\, d\vec{x}} \\ & - \frac{\mu \frac{\partial}{\partial w_i}\int_\Omega H(-\Phi_{\text{glob}}(\vec{w}))\, d\vec{x}}{\int_\Omega H(-\Phi_{\text{glob}}(\vec{w}))\, d\vec{x}} \Bigg) \\ & - \Bigg(\frac{\frac{\partial}{\partial w_i}\int_\Omega I H(\Phi_{\text{glob}}(\vec{w}))\, d\vec{x}}{\int_\Omega H(\Phi_{\text{glob}}(\vec{w}))\, d\vec{x}} \\ & - \frac{\nu \frac{\partial}{\partial w_i}\int_\Omega H(\Phi_{\text{glob}}(\vec{w}))\, d\vec{x}}{\int_\Omega H(\Phi_{\text{glob}}(\vec{w}))\, d\vec{x}} \Bigg) \Bigg].\end{aligned} \tag{E.7}$$

This can be simplified with the Leibniz integral rule and Eq. (5.42) to

$$\begin{aligned}\frac{\partial E_{\text{binary}}}{\partial w_i} = (\nu - \mu)\Bigg(& \frac{-\int_\Omega \delta_{\Phi_{\text{glob}}(\vec{w})} I \tilde{\Phi}_i\, d\vec{x} + \mu \int_\Omega \delta_{\Phi_{\text{glob}}(\vec{w})} \tilde{\Phi}_i\, d\vec{x}}{\int_\Omega H(-\Phi_{\text{glob}}(\vec{w}))\, d\vec{x}} \\ & - \frac{\int_\Omega \delta_{\Phi_{\text{glob}}(\vec{w})} I \tilde{\Phi}_i\, d\vec{x} - \nu \int_\Omega \delta_{\Phi_{\text{glob}}(\vec{w})} \tilde{\Phi}_i\, d\vec{x}}{\int_\Omega H(\Phi_{\text{glob}}(\vec{w}))\, d\vec{x}} \Bigg).\end{aligned} \tag{E.8}$$

The gradient of *Binary Variance Model* is given as

$$\begin{aligned}\frac{\partial E_{\text{variance}}}{\partial w_i} = (\sigma_\nu^2 - \sigma_\mu^2)\Bigg[& \Bigg(\frac{(\mu^2 - \sigma_\mu^2)\frac{\partial}{\partial w_i}\int_\Omega H(-\Phi_{\text{glob}}(\vec{w}))\, d\vec{x}}{\int_\Omega H(-\Phi_{\text{glob}}(\vec{w}))\, d\vec{x}} \\ & - \frac{2\mu \frac{\partial}{\partial w_i}\int_\Omega I H(-\Phi_{\text{glob}}(\vec{w}))\, d\vec{x}}{\int_\Omega H(-\Phi_{\text{glob}}(\vec{w}))\, d\vec{x}} \\ & + \frac{\frac{\partial}{\partial w_i}\int_\Omega I^2 H(-\Phi_{\text{glob}}(\vec{w}))\, d\vec{x}}{\int_\Omega H(-\Phi_{\text{glob}}(\vec{w}))\, d\vec{x}} \Bigg) \\ & - \Bigg(\frac{(\nu^2 - \sigma_\nu^2)\frac{\partial}{\partial w_i}\int_\Omega H(\Phi_{\text{glob}}(\vec{w}))\, d\vec{x}}{\int_\Omega H(\Phi_{\text{glob}}(\vec{w}))\, d\vec{x}} \\ & - \frac{2\nu \frac{\partial}{\partial w_i}\int_\Omega I H(\Phi_{\text{glob}}(\vec{w}))\, d\vec{x}}{\int_\Omega H(\Phi_{\text{glob}}(\vec{w}))\, d\vec{x}} \\ & + \frac{\frac{\partial}{\partial w_i}\int_\Omega I^2 H(\Phi_{\text{glob}}(\vec{w}))\, d\vec{x}}{\int_\Omega H(\Phi_{\text{glob}}(\vec{w}))\, d\vec{x}} \Bigg) \Bigg],\end{aligned} \tag{E.9}$$

which simplifies to

$$\frac{\partial E_{\text{variance}}}{\partial w_i} = (\sigma_\nu^2 - \sigma_\mu^2)\Bigg[\Bigg(-\frac{(\mu^2 - \sigma_\mu^2)\int_\Omega \delta_{\Phi_{\text{glob}}(\vec{w})}\tilde{\Phi}_i \, d\vec{x}}{\int_\Omega H(-\Phi_{\text{glob}}(\vec{w}))\, d\vec{x}} + \frac{2\mu \int_\Omega \delta_{\Phi_{\text{glob}}(\vec{w})} I \tilde{\Phi}_i \, d\vec{x}}{\int_\Omega H(-\Phi_{\text{glob}}(\vec{w}))\, d\vec{x}} - \frac{\int_\Omega \delta_{\Phi_{\text{glob}}(\vec{w})} I^2 \tilde{\Phi}_i \, d\vec{x}}{\int_\Omega H(-\Phi_{\text{glob}}(\vec{w}))\, d\vec{x}} \Bigg) - \Bigg(\frac{(\nu^2 - \sigma_\nu^2)\int_\Omega \delta_{\Phi_{\text{glob}}(\vec{w})}\tilde{\Phi}_i \, d\vec{x}}{\int_\Omega H(\Phi_{\text{glob}}(\vec{w}))\, d\vec{x}} - \frac{2\nu \int_\Omega \delta_{\Phi_{\text{glob}}(\vec{w})} I \tilde{\Phi}_i \, d\vec{x}}{\int_\Omega H(\Phi_{\text{glob}}(\vec{w}))\, d\vec{x}} + \frac{\int_\Omega \delta_{\Phi_{\text{glob}}(\vec{w})} I^2 \tilde{\Phi}_i \, d\vec{x}}{\int_\Omega H(\Phi_{\text{glob}}(\vec{w}))\, d\vec{x}} \Bigg) \Bigg] . \tag{E.10}$$

In Eqs. (E.7) and (E.9) it can be seen that, for the computation of the gradient, the partial derivative with regard to w_i has to be computed for terms that all have the form

$$\frac{\partial}{\partial w_i} \int_\Omega I^j H(\pm\Phi_{\text{glob}}(\vec{w}))\, d\vec{x} \,, \; j = \{0, 1, 2\} . \tag{E.11}$$

They result in terms of the form (c.f. Eqs. (E.8) and (E.10))

$$\pm \int_\Omega \delta_{\Phi_{\text{glob}}(\vec{w})} I^j \tilde{\Phi}_i \, d\vec{x} \,, \; j = \{0, 1, 2\} \tag{E.12}$$

after taking the partial derivative with regard to w_i. This is very similar to the terms which occur in the gradient of the first example from Sect. 5.3.2. So, it can easily be shown that relation (5.44) also holds for the *Binary Mean Model* and the *Binary Variance Model*:

$$\frac{\partial E_{\text{mean}}(\Phi_{\text{glob}}(\vec{w}))}{\partial w_i} = dE_{\text{mean}}(\Phi, \tilde{\Phi}_i) \tag{E.13}$$

$$\frac{\partial E_{\text{variance}}(\Phi_{\text{glob}}(\vec{w}))}{\partial w_i} = dE_{\text{variance}}(\Phi, \tilde{\Phi}_i) . \tag{E.14}$$

This means, likewise to Sect. 5.3.2, the gradients in Eqs. (E.8) and (E.10) can be interpreted as the orthogonal projection of the functional derivatives $\frac{\partial E_{\text{mean}}(\Phi)}{\partial \Phi(\vec{x})}$ and $\frac{\partial E_{\text{variance}}(\Phi)}{\partial \Phi(\vec{x})}$ into the low-dimensional SSM subspace. For a general solution, we refer the reader to Sect. 5.3.3.

Appendix F
Rigid Shape Alignment

As mentioned earlier in this thesis, the rigid shape alignment can be achieved for example by manually defining a few distinctive corresponding points (so-called *landmarks*) on the training data of the SSM and the data under consideration, respectively, and to align the considered data and the SSM by a similarity transformation (as described in Sect. 2.4.1). For surface data, this similarity transformation can also automatically determined by the method that we have described in Sect. 4.3.3.1.

However, an alternative approach to deal with the rigid transformation from the SSM to the data under consideration would be to consider the parameters of the similarity transformation as further unknowns in the segmentation process that have to be identified simultaneously to the determination of the weight vector $\vec{w}$. This approach has been pursued by Tsai et al. in [130]. So, we will briefly outline their approach in the rest of this section for the sake of completeness, but we will not include this determination of rigid transformation parameters in our experimental evaluations.

Tsai et al. proposed to combine the unknown parameters of the similarity transformation that rigidly aligns the SSM to the data in a vector $\vec{p}$ and to perform simultaneous gradient descent with regard to the vector $\vec{p}$ that describes the rigid transformation and the vector $\vec{w}$ which describes the nonrigid transformation. So, Eq. (5.33) modifies to

$$\begin{aligned} \vec{w}^{k+1} &= \vec{w}^k - \gamma_{\vec{w}}^k \, \nabla_{\vec{w}} E(\Phi_{\text{glob}}(\vec{w}^k, \vec{p}^k)) \\ \vec{p}^{k+1} &= \vec{p}^k - \gamma_{\vec{p}}^k \, \nabla_{\vec{p}} E(\Phi_{\text{glob}}(\vec{w}^k, \vec{p}^k)), \end{aligned} \tag{F.1}$$

where $\nabla_{\vec{w}} E(\Phi_{\text{glob}}(\vec{w}, \vec{p}))$ denotes the gradient of the energy function with regard to the weight vector $\vec{w}$ of the SSM (as defined in Sect. 5.3.3) and $\nabla_{\vec{p}} E(\Phi_{\text{glob}}(\vec{w}, \vec{p}))$ denotes the gradient of the energy function with regard to the vector of rigid transformation parameters $\vec{p}$. The respective step sizes in each iteration k are indicated by $\gamma_{\vec{w}}^k$ and $\gamma_{\vec{p}}^k$.

C. Last, *From Global to Local Statistical Shape Priors*, Studies in Systems, Decision and Control 98, DOI 10.1007/978-3-319-53508-1

In Eq. (F.1), $\Phi_{\text{glob}}(\vec{w}, \vec{p})$ is the global SSM $\Phi_{\text{glob}}(\vec{w})$ that is additionally transformed by the above mentioned similarity transformation $T(\vec{p})$ with parameters $\vec{p}$. For the 2D case, this similarity transformation can be written as the product of the following three homogeneous transformation matrices:

$$T(\vec{p}) = \begin{pmatrix} 1 & 0 & a \\ 0 & 1 & b \\ 0 & 0 & 1 \end{pmatrix} \begin{pmatrix} h & 0 & 0 \\ 0 & h & 0 \\ 0 & 0 & 1 \end{pmatrix} \begin{pmatrix} \cos(\theta) & -\sin(\theta) & 0 \\ \sin(\theta) & \cos(\theta) & 0 \\ 0 & 0 & 1 \end{pmatrix}. \tag{F.2}$$

Here, a and b are the translations in x- and y-direction, h is the isotropic scale, and θ is the rotation angle, so that the parameter vector $\vec{p}$ contains 4 unknowns:

$$\vec{p} = (a, b, h, \theta). \tag{F.3}$$

The connection between $\Phi_{\text{glob}}(\vec{w}, \vec{p})$ and $\Phi_{\text{glob}}(\vec{w})$ can be expressed as

$$\Phi_{\text{glob}}(\vec{w}, \vec{p})(x, y) = \Phi_{\text{glob}}(\vec{w})(\tilde{x}, \tilde{y}) = \bar{\Phi}(\tilde{x}, \tilde{y}) + \sum_{i=1}^{m} w_i\, \tilde{\Phi}_i(\tilde{x}, \tilde{y}), \tag{F.4}$$

where the relation between the coordinates $(\tilde{x}, \tilde{y})$ of the untransformed SSM and the coordinates (x, y) of the transformed SSM is given by

$$\begin{pmatrix} \tilde{x} \\ \tilde{y} \\ 1 \end{pmatrix} = T(\vec{p}) \begin{pmatrix} x \\ y \\ 1 \end{pmatrix}, \tag{F.5}$$

so that for $a = b = 0$, $h = 1$, and $\theta = 0$ the untransformed and the transformed SSM are identical. The similarity transformation for the 3D case can be described in a similar way. However, for the 3D case the parameter vector $\vec{p}$ contains 7 unknowns, as also the translation in z-direction and the rotations around the other two spatial axes have to be determined.

Similar to Eq. (5.47), the partial derivative of the energy function $E(\Phi_{\text{glob}}(\vec{w}, \vec{p}))$ with regard to the i-th pose parameter p_i can be obtained with the help of the chain rule for the functional derivative as

$$\frac{\partial E(\Phi_{\text{glob}}(\vec{w}, \vec{p}))}{\partial p_i} = \int_{\Omega} \frac{\partial E(\Phi_{\text{glob}}(\vec{w}, \vec{p}))}{\partial \Phi_{\text{glob}}(\vec{w}, \vec{p})(\vec{x})} \frac{\partial \Phi_{\text{glob}}(\vec{w}, \vec{p})(\vec{x})}{\partial p_i}\, d\vec{x}. \tag{F.6}$$

In Eq. (F.6), the partial derivative of the transformed SSM $\Phi_{\text{glob}}(\vec{w}, \vec{p})$ with regard to the i-th pose parameter p_i can be calculated by considering that the coordinates $(\tilde{x}, \tilde{y})$ of the untransformed SSM can be expressed as a function of the coordinates (x, y) of the transformed SSM (c.f. Eqs. (F.4) and (F.5)).

So, one can use once more the chain rule in order to obtain the desired derivative. For the 2D case, the partial derivative of the transformed SSM $\Phi_{\text{glob}}(\vec{w}, \vec{p})$ with regard to the i-th pose parameter p_i reads as (c.f [130])

$$\frac{\partial \Phi_{\text{glob}}(\vec{w}, \vec{p})(x, y)}{\partial p_i} = \frac{\partial \Phi_{\text{glob}}(\vec{w})(\tilde{x}, \tilde{y})}{\partial p_i} = \left(\frac{\partial \Phi_{\text{glob}}(\vec{w})(\tilde{x}, \tilde{y})}{\partial \tilde{x}}, \frac{\partial \Phi_{\text{glob}}(\vec{w})(\tilde{x}, \tilde{y})}{\partial \tilde{y}}, 0 \right) \frac{\partial T(\vec{p})}{\partial p_i} \begin{pmatrix} x \\ y \\ 1 \end{pmatrix}, \tag{F.7}$$

where the partial derivatives of the transformation matrix $T(\vec{p})$ with regard to the individual transformation parameters are given as

$$\begin{aligned} \frac{\partial T(\vec{p})}{\partial a} &= \begin{pmatrix} 0 & 0 & 1 \\ 0 & 0 & 0 \\ 0 & 0 & 0 \end{pmatrix}, & \frac{\partial T(\vec{p})}{\partial h} &= \begin{pmatrix} \cos(\theta) & -\sin(\theta) & 0 \\ \sin(\theta) & \cos(\theta) & 0 \\ 0 & 0 & 0 \end{pmatrix} \\ \frac{\partial T(\vec{p})}{\partial b} &= \begin{pmatrix} 0 & 0 & 0 \\ 0 & 0 & 1 \\ 0 & 0 & 0 \end{pmatrix}, & \frac{\partial T(\vec{p})}{\partial \theta} &= \begin{pmatrix} -h\sin(\theta) & -h\cos(\theta) & 0 \\ h\cos(\theta) & -h\sin(\theta) & 0 \\ 0 & 0 & 0 \end{pmatrix}. \end{aligned} \tag{F.8}$$

By inserting this into Eq. (F.7), we obtain

$$\begin{aligned} \frac{\partial \Phi_{\text{glob}}(\vec{w}, \vec{p})(x, y)}{\partial a} &= \frac{\partial \Phi_{\text{glob}}(\vec{w})(\tilde{x}, \tilde{y})}{\partial \tilde{x}} \\ \frac{\partial \Phi_{\text{glob}}(\vec{w}, \vec{p})(x, y)}{\partial b} &= \frac{\partial \Phi_{\text{glob}}(\vec{w})(\tilde{x}, \tilde{y})}{\partial \tilde{y}} \end{aligned} \tag{F.9}$$

$$\frac{\partial \Phi_{\text{glob}}(\vec{w}, \vec{p})(x, y)}{\partial h} = \left(\frac{\partial \Phi_{\text{glob}}(\vec{w})(\tilde{x}, \tilde{y})}{\partial \tilde{x}}, \frac{\partial \Phi_{\text{glob}}(\vec{w})(\tilde{x}, \tilde{y})}{\partial \tilde{y}} \right) \times \begin{pmatrix} \cos(\theta) & -\sin(\theta) \\ \sin(\theta) & \cos(\theta) \end{pmatrix} \begin{pmatrix} x \\ y \end{pmatrix} \tag{F.10}$$

$$\frac{\partial \Phi_{\text{glob}}(\vec{w}, \vec{p})(x, y)}{\partial \theta} = \left(\frac{\partial \Phi_{\text{glob}}(\vec{w})(\tilde{x}, \tilde{y})}{\partial \tilde{x}}, \frac{\partial \Phi_{\text{glob}}(\vec{w})(\tilde{x}, \tilde{y})}{\partial \tilde{y}} \right) \times \begin{pmatrix} -h\sin(\theta) & -h\cos(\theta) \\ h\cos(\theta) & -h\sin(\theta) \end{pmatrix} \begin{pmatrix} x \\ y \end{pmatrix}. \tag{F.11}$$

The extension of Eq. (F.7) to the 3D case is straightforward.

Own Publications and References

Own Publications

1. Eichhorn KWG, Westphal R, Last C, Rilk M, Bootz F, Wahl FM, Jakob M (2015) Workspace and pivot point for robot-assisted endoscope guidance in functional endonasal sinus surgery (FESS). Int J Med Robot Comput Assist Surg 11:30–37
2. Last C, Winkelbach S, Wahl FM (2014) Global-to-local shape priors for variational image segmentation. In: IEEE international conference on image processing - ICIP 2014. IEEE, pp 6056–6060
3. Last C, Winkelbach S (2014) Global-to-local statistical shape priors. In: Symposium on statistical shape models & applications - SHAPE 2014, SICAS, p 30
4. Westphal R, Eichhorn KWG, Last C, Rilk M, Bootz F, Wahl FM (2013) Der Einsatz chirurgischer Navigation zur Beschreibung von Arbeitsräumen bei FESS Operationen. 12. Jahrestagung der Deutschen Gesellschaft für Computer- und Roboterassistierte Chirurgie, Innsbruck Medical University, pp 37–40
5. Last C, Winkelbach S, Wahl FM (2013) A new framework for fitting shape models to range scans: local statistical shape priors without correspondences. In: 18th international workshop on vision, modeling & visualization - VMV 2013. The Eurographics Association, pp 153–160
6. Last C, Namueangrak T, Westphal R, Rilk M, Eichhorn K, Bootz F, Wahl F (2013) An approach towards the automatic extraction of critical structures from CT-Data for endoscopic sinus surgeries. 17th annual conference of the international society for computer aided surgery - CARS 2013. Springer, New York, pp 359–360
7. Last C, Sommerkorn A, Westphal R, Wiebking U, Krettek C, Wahl F (2012) Fully automatic knee CT image segmentation with locally adaptive shape priors. 16th annual conference of the international society for computer aided surgery - CARS 2012. Springer, New York, pp 406–408
8. Dorda P, Westphal R, Last C, Bredow J, Schlüter-Brust K, Eysel P, Wahl F (2012) X-ray based identification of implanted knee prostheses. 16th annual conference of the international society for computer aided surgery - CARS 2012. Springer, New York, pp 379–380
9. Last C, Winkelbach S, Wahl FM, Eichhorn KWG, Bootz F (2011) A locally deformable statistical shape model. Machine learning in medical imaging, 2nd international workshop - MLMI 2011, LNCS, vol 7009. Springer, Berlin, pp 51–58
10. Last C, Winkelbach S, Wahl FM, Eichhorn KWG, Bootz F (2010) A model-based approach to the segmentation of nasal cavity and paranasal sinus boundaries. In: Pattern recognition - DAGM 2010, 32nd DAGM symposium, LNCS, vol 6376. Springer, Berlin, pp 333–342

C. Last, *From Global to Local Statistical Shape Priors*, Studies in Systems, Decision and Control 98, DOI 10.1007/978-3-319-53508-1

References

11. Amberg M, Lüthi M, Vetter T (2010) Local regression based statistical model fitting. Pattern recognition - DAGM 2010, 32nd DAGM symposium, LNCS, vol 6376. Springer, Berlin, pp 452–461
12. Apelt D, Preim B, Hahn HK, Strauß G (2004) Bildanalyse und Visualisierung für die Planung von Nasennebenhöhlen-Operationen. Bildverarbeitung für die Medizin, Informatik aktuell. Springer, Berlin, pp 194–198
13. Bagnell JA (2012) Functional gradient descent. In: Course 16–831: statistical techniques in robotics, Lecture 21, Class notes, Carnegie Mellon University. https://www.cs.cmu.edu/afs/cs.cmu.edu/Web/People/16831-f12/notes/F12/16831_lecture21_danielsm.pdf
14. Berlit PD, Grams AE (2010) Computertomografie. In: Bildgebende Diagnostik in Neurologie und Neurochirurgie, Georg Thieme Verlag KG, pp 3–6
15. Besl PJ, McKay ND (1992) A method for registration of 3-D shapes. IEEE Trans Pattern Anal Mach Intell 14(2):239–256
16. Bird RB, Stewart WE, Lightfoot EN (2007) Transport phenomena, 2nd edn. Wiley, New York
17. Blais F (2004) Review of 20 years of range sensor development. J Electr Imaging 13(1):231–240
18. Blake A, Isard M (1998) Active contours. Springer, London
19. Blanz V, Vetter T (1999) A morphable model for the synthesis of 3D faces. In: Computer graphics and interactive techniques - SIGGRAPH 2013, 26th International conference. ACM, pp 187–194
20. Bowyer KW, Chang K, Flynn P (2006) A survey of approaches and challenges in 3d and multi-modal 3d + 2d face recognition. Comput Vis Image Underst 101(1):1–15
21. Boykov Y, Funka-Lea G (2006) Graph cuts and efficient N-D image segmentation. Int J Comput Vision 70(2):109–131
22. Bresson X, Vandergheynst P, Thiran JP (2003) A priori information in image segmentation: energy functional based on shape statistical model and image information. In: International conference on image processing - ICIP 2003. IEEE, pp 425–428
23. Bronshtein I, Semendyayev K, Musiol G, Mühlig H (2015) Handbook of mathematics, 6th edn. Springer, Berlin
24. de Bruijne M, van Ginneken B, Viergever MA, Niessen WJ (2003) Adapting active shape models for 3D segmentation of tubular structures in medical images. Information processing in medical imaging - IPMI 2003, 18th international conference, LNCS, vol 2732. Springer, Berlin, pp 136–147
25. Caselles V, Kimmel R, Sapiro G (1997) Geodesic active contours. Int J Comput Vis 22(1):61–79
26. Chan TF, Vese LA (2001) Active contours without edges. IEEE Trans Image process 10(2):266–277
27. Chen Y, Tagare HD, Thiruvenkadam S, Huang F, Wilson D, Gopinath KS, Briggs RW, Geiser EA (2002) Using prior shapes in geometric active contours in a variational framework. Int J Comput Vis 50(3):315–328
28. Cootes T, Taylor C (1995) Combining point distribution models with shape models based on finite element analysis. Image Vis Comput 13(5):403–409
29. Cootes T, Taylor C (1996) Data driven refinement of active shape model search. In: Proceedings of the British machine vision conference - BMVC 1996. BMVA Press, pp 10.1-10.10
30. Cootes T, Taylor C (2000) Combining elastic and statistical models of appearance variation. 6th European conference on computer vision - ECCV 2000, LNCS, vol 1842. Springer, Berlin, pp 149–163
31. Cootes T, Taylor C (2004) Statistical models of appearance for computer vision. Technical report, University of Manchester. http://personalpages.manchester.ac.uk/staff/timothy.f.cootes/Models/app_models.pdf
32. Cootes T, Taylor C, Cooper D, Graham J (1995) Active shape models - their training and application. Comput Vis Image Underst 61(1):28–59

33. Cootes T, Roberts M, Babalola K, Taylor C (2015) Active shape and appearance models. In: Paragios N, Duncan J, Ayache N (eds) Handbook of biomedical imaging. Springer, Berlin, pp 105–122
34. Cootes TF, Taylor CJ, Cooper DH, Graham J (1992) Training models of shape from sets of examples. Proceedings of the british machine vision conference - BMVC 1992. Springer, London, pp 9–18
35. Cremers D (2006) Dynamical statistical shape priors for level set-based tracking. IEEE Trans Pattern Anal Mach Intell 28(8):1262–1273
36. Cremers D (2008) Nonlinear dynamical shape priors for level set segmentation. J Sci Comput 35(2–3):132–143
37. Cremers D (2013) Shape priors for image segmentation. In: Shape perception in human and computer vision, Springer, London
38. Cremers D, Kohlberger T, Schnörr C (2003) Shape statistics in kernel space for variational image segmentation. Pattern Recogn 36(9):1929–1943
39. Cremers D, Osher SJ, Soatto S (2006) Kernel density estimation and intrinsic alignment for shape priors in level set segmentation. Int J Comput Vis 69(3):335–351
40. Cremers D, Rousson M, Deriche R (2007) A review of statistical approaches to level set segmentation: integrating color, texture, motion and shape. Int J Comput Vis 72(2):195–215
41. Cremers D, Schmidt FR, Barthel F (2008) Shape priors in variational image segmentation: convexity, lipschitz continuity and globally optimal solutions. In: IEEE conference on computer vision and pattern recognition - CVPR 2008. IEEE, pp 1–6
42. Curless B, Levoy M (1996) A volumetric method for building complex models from range images. In: Computer Graphics and Interactive Techniques - SIGGRAPH 1996, 23rd international conference. ACM, pp 303–312
43. Davatzikos C, Tao X, Shen D (2003) Hierarchical active shape models, using the wavelet transform. IEEE Trans Med Imaging 22(3):414–423
44. DAVID Vision Systems GmbH (2013) DAVID-SLS-2 Structured Light 3D Scanner. http://www.david-3d.com/de/products/sls-2
45. Deserno TM (ed) (2011a) Biomedical image processing. Springer, Berlin
46. Deserno TM (2011b) Fundamentals of biomedical image processing. In: Biomedical image processing. Springer, Berlin
47. Dhawan AP (2003) Medical image analysis. Wiley-IEEE Press, New York
48. Dubuisson MP, Jain AK (1994) A modified hausdorff distance for object matching. In: Proceedings of the 12th IAPR international conference on pattern recognition, vol 1 - Conference A: computer vision & image processing. IEEE, pp 566–568
49. Engel E, Dreizler RM (2011) Appendix A - functionals and the functional derivative. In: Density functional theory: an advanced course. Springer, Berlin
50. FEI Visualization Sciences Group (2011) Amira 3D Software for Life Sciences. http://www.fei.com/software/amira-3d-for-life-sciences/
51. Felzenszwalb P (2005) Representation and detection of deformable shapes. IEEE Trans Pattern Anal Mach Intell 27(2):208–220
52. Felzenszwalb PF, Huttenlocher DP (2012) Distance transforms of sampled functions. Theory Comput 8(19):415–428
53. Fischler MA, Bolles RC (1981) Random sample consensus: a paradigm for model fitting with applications to image analysis and automated cartography. Commun ACM 24(6):381–395
54. Franke D (2014) Graph cut segmentation with statistical shape priors. Master's thesis, Technische Universität Braunschweig, supervisor: C. Last
55. Frigyik BA, Srivastava S, Gupta MR (2008) An introduction to functional derivatives. Technical report. UWEETR-2008-0001, University of Washington
56. Gonzalez RC, Woods RE (2008) Digital image processing, 3rd edn. Prentice Hall, Upper Saddle River
57. Graphics and Vision Research Group, University of Basel (2009) The Basel Face Model. http://faces.cs.unibas.ch

58. Greiner W, Reinhardt J (1996) Chapter 2.3 - Functional Derivatives. In: Field quantization. Springer, Berlin, pp 37–39
59. Heimann T, Delingette H (2011) Model-based segmentation. In: Biomedical image processing. Springer, Berlin
60. Heimann T, Meinzer HP (2009) Statistical shape models for 3D medical image segmentation: A review. Med Image Anal 13:543–563
61. Himmelblau DM (1972) Applied nonlinear programming. McGraw-Hill, New York
62. Honrado CP, Larrabee WFJ (2004) Update in three-dimensional imaging in facial plastic surgery. Curr Opin Otolaryngol Head Neck Surg 12(4):327–331
63. Horn B, Schunck B (1981) Determining optical flow. Artif Intell 17:185–203
64. Dryden IL, Mardia KV (2016) Statistical shape analysis, with applications in R, 2nd edn. Wiley, Chichester
65. Johnson HJ, McCormick MM, Ibanez L (2015) The ITK software guide book 2: design and functionality, 4th edn. Kitware, Incorporated
66. Jolliffe IT (2002) Principal component analysis, 2nd edn. Springer series in statistics, Springer, New York
67. Jud C (2014) Object segmentation by fitting statistical shape models: a kernel-based approach with application to wisdom tooth segmentation from CBCT images. Phd thesis, Universität Basel. http://edoc.unibas.ch/diss/DissB_10884
68. Jud C, Vetter T (2014) Geodesically damped shape models. In: Symposium on statistical shape models & applications - SHAPE 2014, SICAS, p 13
69. Kalpathy-Cramer J, Müller H (2011) Systematic evaluations and ground truth. In: Biomedical image processing. Springer, Berlin
70. Kendall DG (1977) The diffusion of shape. Adv Appl Probab 9(3):428–430
71. Knothe R (2009) A global-to-local model for the representation of human faces. Phd thesis, Universität Basel. http://edoc.unibas.ch/diss/DissB_8817
72. Koikkalainen J, Tölli T, Lauerma K, Antila K, Mattila E, Lilja M, Lötjönen J (2008) Methods of artificial enlargement of the training set for statistical shape models. IEEE Trans Med Imaging 27(11):1643–1654
73. Kovesi PD (2000) MATLAB and octave functions for computer vision and image processing. Centre for Exploration Targeting, School of Earth and Environment, The University of Western Australia, Perth, Australia. http://www.peterkovesi.com/matlabfns/
74. Lagarias JC, Reeds JA, Wright MH, Wright PE (1998) Convergence properties of the nelder-mead simplex method in low dimensions. SIAM J Optim 9(1):112–147
75. Leont'ev AF (2002) Finite-difference calculus. In: Encyclopaedia of mathematics. Kluwer Academic Publishers, Dordrecht
76. Lester H, Arridge SR (1999) A survey of hierarchical non-linear medical image registration. Pattern Recogn 32(1):129–149
77. Leventon ME, Grimson WEL, Faugeras O (2000) Statistical Shape Influence in Geodesic Active Contours. In: IEEE conference on computer vision and pattern recognition - CVPR 2000. vol 1. IEEE, pp 316–323
78. Li C, Xu C, Gui C, Fox MD (2005) Level set evolution without re-initialization: a new variational formulation. In: IEEE computer society conference on computer vision and pattern recognition - CVPR, 2005, vol 1, pp 430–436
79. Li H, Yu J, Ye Y, Bregler C (2013) Realtime facial animation with on-the-fly correctives. In: Computer graphics and interactive techniques - SIGGRAPH 2013, 40th international conference, ACM, pp 42:1–42:10
80. Loog M (2007) Localized maximum entropy shape modelling. Information processing in medical imaging - IPMI 2007, 20th international conference, LNCS, vol 4584. Springer, Berlin, pp 619–629
81. Lorensen WE, Cline HE (1987) Marching cubes: a high resolution 3d surface construction algorithm. Computer graphics and interactive techniques - SIGGRAPH 1987, 14th International conference. ACM, pp 163–169

82. Lüthi M (2010) A machine learning approach to statistical shape models with applications to medical image analysis. Phd thesis, Universität Basel. http://edoc.unibas.ch/diss/DissB_9149
83. Lüthi M, Jud C, Vetter T (2013) A unified approach to shape model fitting and non-rigid registration. In: Machine learning in medical imaging, 4th international workshop - MLMI 2013, LNCS, vol 8184. Springer International Publishing, pp 66–73
84. Lötjönen J, Mäkelä T (2001) Elastic matching using a deformation sphere. In: Medical image computing and computer-assisted intervention - MICCAI 2001, 4th International Conference, LNCS, vol 2208. Springer, pp 541–548
85. Malcolm J, Rathi Y, Tannenbaum A (2007) Graph cut segmentation with nonlinear shape priors. IEEE international conference on image processing - ICIP 2007:365–368
86. Malladi R, Sethian J, Vemuri B (1995) Shape modeling with front propagation: a level set approach. IEEE Trans Pattern Anal Mach Intell 17(2):158–175
87. Mertens S (2008) Variationsrechnung. In: Kursvorlesung Theoretische Physik I, Otto-von-Guericke-Universität Magdeburg. http://www-e.uni-magdeburg.de/mertens/teaching/mech/skript/variationen.pdf
88. Meyer CD (2000) Matrix analysis and applied linear algebra. Siam
89. Mian A, Pears N (2012) Chapter 8–3D face recognition. 3D imaging, analysis and applications. Springer, London, pp 311–366
90. Müller M, Röder T, Clausen M, Eberhardt B, Krüger B, Weber A (2007) Documentation: mocap database HDM05. Technical report. CG-2007-2, Universität Bonn
91. Modersitzki J (2004) Numerical methods for image registration. Oxford University Press, Oxford
92. Nain D, Haker S, Bobick A, Tannenbaum A (2007) Multiscale 3-D shape representation and segmentation using spherical wavelets. IEEE Trans Med Imaging 26(4):598–618
93. Osher S, Sethian JA (1988) Fronts propagating with curvature-dependent speed: algorithms based on Hamilton-Jacobi formulations. J Comput Phys 79(1):12–49
94. Paragios N, Deriche R (2002) Geodesic active regions and level set methods for supervised texture segmentation. Int J Comput Vis 46(3):223–247
95. Paragios N, Rousson M, Ramesh V (2002) Matching distance functions: a shape-to-area variational approach for global-to-local registration. 7th European conference on computer vision - ECCV 2002. Springer, Berlin, pp 775–789
96. Parr RG, Yang W (1989) Appendix A - functionals. In: Density-functional theory of atoms and molecules. Oxford University Press, New York
97. Parzen E (1962) On estimation of a probability density function and mode. Ann Math Stat 33(3):1065–1076
98. Paysan P, Knothe R, Amberg B, Romdhani S, Vetter T, (2009) A 3D Face Model for Pose and Illumination Invariant Face Recognition. Advanced Video and Signal Based Surveillance - AVSS 2009, 6th IEEE International Conference. IEEE, pp 296–301
99. Pirner S, Tingelhoff K, Wagner I, Westphal R, Rilk M, Wahl F, Bootz F, Eichhorn K (2009) CT-based manual segmentation and evaluation of paranasal sinuses. Eur Arch Otorhinolaryngol 266(4):507–518
100. Press WH, ATeukolsky S, Vettering WT, Flannery BP (2002) Numerical recipes in C++: the art of scientific computing, 2nd edn. Cambridge University Press, Cambridge
101. Recktenwald G (2014) FTCS solution to the heat equation. In: Course ME 448/548: applied computational fluid dynamics, Portland State University. http://web.cecs.pdx.edu/gerry/class/ME448/notes/pdf/FTCS_slides.pdf
102. Rilk M, Wahl FM, Eichhorn KWG, Wagner I, Bootz F (2009) Path planning for robot-guided endoscopes in deformable environments. In: Advances in robotics research. Springer, Berlin
103. Rilk M, Kubus D, Wahl FM, Eichhorn KWG, Wagner I, Bootz F (2010) Demonstration of a prototype for robot assisted endoscopic sinus surgery. In: IEEE international conference on robotics and automation - ICRA 2010. IEEE, pp 1090–1091
104. Roberts MG, Cootes TF, Adams JE (2003) Linking sequences of active appearance sub-models via constraints: an application in automated vertebral morphometry. In: Proceedings of the British Machine Vision Conference - BMVC 2003. BMVA Press, pp 38.1–38.10

105. Rosenblatt M (1956) Remarks on some nonparametric estimates of a density function. Ann Math Stat 27(3):832–837
106. Rousson M, Cremers D (2005) Efficient kernel density estimation of shape and intensity priors for level set segmentation. Medical image computing and computer-assisted intervention - MICCAI 2005, 8th international conference, LNCS, vol 3750. Springer, Berlin, pp 757–764
107. Rousson M, Paragios N (2002) Shape priors for level set representations. 7th European conference on computer vision - ECCV 2002. Springer, Berlin, pp 78–92
108. Rousson M, Paragios N (2008) Prior knowledge, level set representations & visual grouping. Int J Comput Vis 76(3):231–243
109. Rousson M, Paragios N, Deriche R (2004) Implicit Active Shape Models for 3D Segmentation in MR Imaging. In: Medical Image Computing and Computer-Assisted Intervention - MICCAI 2004, 7th Int. Conf., Springer, Berlin Heidelberg, LNCS, vol 3216, pp 209–216
110. Rueckert D, Frangi A, Schnabel J (2003) Automatic construction of 3-d statistical deformation models of the brain using nonrigid registration. IEEE Trans Med Imaging 22(8):1014–1025
111. Salah Z, Bartz D, Dammann F, Schwaderer E, Maassen M, Straßer W (2005) A fast and accurate approach for the segmentation of the paranasal sinus. Bildverarbeitung für die Medizin, Informatik aktuell. Springer, Berlin, pp 93–97
112. Schmidt FR, Cremers D (2009) A closed-form solution for image sequence segmentation with dynamical shape priors. Pattern recognition, 31st DAGM symposium, LNCS, vol 5748. Springer, Berlin, pp 31–40
113. Schoenemann T, Cremers D (2010) A combinatorial solution for model-based image segmentation and real-time tracking. IEEE Trans Pattern Anal Mach Intell 32(7):1153–1164
114. Seo A, Chung S, Lee J, Kim JI, Kim H (2010) Semiautomatic segmentation of nasal airway based on collaborative environment. In: Proceedings of the international symposium on ubiquitous virtual reality (ISUVR 2010). IEEE, pp 56–59
115. Sethian J (1999) Level set methods and fast marching methods: evolving interfaces in computational geometry, fluid mechanics, computer vision, and materials science. Cambridge University Press, Cambridge
116. Shang Y, Dössel O (2004) Statistical 3D shape-model guided segmentation of cardiac images. In: Computers in cardiology - CinC 2004. IEEE, pp 553–556
117. Shen D, Davatzikos C (2000) An adaptive-focus deformable model using statistical and geometric information. IEEE Trans Pattern Anal Mach Intell 22(8):906–913
118. Shen D, Herskovits EH, Davatzikos C (2001) An adaptive-focus statistical shape model for segmentation and shape modeling of 3-d brain structures. IEEE Trans Med Imaging 20(4):257–270
119. Sklyarenko E (2002) Barycentric coordinates. In: Encyclopaedia of mathematics. Kluwer Academic Publishers, Dordrecht
120. Sommerkorn A, Westphal R, Wiebking U, Liodakis E, Krettek C, Wahl FM (2012) Berechnung von Korrekturwinkeln für Hohe Tibia Osteotomie anhand von 3d Oberflächendruckverteilungen im Knie. In: 11. Jahrestagung der Gesellschaft für Computer- und Roboterassistierte Chirurgie (CURAC 2012)
121. Sotiras A, Davatzikos C, Paragios N (2013) Deformable medical image registration: a survey. IEEE Trans Med Imaging 32(7):1153–1190
122. Staib LH, Duncan JS (1992) Boundary finding with parametrically deformable models. IEEE Trans Pattern Anal Mach Intell 14(11):1061–1075
123. Stegmann MB, Gomez DD (2002) A brief introduction to statistical shape analysis. Informatics and mathematical modelling, Technical University of Denmark, DTU. http://www2.imm.dtu.dk/pubdb/p.php?403
124. Tanguay A (2013) Directional derivatives and the gradient vector. In: Course Math 233: multivariate calculus, Lecture 14.6, Class notes, University of Massachusetts. http://www.math.umass.edu/tanguay/233/Section14-6.pdf
125. Taron M, Paragios N, Jolly MP (2007) From uncertainties to statistical model building and segmentation of the left ventricle. In: IEEE 11th international conference on computer vision - ICCV 2007. IEEE, pp 1–8

126. The OpenMP Architecture Review Board (2002) OpenMP C and C++ Application Program Interface Version 2.0. http://www.openmp.org/mp-documents/cspec20.pdf
127. Thirion JP (1998) Image matching as a diffusion process: an analogy with Maxwell's demons. Med Image Anal 2(3):243–260
128. Tingelhoff K, Moral A, Kunkel M, Rilk M, Wagner I, Eichhorn K, Wahl F, Bootz F (2007) Comparison between manual and semi-automatic segmentation of nasal cavity and paranasal sinuses from CT images. In: Proceedings of the 29th annual international conference of the IEEE engineering in medicine and biology society (EMBS 2007). IEEE, pp 5505–5508
129. Tingelhoff K, Eichhorn K, Wagner I, Kunkel M, Moral A, Rilk M, Wahl F, Bootz F (2008) Analysis of manual segmentation in paranasal ct images. Eur Arch Otorhinolaryngol 265(9):1061–1070
130. Tsai A, Yezzi A Jr, Wells W, Tempany C, Tucker D, Fan A, Grimson WE, Willsky A (2003) A shape-based approach to the segmentation of medical imagery using level sets. IEEE Trans Med Imaging 22(2):137–154
131. Wahl FM (1984) Digitale Bildsignalverarbeitung: Grundlagen, Verfahren. Beispiele, Springer, Berlin
132. Wahl FM (1986) A coded light approach for depth map acquisition. Mustererkennung, 8. DAGM symposium. Springer, Berlin, pp 12–17
133. Wang Y, Staib LH (1998) Elastic model based non-rigid registration incorporating statistical shape information. Medical image computing and computer-assisted interventation - MICCAI'98, 1st International Conference, LNCS, vol 1496. Springer, Berlin, pp 1162–1173
134. Wang Y, Staib LH (2000) Boundary finding with prior shape and smoothness models. IEEE Trans Pattern Anal Mach Intell 22(7):738–743
135. Wang Y, Staib LH (2000) Physical model-based non-rigid registration incorporating statistical shape information. Med Image Anal 4(1):7–20
136. Weese J, Kaus M, Lorenz C, Lobregt S, Truyen R, Pekar V (2001) Shape constrained deformable models for 3D medical image segmentation. Information processing in medical imaging - IPMI 2001, 17th International Conference, LNCS, vol 2082. Springer, Berlin, pp 380–387
137. Westphal R, Zaremba D, Hassel T, Liodakis E, Suero E, Krettek C, Bach FW, Wahl F (2013) Roboterassistierte Umstellungsosteotomie mittels Wasserabrasivstrahltechnik. In: 12. Jahrestagung der Gesellschaft für Computer- und Roboterassistierte Chirurgie (CURAC 2013), pp 244–247
138. Wilcoxon F (1945) Individual comparisons by ranking methods. Biom Bull 1(6):80–83
139. Winkelbach S, Molkenstruck S, Wahl FM (2006) Low-cost laser range scanner and fast surface registration approach. Pattern recognition, 28th DAGM symposium, LNCS, vol 4174. Springer, Heidelberg, pp 718–728
140. Yezzi Jr A, Tsai A, Willsky A (1999) A statistical approach to snakes for bimodal and trimodal imagery. In: Computer vision - ICIP 1999, 7th IEEE International Conference, vol 2. IEEE, pp 898–903
141. Zhao Z, Aylward SR, Teoh EK (2005) A novel 3D partitioned active shape model for segmentation of brain MR images. Medical image computing and computer-assisted intervention - MICCAI 2005, 8th International Conference, LNCS, vol 3749. Springer, Berlin, pp 221–228
142. Zitova B, Flusser J (2003) Image registration methods: a survey. Image Vis Comput 21(11):977–1000

Printed in the United States
By Bookmasters